Solid State Chemistry in Catalysis

ACS SYMPOSIUM SERIES **279**

Solid State Chemistry in Catalysis

Robert K. Grasselli, EDITOR
The Standard Oil Company

James F. Brazdil, EDITOR
The Standard Oil Company

Based on a symposium sponsored by
the Division of Petroleum Chemistry, Inc.
at the 186th Meeting
of the American Chemical Society,
Washington, D.C.,
August 28–September 2, 1983

American Chemical Society, Washington, D.C. 1985

Library of Congress Cataloging in Publication Data
Solid state chemistry in catalysis.
  (ACS symposium series, ISSN 0097-6156; 279)

  "Based on a symposium sponsored by the Division of
Petroleum Chemistry, Inc. at the 186th Meeting of the
American Chemical Society, Washington, D.C.,
August 28–September 2, 1983."

  Includes bibliographies and indexes.

  1. Catalysis—Congresses. 2. Solid state chemistry—
Congresses.

  I. Grasselli, Robert K., 1930–   . II. Brazdil,
James F., 1953–   . III. American Chemical Society.
Division of Petroleum Chemistry. IV. American
Chemical Society. Meeting (186th: 1983: Washington,
D.C.) V. Series.

QD505.S65   1985   541.3'95   85-9190
ISBN 0-8412-0915-4

ACS Symposium Series

M. Joan Comstock, *Series Editor*

Advisory Board

FOREWORD

The ACS SYMPOSIUM SERIES was founded in 1974 to provide a medium for publishing symposia quickly in book form. The format of the Series parallels that of the continuing ADVANCES IN CHEMISTRY SERIES except that, in order to save time, the papers are not typeset but are reproduced as they are submitted by the authors in camera-ready form. Papers are reviewed under the supervision of the Editors with the assistance of the Series Advisory Board and are selected to maintain the integrity of the symposia; however, verbatim reproductions of previously published papers are not accepted. Both reviews and reports of research are acceptable, because symposia may embrace both types of presentation.

CONTENTS

PREFACE

HETEROGENEOUS CATALYSIS has been synonymous with industrial catalysis and the chemical industry since the time of Berzelius, Sabatier, Ostwald, Haber, Bosch, Mittasch, Fischer, and Hüttig, as well as many others. To this day, virtually all chemical and refining processes are based on the use of solid catalysts that effect selective transformations of hydrocarbon molecules to desired products in the vapor phase. Nonetheless, correlations between structural aspects of solid materials and their behavior as catalysts are relatively recent developments. Probably first to recognize the importance of structure in catalysis were researchers in the field of catalytic cracking who investigated the catalytic activity of natural clays and minerals in the early 1930s. The culmination of that work was the discovery of zeolitic cracking catalysts in the early 1960s and the subsequent development of the concept of shape selective catalysis.

With the advent of highly sophisticated instrumentation for precise structure determination, key catalytic roles are being recognized for many subtle features of solid state materials, such as point and extended defects, surface structure and surface composition, atomic coordination, phase boundaries, and intergrowths. Today, X-ray structure analysis is routine in heterogeneous catalysis research and has become as common and necessary as BET surface area analysis for characterizing solid catalysts. In addition, high resolution electron microscopy; photoelectron, IR, and Raman spectroscopies; solid state NMR; and even neutron diffraction are assuming increasingly important roles in both applied and fundamental catalysis research. Information about solids on the atomic and molecular level, which these techniques provide when combined with traditional catalytic studies (e.g., reaction kinetics, tracer studies, and molecular probes), gives a better fundamental understanding of complex catalytic phenomena. Correlations between the solid state and catalytic properties assessed through the application of sophisticated instrumentation and classical mechanistic approaches are the central theme of this book.

The book comprises 20 chapters that focus on state-of-the-art understanding of solid state mechanisms in heterogeneous catalysis and the relationship between catalytic behavior and solid state structure. The volume contains expanded and updated versions of papers presented on this subject at the ACS symposia in Washington, D.C. (1983) and Las Vegas (1982), and of written contributions from invited participants who could not attend these meetings. It emphasizes catalysis with oxides, sulfides, and zeolites. Although by no means an exhaustive treatise, we hope that it provides the reader with an understanding of the role the solid state plays in heterogeneous catalysis and gives an appreciation for the contributions solid state chemistry has made to the advancement of catalytic science and technology.

We should like to thank all the contributors for their excellent cooperation and patience during the process of editing this book and to the ACS for making this publication possible.

ROBERT K. GRASSELLI
JAMES F. BRAZDIL
The Standard Oil Company (Ohio)
Cleveland, OH 44128

December 7, 1984

OXIDES

Catalysis by Transition Metal Oxides

JERZY HABER

Institute of Catalysis and Surface Chemistry, Polish Academy of Sciences, ul.Niezapominajek, 30-239 Krakow, Poland

Catalytic oxidation reactions are divided into two groups: electrophilic oxidation proceeding through activation of oxygen and nucleophilic oxidation in which activation of the hydrocarbon molecule is the first step, followed by consecutive hydrogen abstraction and nucleophilic oxygen insertion. Properties of individual cations and their coordination polyhedra determine their behaviour as active centers responsible for ·activation of hydrocarbon molecules. A facile route for nucleophilic insertion of oxygen into such molecules by group V, VI and VII transition metal oxides is provided by the crystallographic shear mechanism, catalytic properties are thus dependent upon the geometry of the surface. The catalyst surface is in dynamic interaction with the gas phase, and changes of the latter may thus result in surface transformations and appearance of surface phases, which influence the selectivity of catalytic reactions.

The vast majority of catalysts used in modern chemical industry are oxides. Because of their ability to take part in the exchange of electrons, as well as in the exchange of protons or oxide ions, oxides are used as catalysts in both redox and acid-base reactions. They constitute the active phase not only in oxide catalysts but also in the case of many metal catalysts, which in the conditions of catalytic reaction are covered by a surface layer of a reactive oxide. Properties of oxides are also important in the case of preparation of many metal and sulphide catalysts, which are obtained from an oxide precursor. Very often, highly dispersed metals are prepared by reduction of an appropriate oxide phase, and sulphide catalysts are formed from the oxide precursor in the course of the hydrodesulphurization by interaction with the reaction medium. Finally, oxides play an important role in carriers for active metal or oxide phases, very often modifying strongly their catalytic properties. The present paper concerns

0097-6156/85/0279-0003$06.00/0

only one aspect of the vast field of chemistry of oxides, namely
the catalysis by transition metal oxides, which is the basis of the
selective oxidation of hydrocarbons.
    Catalytic oxidation is one of the most important types of
processes, both from a theoretical and a practical point of view.
As early as 1918, the production of phthalic anhydride by oxidation
of naphthalene over V_2O_5 was introduced. The milestone in the
development of modern petrochemical industry was the introduction
of the gas phase oxidation of propylene to acrolein and
ammoxidation to acrylonitrile over bismuth molybdate catalysts,
which provided in the early sixties, an abundant supply of new,
inexpensive, and useful chemical intermediates(1). Today,
catalytic oxidation is the basis of the production of almost all
monomers used in the manufacturing of synthetic fibers, plastics,
and many other products. With the increasing cost of energy and
shrinking supply of cheap hydrocarbons, much effort is now being
expended on the development of new oxidation processes of higher
selectivity and lower energy consumption. Substitution of the
dehydrogenation by oxidative processes, as in the production of
styrene from ethylbenzene, may be quoted as an example. Another
increasingly important field of catalytic application is the
selective oxidation of paraffins.

Discussion

Electrophilic and nucleophilic oxidation

In every oxidation reaction two reactants always take part: oxygen
and the molecule to be oxidized. The reaction may thus start
either by the activation of the dioxygen or by the activation of
the hydrocarbon molecule.
    At ambient or moderate temperatures, an oxygen molecule may be
activated by bonding into an organometallic complex in the liquid
phase. Depending on the type of the central metal atom and on the
properties of the ligands, superoxo-, peroxo- or oxo-complexes may
be formed:

$$M \xrightarrow{+O_2} MO_2 \xrightarrow{+M} MO_2M \longrightarrow 2MO \longrightarrow MOM$$

   perox   superoxo μ-peroxo   oxo μ-oxo

    In the case of the superoxo-complexes, an electrophilic attack
of a terminal oxygen atom on the organic reactant occurs resulting
in the formation of a μ-peroxo-complex, which decomposes into the
oxygenated product (lower left part of Figure 1). In the case of
perox complexes of group IV, V and VI transition metals, a stoichiometric oxidation takes place if a vacant coordination site exists
adjacent to side bonded oxygen and is capable of being occupied by
the organic reactant. Its olefin bond is then inserted into the

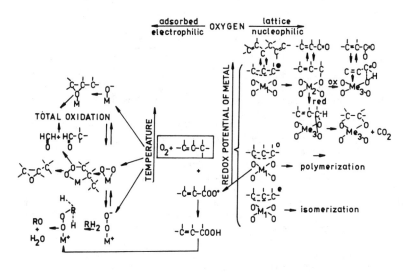

Figure 1. Mechanism of the catalytic oxidation of hydrocarbons. Reproduced with permission from Ref. 31.

metal-oxygen bond forming a peroxometallocycle which is then
decomposed into the oxygenated hydrocarbon molecule and oxo metal
complex (2). In order to transform the latter back into the
reduced metal which could again form the peroxocomplexes with a new
oxygen molecule, a coreducing agent is required, which may be a
hydrogen donor or a reactant itself. It should be born in mind
that in liquid phase oxidation reactions, the original oxygen
complex may be transformed into other reactive species which play
the role of active intermediates. A superoxo-complex may be
transformed into an alkyl peroxide or peracid complex, which is the
oxygen inserting intermediate (3-5).
     At higher temperatures the peroxide O_2^{2-} and superoxide O_2^-
species may appear at the surface of an oxide. Under these
conditions, the peroxide ion is unstable and dissociates forming
the ion radical O^-. Both O_2^- and O^- species are strongly
electrophilic reactants which attack the organic molecule in the
region of its highest electron density. At variance with their
behaviour in the liquid phase, the peroxy- and epoxy-complexes
formed as the result of an electrophilic attack of O_2^- or O^- species
on the olefin molecule are intermediates which lead to the
degradation of the carbon skeleton under heterogeneous catalytic
reaction conditions (6). Saturated aldehydes are formed in the
first stage (upper left part of Figure 1). These are usually much
more reactive than unsaturated aldehydes and at higher temperature
undergo rapidly total oxidation. Indeed, experimental data
collected in recent years clearly show that electrophilic oxygen
species in heterogeneous processes are responsible for total
oxidation (7).
     When hydrocarbon molecules are activated, a variety of
reaction paths may be initiated, consisting of a series of
consecutive oxidative steps, each of them requiring a different
active center to be present at the catalyst surface (8-10). It
should be emphasized at this point that it is the cations of the
catalyst which act as oxidizing agents in some of the consecutive
steps of the reaction sequence, forming the activated hydrocarbon
species. These undergo in subsequent steps a nucleophilic attack
by lattice oxygen ions O^{2-}, which are nucleophilic reagents with no
oxidizing properties. They are inserted into the activated
hydrocarbon molecule by nucleophilic addition forming an oxygenated
product, which after desorption leaves an oxygen vacancy at the
surface of the catalyst. Such vacancies are then filled with
oxygen from the gas phase, simultaneously reoxidizing the reduced
cations. It should be noted that incorporation of oxygen from the
gas phase into the oxide surface does not necessarily take place at
the same site from where surface oxygen is inserted into the
hydrocarbon molecule after being transported through the lattice.
     In the case of complex hydrocarbon molecules, the nucleophilic
addition of oxygen may take place at different sites of the
molecule. It will take place at a site which is made most
electropositive by appropriate bonding of the molecule at the
active center of the catalyst. When adsorption of the hydrocarbon
molecule results in the formation of a radical, interaction between
adsorbed molecules is favoured and dimerization or polymerization
occurs. When the adsorbed species are negatively charged,

isomerization may be favoured. This type of product obtained depends on the type and proportion of different active centers at the catalyst surface as well as on the ratio of the rate of desorption of the particular intermediate product to the rate of its transformation into the intermediate complex next in the series (upper right part of scheme in Figure 1). These rates may strongly depend on the degree of surface reduction attained in the course of the reaction, as is the case with the carboxylate complex, which is an intermediate in the oxidation of aldehydes to carboxylic acids. On oxidized surfaces, this complex desorbs in the form of an acid, whereas on a reduced surface it undergoes decarboxylation, resulting in the deposition of coke (11).

Reactions of catalytic oxidation may be thus divided into two groups: electrophilic oxidation, proceeding through the activation of oxygen, and nucleophilic oxidation, in which activation of the hydrocarbon molecule is the first step, followed by consecutive steps of nucleophilic oxygen insertion and hydrogen abstraction. They may be conveniently systematized according to the number of elementary structural transformations introduced into the reacting molecule (Table I).

An active and selective catalyst for oxidation of hydrocarbons to oxygenated products with retention of double bonds or aromaticity should thus have the following properties:
- activation of the hydrocarbon molecule by modifying its bonds and generating at appropriate sites the electron distribution favouring the nucleophilic attack of oxygen;
- efficient insertion of the nucleophilic lattice oxygen into the activated hydrocarbon molecule;
- rapid interaction with gas phase oxygen to replenish the lattice oxygen and transport it through the lattice to active sites, where the insertion takes place;
- should not generate electrophilic oxygen species.
The fundamental question arises as to how these properties are related to the solid state chemistry of oxides.

Activation of organic molecule

Classical studies of Adams (12, 13), using deuterated propylene and C_4-C_8 olefins, and of Sachtler and de Boer (14), with C^{14}-labelled propylenes, showed that activation of the olefin molecule consists of the abstraction of α-hydrogen and the formation of a symmetric allylic intermediate. Conclusions concerning the role of the cationic and anionic sublattices of complex oxide catalysts having an oxysalt character, such as molybdates, tungstates, etc., in the initial α-hydrogen abstraction and the subsequent steps of the oxidation process, were drawn by comparing the behaviour of Bi_2O_3 and MoO_3 for the reaction of propylene and allyl iodide (15). When allyl iodide was passed over MoO_3, practically total conversion was observed already at 310°C with 98% selectivity to acrolein. Under the same conditions, MoO_3 was completely inactive with respect to propylene. On contacting allyl iodide with Bi_2O_3, total conversion at 310°C was also observed. However, in this case 70% of the products formed were 1,5-hexadiene with practically no acrolein being detected. 1,5-hexadiene was also the main product

Table I. Heterogeneous Oxidation of Hydrocarbons

Electrophilic Oxidation		Nucleophilic Oxidation	
Reaction Type	Catalyst	Reaction Type	Catalyst
1. With double bond fission		1. Without introduction of hetero-atom	
1.1. oxidation of olefins to oxides	Ag_2O	1.1. oxidative dehydrogenation of alkanes and alkenes to dienes	$Bi_2O_3-MoO_3$ P_2O_5
1.2. oxyhydration of olefins to saturated ketones	SnO_2-MoO_3	1.2. oxidative dehydrodimerization and dehydrocyclization of alkenes	$MoO_3-Al_2O_3$
2. With C-C bond fission		2. With introduction of heteroatom	
2.1. oxidation of olefins to saturated aldehydes	V_2O_5	2.1. introduction of heteroatom into hydrocarbon chain	
2.2. oxidation of aromatics to anhydrides and acids with ring rupture	$V_2O_5-MoO_3$	2.1.1. introduction of oxygen a. oxidation of olefins to unsaturated aldehydes and ketones	$Bi_2O_3-MoO_3$
3. Total oxidation to CO_2+H_2O	Co_3O_4 $CuCo_2O_4$ $CuCr_2O_4$	b. oxidation of alkylaromatics to aldehydes	$Bi_2O_3-MoO_3$
		2.1.2. introduction of nitrogen a. ammoxidation of olefins to nitriles	$UO_3-Sb_2O_4$
		2.2. introduction of heteroatom into acyl group a. oxidation of aldehydes to acids	$NiO-MoO_3$
		b. oxidation of alkylaromatics to anhydrides	$V_2O_5-TiO_2$

in the reaction of propylene over Bi_2O_3. These results clearly
indicated that activation of the olefin molecule, which consists of
the abstraction of α-hydrogen and the formation of an allylic
specie, takes place on cationic active centers, whereas the MoO_3 or
molybdate anionic sublattice is responsible for the insertion of
oxygen into the hydrocarbon molecule. Indeed, quantum chemical
calculations of the system, composed of cobalt ion in octahedral
coordination of five oxygen atoms and propylene molecule as the
sixth ligand, have shown (16, 17) that on approaching the propylene
molecule to the plane of the active center, the C-H bond is
continuously destabilized with an interaction appearing
simultaneously between this hydrogen and the oxygen of the active
center and increasing until the total energy attained a minimum and
an intermediate complex is formed. When the allyl specie is
removed, the energy required for this process is much smaller than
that needed to remove the whole propylene molecule. The OH bond is
further strengthened, and its energy attains the value
characteristic of a normal the OH group. Thus, it may be concluded
that on contacting propylene with the surface of the transitition
metal oxide, reactive chemisorption takes place, in which the C-H
bond is broken and an absorption complex with allyl species as one
of the ligands is formed. Considerable charge transfer takes place
from the allyl species onto the transition metal orbitals,
rendering the species positive, the charge distribution depending
on the type of metal, its valence state and the ligand field
strength (18). Experiments, with azopropene and allyl alcohol,
carried out by Grasselli et al. (19-22), demonstrated that after
the first hydrogen abstraction, insertion of oxygen takes place,
and only then the second hydrogen is abstracted resulting in the
formation of the acrolein precursor.

    An important question may be raised at this point as to what
is the structure of the cationic active center activating the
hydrocarbon molecule. Can every cation situated at the surface of
the given oxide perform the role of the active center, or must this
cation be localized at some special site of the surface, and how
does its activity depends on this location? In order to obtain
some relevant information about this question, isolated bismuth
ions were supported at the surface of MoO_3 (23). Taking into
account the very high efficiency of MoO_3 for inserting oxygen into
activated hydrocarbon molecules, it might be assumed that every
propylene molecule activated at the isolated bismuth ion would pick
up oxygen and be converted to acrolein. The number of acrolein
molecules would thus be a measure of the number of propylene
molecules activated by the known number of bismuth ions.
Measurements of the propylene oxidation activity as a function of
the surface concentration of bismuth ions, expressed as their
number per surface molybdenum atom, are shown in Figure 2. When
allyl iodide was introduced, the yield of acrolein was constant and
independent of the bismuth coverage confirming the assumption that
once allyl radicals have been generated they rapidly undergo a
nucleophilic attack by oxide ions from the lattice of MoO_3. On
introducing a mixture on propylene and oxygen, the activity at low
bismuth surface coverages increased proportionally to the surface
concentration of bismuth, the turnover frequency per bismuth ion

being thus constant. At higher bismuth coverages, the activity
leveled off because bismuth ions became located too close to each
other to operate simultaneously in the reaction. It is noteworthy
that the yield of acrolein observed at the plateau is similar to
that observed in the case of the $Bi_2(MoO_4)_3$ phase. This clearly
demonstrates that the ability to activate the hydrocarbon molecule
is related to the individual bismuth cations and their nearest
neighbors. These active centers function independent of whether
they are distributed randomly as a monolayer at the surface of MoO_3
or form the surface of the bismuth molybdate phase with long range
order. It should also be noted that the amount of CO_2 formed
remains constant indicating that the side reaction of total
oxidation proceeds at some other sites, resulting from the
properties of MoO_3 itself.

Insertion of Oxygen

As already mentioned, experiments in which allyl compounds were
reacted with complex oxides, such as molybdatess or tungstates,
showed that it is MoO_3, WO_3, or the corresponding anionic
sublattices which perform the insertion of oxygen into the
hydrocarbon molecule. The question may thus be raised as to which
properties of these oxides are responsible for the very high
activity and selectivity in the insertion of oxygen. One of the
features common to all group V, VI, and VII transition metal oxide
lattices, known to be good catalysts for selective oxidation of
hydrocarbons, is their ability to form shear structures which
relates to the facile planar rearrangement of coordination
polyhedra and their particular
spacial arrangement. In the octahedral coordination of oxide ions,
in which d^2sp^3 hybridized orbitals are used by the metal to form
σ-bonds, the remaining d_{xy}, d_{xz} and d_{yz} orbitals of group V, VI, and
VII metals extend far enough to considerably overlap with πp
orbitals of oxygen, and the position of their redox potential
relative to the anion valence band edge are favourable for the bond
formation (24). As a result, π-bonds with oxygen ions are formed
and the cations become displaced from the centers of octahedra
towards terminal oxygen atoms. Large displacement polarizabilities
give rise to high relaxation energy, which decreases the cation-
cation repulsions opposing the formation of a structure with
shorter metal-metal distance (25). Thus, removal of oxygen ions
from the lattice of these oxides results in the formation of
ordered arrays of oxygen vacancies, followed by a very facile
rearrangement of the layers of initially corner-linked metal oxygen
octahedra into an arrangement of edge-linked octahedra, resulting
in the formation of a shear plane (Figure 3). A hypothesis was
advanced (26) that the easy evolution of one oxygen ion on the
transformation from corner-linked to edge-linked arrangement of
metal-oxygen octahedra may be one of the factors creating the
ability of these structures to insert oxygen into the organic
molecule in processes of selective oxidation of hydrocarbons.
Studies of allyl iodide activity on different tungsten oxides seem
to confirm this hypothesis (27). The experiments were carried out
with two groups of tungsten oxides: those in which shear

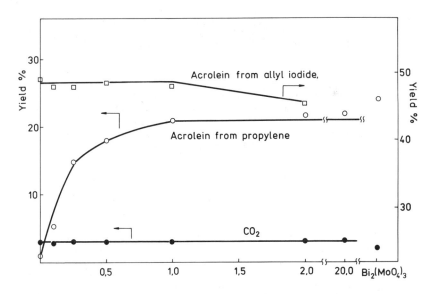

Figure 2. Yield of acrolein and CO_2 in oxidation of propylene and allyl iodide as function of the coverage of MoO_3 with bismuth ions. (Reproduced with permission from Ref. 23.)

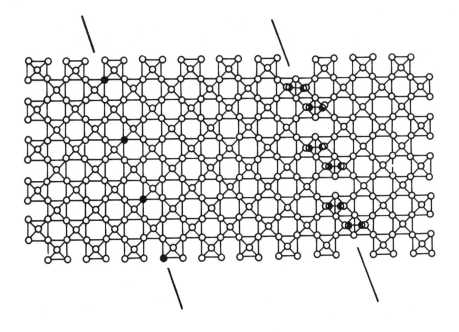

Figure 3. Formation of a shear plane in ReO_3-type oxide structure.

structures are formed on reduction (WO_3, $W_{20}O_{59}$) and those which do not show this phenomenon ($W_{18}O_{49}$,WO_2), as may be seen from their structures illustrated in Figure 4. The selectivity of the allyl iodide reaction to acrolein as a function of the number of pulses introduced into the reactor are seen in Figure 5. For comparison, results obtained with MoO_3 are also shown. In the case of $W_{20}O_{59}$ which, similarly to MoO_3, is able to generate shear planes on interaction with the reducing atmosphere, acrolein appears in the products after the first few pulses. On the other hand, in the case of samples of $W_{18}O_{49}$ and WO_2, which are unable to form shear structures, practically no acrolein was present in the products of the reaction.

Let us now return to the surface of a molybdate catalyst. Due to the displacement of Mo^{6+} ions towards terminal oxygens the bridging oxygens become more basic and thus more reactive in the nucleophilic attack on the activated hydrocarbon molecule. On raising the temperature, the rearrangement of the corner-linked metal-oxygen polyhedra into edge linked arrays proceeds more and more readily, and provides a facile and efficient route for the addition on a nucleophilic lattice oxygen to the hydrocarbon molecule without the generation of point defects, which could be involved in the formation of electrophilic oxygen species.

Strong evidence supporting this model is provided by the ESR studies of MoO_3 in the course of its interaction with different atmospheres (28). As an example, Figure 6 shows the ESR spectra of MoO_3 after outgassing the 430°C for 5 min. (curve A) and for 35 min. (curve B). Analysis of the values of the g-tensor reveals the appearance of two different Mo^{5+} centers: type A, formed at an early stage of reduction, and characterized by rhombically distorted square pyramidal surrounding of non-axial symmetry along the -double bonded oxygen, and type B, of distorted octahedral coordination, and appearing in strongly reduced samples. Comparison of these results with the situation at the surface of MoO_3 crystallites (Figure 7) leads to the conclusion that the only surface oxygen ion, which can be removed leaving reduced molybdenum cation in square pyramidal surrounding with double-bonded oxygen in the opposite apex, is the surface oxygen bridging two adjacent octahedra in the double string of edge-linked Mo-O octahedra. When concentration of vacancies increases, crystallographic shear takes place (Figure 7b), and Mo^{5+} cations assume the octahedral coordination along the shear planes. It is noteworthy that on exposing MoO_3 to allyl compounds only the ESR spectrum of type B centers appears. This indicates that insertion of oxygen into the organic molecule is accompanied by a simultaneous rearrangement of the coordination octahedra at the surface of MoO_3.

Mo^{5+} ions registered in the ESR measurement constitute only a small fraction of the reduced species, the majority being Mo^{4+} ions, which as non-Kramers ions are not expected to give an ESR signal. As these ions are located in the shear planes in edge-linked octahedra, Mo-Mo bonds are formed as revealed by the XPS studies (26, 29). UV photoelectron spectra shown in Figure 8 indicate that these clusters of tetravalent molybdenum ions constitute energy levels situated in the forbidden energy gap of the oxide (30). MoO_3, which has been oxidized at 400°C, shows the

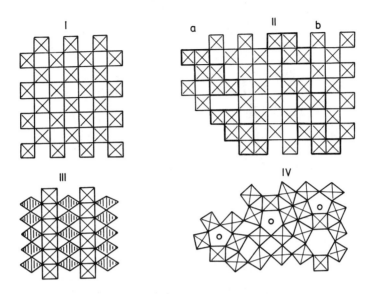

Figure 4. Structures of different tungsten oxides:
I – WO_3, II – $W_{20}O_{58}$, III – WO_2, IV – $W_{18}O_{49}$. Reproduced with
permission from Ref. 27. Copyright 1983, Academic Press.

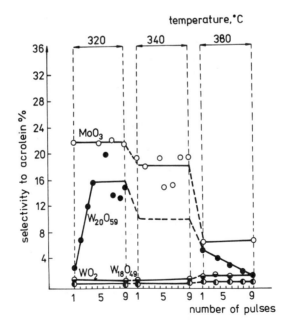

Figure 5. Selectivity to acrolein on interaction of allyl
iodide with MoO_3 and different tungsten oxides (17).
Reproduced with permission from Ref. 27. Copyright 1983,
Academic Press.

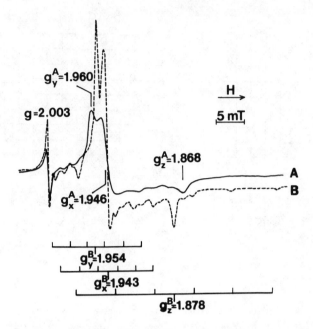

Figure 6.   ESR spectra of MoO$_3$ after outgassing at 430RC for 5 min (curve A) and 35 min (curve B). (Reproduced with permission from Ref. 28.)

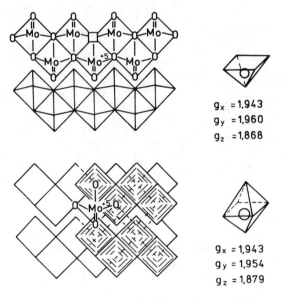

Figure 7.   Generation of oxygen vacancies   and shear planes in MoO$_3$ structure.

valence band with two maxima at 5.0 and 7.6 eV (curve IA). Its
edge is observable at a binding energy of about 2.8 eV. After
outgassing (curve IB) local energy levels appear at about 0.9 eV
and 2.0 eV. Similar levels are also formed on outgassing the
Bi_2MoO_6, as indicated by curve IIB. These levels have a donor
character as the substitution of higher valent ions in the lattice
sites of an oxide by lower valent ions results in the appearance of
n-type semiconductor.

A general conclusion may thus be formulated that the ability
of group V, VI, and VII transition metal oxides to form different
types of bonding between coordination polyhedra plays an important
role in determining their catalytic properties by providing a
facile route for insertion of oxygen into an organic molecule. The
mechanism of such insertion is shown in Figure 9 (31). Here in
contrast to those oxides in which the desorption of an oxygenated
product results in the generation of an oxygen vacancy at
considerable expenditure of energy, the desorption is accompanied
by the simultaneous facile rearrangement of octahedra.

After the given elementary step of the catalytic reaction has
taken place with the participation of the lattice oxygen and
formation of a nucleus of the shear plane, the active center at the
surface is left in the reduced state. Before such an elementary
step can be repeated, the active center must be reoxidized. This
reoxidation can be realized either by incorporation of oxygen from
the gas phase or by diffusion of oxygen ions from the bulk.
Depending on the rate of such regeneration of the active center,
its "dead time" may be short or long. In the case of oxides such
as bismuth molybdate, the mobility of lattice oxygen is high (32,
33), and regeneration by diffusion from the bulk operates very
efficiently because reoxidation of the lattice may take place at
centers different from those participating in the reaction. Under
such conditions, the "dead time" of active centers is very short,
the turn-over frequency very large, and the catalyst activity very
high. Thus, it may be concluded that parameters modifying the
mobility of oxide ions of a solid lattice may strongly influence
the catalytic activity.

The Role of Surface Geometry

As already mentioned in transition metal oxides of group V, VI, and
VII, the appearance of a π-bond component of the metal-anion
bonding results in the displacement of the cation from the center
of site symmetry which stabilizes the layered arrangement of
coordination polyhedra and extended defects - shear and block
structures. As the charge density and hence the acid-base
character of oxide ions is strongly influenced by the metal-oxygen
separation, the displacement causes the differentiation of oxide
ions in various crystallographic positions in respect to their
acid-base properties. Simultaneously, redox properties are
modified as manifested by the dependence of the work function on
the type of the crystallographic plane. It may be concluded that
in the case of such oxides the catalytic properties will depend on
the geometrical structure of the surface. Indeed, experimental
data collected in the last few years clearly indicate that

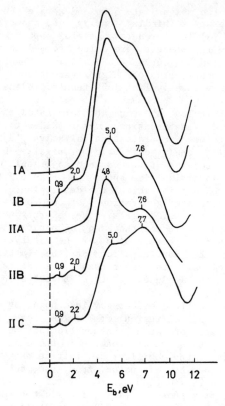

Figure 8. UPS spectra of MoO$_3$ (I) and Bi$_2$MoO$_6$ (II):
A - oxidized at 470RC for 1 hr at 1 atm of oxygen;
B -outgassed for 10 hr at 470RC; C - contacted with propylene
at 440RC (20). Reproduced with permission from Ref. 30.
Copyright 1976, Academic Press.

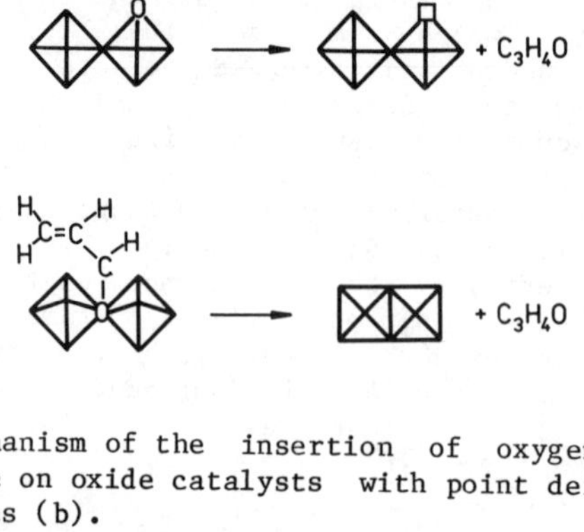

Figure 9. Mechanism of the insertion of oxygen into hydrocarbon molecule on oxide catalysts with point defects (a) and
shear structures (b).

different crystal modifications of oxide systems or different
crystal planes of given oxide crystallites may differ considerbly
in their catalytic properties.   Tatibouet and Germain (34), using
crystallites of MoO_3, prepared by sublimation and appropriate
seiving showed that on the basal (010) plane of MoO_3, methanol, in
presence of oxygen, becomes dehydrogenated to formaldehyde, whereas
on the (001) and (101) faces it is dehydrated to dimethylether.
This suggests that the (010) plane shows more pronounced redox
properties, whereas the (001) and (101) planes behave as acid-base
surfaces. This is in line with the result of extensive studies of
V_2O_5 and V_2O_5-TiO_2 catalysts (35).   Namely, Gasior and Grzybowska
(36) observed a drastic decrease of acidity of V_2O_5 when it was
supported on anatase at small coverages not exceeding a monolayer.
At higher loadings, the acidity increased with V_2O_5 content,
finally attaining that of the pure phase.   Crystal structure
analysis of V_2O_5 and anatase reveals a good crystallographic fit
between the (001) cleavage plane of anatase and the (001) basal
plane of V_2O_5 containing the V=O groups sticking out perpendicular
to the surface. This structure may be assumed to prevail in the
monolayer of V_2O_5 on anatase.   However, such catalysts show no
acidity. Therefore, it may be concluded that on pure V_2O_5 the
acid-base properties are located mainly at the side planes (110)
and (100). This has a direct bearing on the catalytic properties.
Thus, Gasior and Machej (37) studied the catalytic activity of V_2O_5
samples of different crystal habit and found that plate-like
crystallites exposing mainly the basal (001) planes with V=O groups
show very high selectivity in the oxidation of o-xylene to phthalic
anhydride, whereas in the case of needle-like crystallites with
predominance of (110) and (100) side planes mainly total oxidation
to CO_2 is observed.
    It is interesting that at variance with these conclusions,
Volta et. al. (38, 39), studying oriented samples of MoO_3, obtained
from intercalation compounds of $MoCl_5$ - graphite, concluded that
selective oxidation of propylene to acrolein or isobutene to
methacrolein is mainly catalysed by the (100) side faces, whereas
the (010) basal race is responsible for total oxidation. It should
be remembered, however, that selective oxidation is a multistep
process, consisting of the consecutive abstractions of hydrogen
atoms and insertion of oxygen atoms.   As already described,
experiments with allyl compounds (15) showed that MoO_3 lattice
efficiently inserts nucleophilic lattice oxygen ions into the
activated hydrocarbon molecule but has only a limited ability to
activate the hydrocarbon, this step being rate determining in
selective oxidation of olefins.   Therefore, determination of
catalytic activity in the oxidation of olefins can only yield
limited information about the mechanism of the reaction. Indeed,
experiments on the interaction of MoO_3 crystallite of different
crystal habit with allyl iodide seems to indicate (40) that it is
the (010) basal plane which is responsible for the insertion of
oxygen into the organic molecule by the shearing mechanism
described in the preceding section.

Dynamics of the Catalyst Surface

Transition metal oxides belong to non-stoichiometric compounds, their composition depending on the equilibrium between the lattice and its consitutents in the gas phase, i.e. it is a function of oxygen pressure. Due to the contribution of the surface free energy, the compostion of the surface of the solid differs from that of the bulk, and the system in equilibrium is composed of bulk crystallites, their surface and the gas phase. However, equilibration of the gas phase with the bulk of crystallites takes place only after annealing at high temperatures, when diffusion of lattice constitutents becomes sufficiently rapid. When an oxide, which has been equilibriated at a high temperature in oxygen at a given pressure, is then heated under a different pressure at a low temperature at which the diffusion of defects in the lattice is slow, the new equilibrium comprises only the surface layer. When hydrocarbon molecules, which have reducing properties, are also present in the gas phase, a certain degree of reduction of the surface is reached corresponding to a steady state in which the rate of reduction of the surface by hydrocarbon molecules becomes equal to the rate of its reoxidation by gas phase oxygen. When the composition of the gas phase is changed, a corresponding change of the surface compostion occurs which in turn may result in changes of catalytic activity (40, 41). However, several other phenomena may also take place, such as ordering of defects at the surface, surface transformations, and precipitation of new bidimensional surface phases. They may result in the appearance of new types of active centers at the surface of the catalyst, directing the catalytic reaction along a new pathway and thus profoundly influencing the selectity (31, 41-44).

The catalyst surface is in a dynamic interaction with the gas phase. Depending on the properties of the mixture of reactants of the catalytic reaction, different surface phases may be formed at the surface of the catalyst, directing the rection along different reaction paths. Thus, when the steady state conditions of the reaction are changed, the structure of the catalyst surface also may change, modifying the activity and selectivity of the catalyst itself. This means that in the rate equation it is not only the concentration term which depends on the pressure of reactants, but also the rate constant.

Concluding Remark

In the catalytic reaction of organic molecules with gas phase oxidants (e.g. oxygen, sulphur, chlorine) either the oxidant is activated and performs an electrophilic attack, or the organic molecules are activated and the reaction proceeds in consecutive steps of hydrogen abstraction and nucleophilic oxygen insertion. Reactions of catalytic oxidation may be thus divided into two groups: a. electrophilic oxidation, in which epoxides are formed in case of liquid phase reaction and degradation of the carbon skeleton takes place under conditions of heterogeneous reaction, resulting in the total oxidation, and b. nucleophilic oxidation, in which products of the successive nucleophilic insertion of appropriate anionic lattice constituents into the organic molecule are formed. This insertion takes place at the site of the molecule, which by its appropriate bonding at the active center of the catalyst, is made most positive. The structure of the intermediate complex composed of the reacting molecule and the active center thus determines the reaction pathway and consequently the selectivity.

The ability to activate the hydrocarbon molecule is related to the properties of individual cations and their nearest neighbours, constituting active centers.

When discussing the behaviour of an intermediate complex located at the surface of a solid it is necessary to take into account the fact that the occupancy of different orbitals is determined by the chemical potential of electrons in the solid, given by the position of the Fermi level. Shifting of this position e.g., by introduction of additives changes the orbital occupancy, may in turn change the reactivity of bonds and modify the activity and selectivity.

Transition metal oxides are nonstoichiometric compounds. The nonstoichiometry may be introduced either by the generation of point defects or by the change of the mode of linkage between the coordination polyhedra, resulting in the formation of extended defects/shear structures. This latter way of changing the stoichiometry is a characteristic feature of group V, VI, and VII transition metal oxides and is related to the presence of the π-orbital component of the metal-oxygen bonds, resulting in the displacement of the cations from the center of site symmetry, which stabilizes the layered arrangement of coordination polyhedra. As the charge density on oxide ions is strongly influenced by the metal-oxygen separation, the displacement causes the differentation of oxide ions in various crystallographic positions with respect to their redox and acid-base properties. As a result, the catalytic properties strongly depend on the geometrical structure of the surface. Different polymorphic modifications or different crystal planes may differ considerably in their catalytic behaviour. A general conclusion may be formulated that the ability of group V, VI and VII transition metal oxides to form different types of bonding between coordination polyhedra plays an important role in determining their catalytic properties by providing a facile route for insertion of oxygen into an organic molecule. No oxygen vacancies are formed and the generation of electrophilic oxygen, which could initiate the side reaction of total oxidation, is thus eliminated.

The catalyst surface is in dynamic interaction with the gas phase. Depending on the properties of the mixture of reactants of the catalytic reaction, different surface phases may be formed at the surface of the catalyst, directing the reaction along different reaction pathways. A change of the steady state-conditions influences the catalytic reaction, not only directly through the concentration term in the rate equation, but also by modifying the properties of the catalyst itself, i.e. the rate constant k. Thus, heterogeneous catalytic systems should not be treated as two phases, but as three phase systems composed of the gas phase, the solid, and the surface region. The latter is composed of the surface atoms of the catalyst lattice interacting with the adsorbed molecules of the catalytic reaction.

Literature Cited

1. Callahan, J. L.; Grasselli, R. K.; Milberger, E. C.;
    Strecker, H. A. Ind. Eng. Chem., Prod. Res. Develop. 1970,
    134.
2. Sheldon, R. A.; Kochi, J. K. Adv. Catal. 1976, 25, 272.
3. Apostol, I.; Haber, J.; Mlodnicka, T.; Poltowicz, J.
    J. Molecular Catal. 1982, 14, 197.
4. Apostol, I.; Haber, J.; Mlodnicka, T.; Poltowicz, J.
    J. Molecular Catal. 1984, 26, 239.
5. Apostol, I.; Haber, J.; Mlodnicka, T.; Poltowicz, J.
    Proc. 8th Intern. Congr. Catalysis, Berlin 1984, p. IV-497.
6. Bielanski, A.; Haber, J. Catal. Rev. 1979, 19, 1.
7. Libre, J. M.; Barbaux, Y.; Grzybowska, B.; Bonnelle, J. P.
    React. Kinet. Catal. Lett. 1982, 20, 249.
8. Haber, J. Proc. 5th Intern. Congr. Catalysis, Palm Beach 1972,
    p. 1006.
9. Haber, J.; Grzybowska, B. J. Catal. 1973, 28, 489.
10. Haber, J.; Pure and Appl. Chem. 1978, 50, 923.
11. Haber, J.; Marczewski, J.; Stoch, J.; Unger, L. Proc. 6th
    Intern. Congr. Catalysis, London 1976, p. 827.
12. Adams, C. R.; Jennings, I. J. J. Catal. 1963, 2, 63.
13. Adams, C. R. Proc. 3rd Intern. Congr. Catalysis, Amsterdam
    1964, p. 240.
14. Sachtler, W. M. H.; de Boer, N. H. Proc. 3rd Intern. Congr.
    Catalysis, Amsterdam 1964, p. 252.
15. Grzybowska, B.; Haber, J.; Janas, J. J. Catal. 1977, 49, 150.
16. Haber, J.; Witko, M. J. Molecular Catal. 1980, 9, 399.
17. Haber, J.; Witko, M. Acc. Chem. Res. 1981, 14, 1.
18. Haber, J.; Sochacka, M.; Grzybowska, B.; Golebiewski, A.
    J. Molecular Catal. 1975, 1, 35.
19. Burrington, J. D.; Grasselli, R. K. J. Catal. 1979, 59, 79.
20. Grasselli, R. K.; Burrington, J. D.; Brazdil, J. F. Disc.
    Faraday Soc. 1981, 72, 203.
21. Grasselli, R. K.; Burrington, J. D. Adv. Catal. 1981, 30,
    133.
22. Burrington, J. D.; Kartisek, C. T.; Grasselli, R. K.
    J. Catal. 1983, 81, 489.
23. Bruckman, K.; Haber, J.; Wiltowski, T. in preparation.
24. Goodenough, J. Proc. Solid State Chem. 1971, 5, 145.
25. Catlow, C.A. In "Nonstoichiometric Oxides"; Sorenson O. T.,
    Ed.; Academic Press: New York, 1981, Chap. 2.
26. Haber, J. Proc. 2nd Intern. Conf. Chemistry and Uses of
    Molybdenum, Oxford 1976, p. 119.
27. Haber, J.; Janas, J.; Schiavello, M.; Tilley, R. J. D.
    J. Catal. 1983, 82, 395.
28. Haber, J.; Serwicka, E. in preparation.
29. Haber, J.; Marczewski, J.; Stoch, J.; Unger, L.
    Ber. Bunsenges Phys. Chem. 1975, 79, 970.
30. Grzybowska, B.; Haber, J.; Marczewski, W.; Unger, L.
    J. Catal. 1976, 42, 327.
31. Haber, J. Proc. 3rd Intern. Conf. Chemistry and Uses of
    Molybdenum, AnnArbor 1979, p. 114.

32. Keulks, G. W.; Krenzke, L. D.; Notermann, T. M. Adv. Catal. 1978, 27, 183.
33. Moro-oka, Y.; Veda, W.; Tanaka, S. Proc. 7th Intern. Congr. Catalysis, Tokyo 1980.
34. Tatibouet, J. M.; Germain, J. E. J. Catal. 1981, 72, 375.
35. Gasior, M.; Grzybowska, B.; In "Vanadia Catalysts for Oxidation of Aromatic Hydrocarbons"; Haber, J.; Grzybowska, B.; Ed.; Polish Scientific Publ. Krakow, 1984, p. 133.
36. Gasior, M.; Gasior, I.; Grzybowska, B. Appl. Catal. 1984, 10, 87.
37. Gasior, M.; Machej, T. J. Catal. 1983, 83, 472.
38. Volta, J. C.; Forissier, M.; Theobald, F.; Pham, T. P. Disc. Faraday Soc. 1981, 72, 275.
39. Volta, J. C.; Desquesnes, W.; Moraweck, B.; Tatibouet, J. M. Proc. 7th Intern. Congr. Catalysis, Tokyo 1980, p. 1398.
40. Bruckman, K.; Haber, J.; Wiltowski, T. in preparation.
41. Haber, J.; Stoch, J.; Wiltowski, T. Proc. 7th Intern. Congr. Catalysis., Tokyo 1980.
42. Haber, J.; Stoch, J.; Wiltowski, T. React. Kinet. Catal. Lett. 1980, 13, 161.
43. Haber, J. Proc. 4th Intern. Conf. Chemistry and Uses of Molybdenum, Golden, Colorado, 1982, p. 395.
44. Haber, J. In "Surface Properties and Catalysis by Non-metals"; Bonnelle, J.P.; Delmon, B.; Derouane, E., Ed.; Reidel Publ. Co., Dordrecht, 1983, p. 1.

RECEIVED January 14, 1985

2

Active Sites on Molybdate Surfaces, Mechanistic Considerations for Selective Oxidation, and Ammoxidation of Propene

JANET N. ALLISON and WILLIAM A. GODDARD III

Arthur Amos Noyes Laboratory of Chemical Physics, California Institute of Technology, Pasadena, CA 91125

Molybdates involving various metal additives play a dominant role in such industrially important catalytic processes as selective oxidation (propene to acrolein) and ammoxidation (propene to acrylonitrile); however, the details of the reaction mechanism and of the surface sites responsible are yet quite uncertain. In order to establish the thermochemistry and detailed mechanistic steps involved with such reactions, we have performed *ab initio* quantum chemical calculations [generalized valence bond (GVB) and configuration interaction (CI)]. These studies indicate a special importance of multiple surface dioxo Mo sites (possessing two Mo-O double bonds and hence spectator oxo groups) arranged together so as to provide the means for promoting the sequence of transformations.

Various catalysts based on molybdates have been used both for *selective oxidation of propene to acrolein*

$$H_2C{=}CH{-}CH_3 + O_2 \xrightarrow{\quad} H_2C{=}CH{-}HC{=}O + H_2O \qquad (1)$$

$$Bi_2O_3 \cdot nMoO_3 \qquad n = 2, 3$$

$$320°C$$

and *ammoxidation of propene to acrylonitrile*,

$$H_2C{=}CH{-}CH_3 + O_2 + NH_3 \xrightarrow{\quad} H_2C{=}C{\big<}^{CN}_{H} + H_2O \qquad (2)$$

$$Bi_2O_3 \cdot nMoO_3$$

$$430°C$$

Numerous experimental studies have provided mechanistic information about these catalytic reactions; however, there are as yet many uncertainties concerning the character of the active site and its relation to the details of the mechanism. In this paper we will use the results of *ab initio*

0097–6156/85/0279–0023$06.00/0

quantum chemical calculations [generalized valence bond (GVB) and configuration interaction (CI)] to help analyze the details of the reaction mechanisms and the relation of various reaction steps to specific surface sites.

In the following sections we discuss the principle of spectator oxo promotion that we find to play a crucial role in promoting particular reaction steps; we then examine the details of selective oxidation; and, finally, we outline preliminary results on ammoxidation .

Spectator Oxo Effects

Molybdates lead to bulk structures involving either octahedral or tetrahedral coordination of oxygens about each Mo. On various surfaces, the most stable configurations for such molybdates are likely to be

1   **2**   **3**

where **1** and **3** correspond to bulk octahedral sites and **2** to bulk tetrahedral sites. Here there are four (**1**) and (**3**) or two (**2**) single bonds to oxygen atoms that have a single bond to another Mo center, and one (**1**) and (**3**) or two (**2**) double bonds to oxygens that are not bonded to other Mo atoms. Typically the M-O single bond lengths are ~1.95 Å and the Mo-O double bond lengths are 1.67 to 1.73 Å. In addition, the octahedral site **1** would have a sixth oxygen neighbor at 2.2 to 2.4 Å (**1**). All three surface structures are formally Mo^{VI} and all involve Mo-O double bonds. However, we find that these species lead to extremely different chemistry. Thus, in selective oxidation of propene, a critical step is trapping of an allyl radical at an Mo=O bond. However, we find that only for species **2** is this process strongly exothermic.

$$\Delta H \quad \Delta G_{(400°C)}$$
$$(kcal/mol)$$

$$-2 \qquad +21 \quad (3)$$

$$-35 \qquad -12 \quad (4)$$

This remarkable difference arises from the spectator oxo effect (**2,3**), as discussed below.

Figure 1 shows the bonding for **2'**

2'

as a model of **2**. Here we see that each Mo=O bond has the form of a covalent double bond involving spin pairing of two singly-occupied Mo d orbitals and two singly-occupied O p orbitals. Denoting the Mo=O axis as z and the MoO_2 plane as yz, the Mo-O sigma bond involves Mo d_{z^2} and O p_z orbitals, while the Mo-O pi bond involves Mo d_{xz} and O p_x orbitals . Two such double bonds require four electrons in four orthogonal Mo d orbitals. On the other hand, the two Mo-Cl bonds (modeling single bonds to bridging oxygens in molybdates) involve a large amount of ionic character with some 5s-5p character on the Mo. Thus the Mo^{VI} center in **2** should be best visualized in terms of ionic bonds to the two bridging oxygens, while the Mo=O bonds should be considered as covalent double bonds. The requirement of two singly-occupied d orbitals for each Mo=O bond leads to an O=Mo=O angle of 106° (the π bonds would prefer 90°; the σ bonds, 125°) (3). In addition to the two singly-occupied p orbitals involved in the Mo=O bond, each oxygen has four valence electrons in two nonbonding orbitals (mixture of O 2s and O $2p_y$).

The bonding is quite different when there is only one double bond, as indicated in Figure 2 (the orbitals are for **1'**

1'

where each Cl models the bridging oxygens of **1**). Here there are *two* pi bonds between Mo and O. Thus the Mo has a total of four ionic bonds to the four Cl, two singly-occupied $d\pi$ orbitals (d_{xz} and d_{yz} if the Mo=O axis is z) used in the two Mo=O pi bonds and an empty d_{z^2} orbital. The two Mo=O pi bonds require two singly-occupied $p\pi$ orbitals on the oxygen (p_x and p_y), leaving four electrons in the O 2s and O $2p_z$ orbitals. With two electrons in O $2p_z$ and none in Mo d_{z^2}, we obtain a Lewis base-Lewis acid bond in the sigma system, leading to a net bond involving six electrons (four from oxygen and two from Mo). [As indicated in Figure 2, there is some charge transfer from Mo to O in the π bonds and from O to Mo in the σ bond, but the net description remains a *six- electron* bond.] The result is a partial triple bond or a **super double bond** that is much stronger than the double bond of **2**. Why can't species **2** make two such super double bonds? The requirement is *two* singly-occupied metal pi orbitals (π_x and π_y) for *each* super double bond so that there are just not enough Mo pi orbitals to go around (analogous to the difference between O=C=O with two *double* bonds and C=O with a partial triple bond about twice as strong).

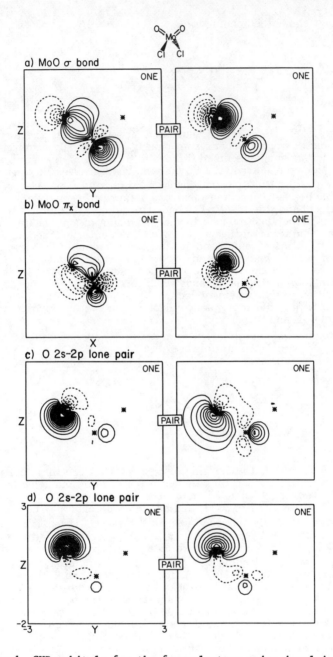

Figure 1. GVB orbitals for the four electron pairs involving the
left Mo=O double in species 2. Dotted contours indicate negative
amplitude. Increments between contours are 0.050 a.u.; the zero
contour is <u>not</u> shown.

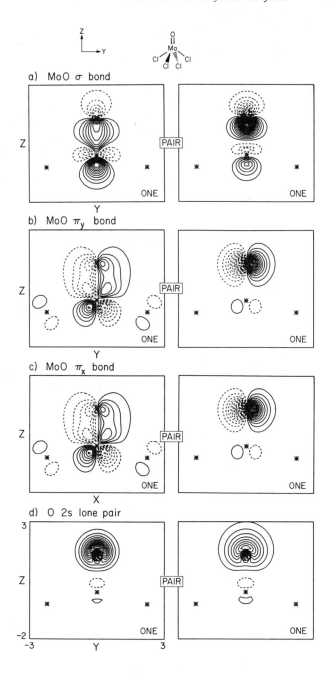

Figure 2. GVB orbitals for the four electron pairs involving the Mo=O super double bond for species 1. (Same plotting conventions as in Figure 1.)

Based on the above arguments, we expect reaction (4) to be much more favorable than reaction (3) because (3) involves attack on a stronger bond. However, there is a *second* equally important factor involved in the difference in reaction enthalpies for (3) and (4). The extra Mo=O bond of **2** would appear to be a spectator to reaction (4), but in fact it helps *promote* the reaction. The reason is that in the product, **5**, this spectator group is free to utilize two Mo dπ orbitals to form a super double bond, whereas in the reactant, **2**, the second Mo=O bond (the one involved directly in the reaction) requires one of these dπ orbitals. Thus the spectator Mo=O bond changes from a double bond to a super double bond when the allyl reacts with the *other* Mo=O bond. The net result then is that reaction (4) is more favored than (3) by 33 kcal.

This spectator oxo promotion is a general effect, so that considering

$$(5)$$

versus

$$(6)$$

the spectator oxo group promotes reaction at the adjacent double-bonded group in (6) by ≈ 33 kcal with respect to (5).

We find a similar but smaller spectator effect of ≈16 kcal for imido groups, promoting reactions such as

$$(7)$$

Thus, as shown in Figure 3, **6** has a (bent) Mo=NH double bond involving two singly-occupied N p orbitals (both perpendicular to the NH bond) paired up to form a sigma bond and a pi bond to the Mo (utilizing an Mo dπ orbital and an Mo dσ orbital). Here the N 2s lone pair is not involved in bonding. However, for **7**, the Mo=NH bond is linear, leading to two Mo-N pi bonds. There is also a Lewis base-Lewis acid sigma bond between N and Mo; however, in this case it involves the N 2s pair (the N $2p_z$ orbital being involved in the NH bond) rather than the pz pair as for O. The result is an Mo-NH super double bond about half as super as the Mo=O super double bond.

In the next section we will use this concept of the *spectator effect* in examining likely pathways for selective oxidation of propene to acrolein.

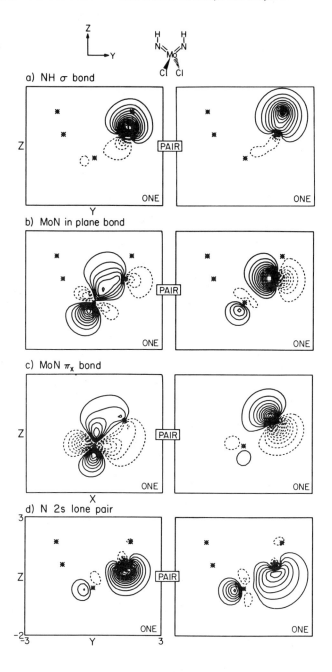

Figure 3. The GVB orbitals for the four electron pairs in the right Mo=NH bond of species 5. (Same plotting conventions as in Figure 1.)

Mechanistic Studies on Selective Oxidation

For selective oxidation (1) and ammoxidation (2) of propene by bismuth molybdates,

i)   it has been well established that the rate-determining step is allyl hydrogen abstraction to yield allyl radical (4,5),

ii)   it is generally acknowledged that the Bi oxide site is involved in this initial step since Bi_2O_3 will abstract the H but does not do oxidation, while MoO_3 will do oxidation of allyl but is ineffective at allyl H abstraction) (6—8), and

iii)   Grasselli, Burrington, and co-workers (9) have proposed a detailed mechanism involving trapping of the allyl radical from (i) at a dioxo Mo site

$$(8)$$

followed by β-hydride elimination to form acrolein

$$(9)$$

As additional evidence for the role of the dioxo site, they carried out experiments using labeled allyl alcohol (rather than propene) that could be most simply interpreted in terms of reaction at a dioxo site followed by interchange of allyl between spectator oxo and alkoxy oxygens (10).

$$(10)$$

In order to explore further the details of these mechanisms, we calculated the energetics for some of the reaction steps.

Concerning trapping of an allyl at a surface molybdate site, we find that the process for the dioxo unit (4) is quite favorable, whereas the process for the monoxo unit (3) is less favorable. This results from the spectator oxo stabilization present in the dioxo unit and agrees with the basic tenet (8) of the Grasselli-Burrington mechanism. However, one-center steps such as (9) for propene and (10) for allyl alcohol involving reaction *with* the spectator oxo group are unfavorable. For example, the step in (9) is endothermic by ~45 kcal and the step in (10) is endothermic by ~25 kcal. As a result, we were led to the idea that the *collections of adjacent dioxo units* are *critical* to the *selective oxidation process.* * *This idea is illustrated in Figure 4 where step b corresponds to (8). Rather than the one- center process (10), we propose that the β- hydride abstraction is by an adjacent dioxo unit* as in Figure 4c. The stabilization due to the spectator oxo group of the second center is critical in keeping the free energy charge negative.

The first question to ask concerning the mechanism in Figure 4 is whether the crystal structure is compatible with adjacent surface dioxo units. Figure 5 shows the (010) surface of α–$Bi_2(MoO_4)_3$ [corresponding to the parent Scheelite structure, $CaWO_4$] (11). Here we see that there are adjacent dioxo centers along the crystal surface.

With these adjacent dioxo units, we would expect allyl transfer steps such as

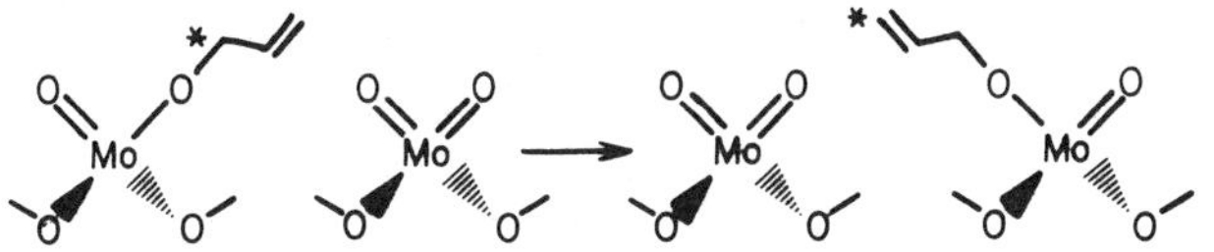

(11)

to be facile. This should lead to equilibration of the C1 and C3 carbons and hence a fixed ratio of

(12)

(a)   (b)

regardless of whether the starting propene is

(13)

or whether the starting allyl alcohol is

---

* This conclusion is buttressed by recent results that were reported after the original theoretical analysis and submission of our paper (10).

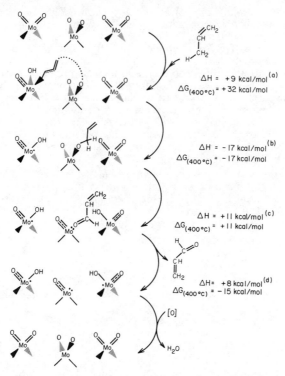

Figure 4.   The multiple dioxo mechanism for selective oxidation.

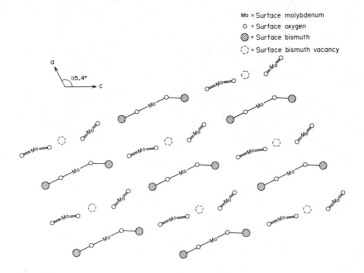

Figure 5. The (010) surface of α-Bi$_2$(MoO$_4$)$_3$ [corresponding to the (001) surface of the parent Scheelite structure, CaWO$_4$].

$$\underset{HO}{\overset{D_2}{\underset{\overset{|}{C}}{C}}}\underset{H}{\overset{}{\diagdown}}\overset{CH_2}{\diagup}\qquad\underset{HO}{\overset{H_2}{\underset{\overset{|}{C}}{C}}}\underset{H}{\overset{}{\diagdown}}\overset{CD_2}{\diagup}\qquad(14)$$

This is consistent with abundant experimental evidence (Grasselli *et al.*) (*9*) *that the ratio of products (12a) to (12b) is 70% to 30%, independent of starting material in (13) or (14).*

It is important to note that $\alpha-Bi_2(MoO_4)_3$ has this active site with adjacent Mo dioxo groups. Such sites do *not* exist in the parent Scheelite structure (*6*) ($CaWO_4$), but with Bi, two adjacent Ca sites out of each six are vacant (indicated by dotted circles in Figure 5), leading to the special active site.

The availability of several adjacent Mo dioxo units suggests the *possibility* of step (a) in Figure 4 in activating the propene. It is generally accepted that this activation occurs on a Bi site since MoO_3 is not effective in abstracting the allyl hydrogen. However, Bi *is essential* to the existence of the chain of Mo dioxo units in Figure 5, and hence it is possible that the *activation* is actually *by a Mo dioxo unit that exists because of the Bi- induced vacancies and not actually on a Bi oxide site.* These ideas are consistent with the fact that allyl iodide is readily converted to acrolein over *bismuth- free* MoO_3, while the conversion of propene to acrolein over MoO_3 is inactive (~2% yield) (*7*). The C-I bond strength as in C_3H_5I is 43.5 kcal/mol, whereas the C-H bond strength as in C_3H_6 is 87.5 kcal/mol (*7*). Therefore it may be postulated that an important differential factor in the breaking of the C-H bond of propene may be the extra stabilization needed that is provided by the Bi-induced configuration of Mo dioxo sites. This is supportive of the idea of nonadditivity of the active sites of the *two* separate oxides, Bi_2O_3 and MoO_3, and the requirement for active sites of Bi, Mo, and O to act together (as a unit) for the conversion of propylene to acrolein (*8*).

The isotope studies *(vide supra)* have been used to establish that a π-allyl species is initially formed and, indeed, scrambling may occur at this stage (*9*). This may well be the case, and we cannot address the issue of σ-allyl versus π-allyl directly since we have not yet studied π-allyl complexes. However, the formation of π-allyl is *not required* in our mechanism. Equilibration via steps as in (*12*) involving only σ-allyls as stable species are mechanistically equivalent to π-allyl. [The energy estimates of Figure 4 do *not* include any special stabilization due to Bi.]

In Figure 4 we quote current estimates of the reaction enthalpies for the various steps in our mechanism. There are numerous uncertainties here (*12*), and with the present uncertainties we certainly cannot say that the mechanism is confirmed. However, each step is approximately thermoneutral and hence the scheme is certainly energetically plausible. Assuming an energy barrier of 10 kcal for the reverse of step (a), Figure 4, leads to allyl H-abstraction as the rate-determining step, in agreement with experiment (the estimated $E_{act} = 19$ kcal would be close to the experimental value of $E_{act} = 22$ kcal).

Summarizing, we agree with Grasselli *et al.* (*9*) that Mo dioxo units are essential in selective oxidation; however, we find that single dioxo units cannot complete the reaction. We propose that selective oxidation requires a collection of at least three adjacent dioxo units to enter into three potentially endothermic reaction steps, facilitating the desired reaction. Indeed, we find for $Bi_2(MoO_4)_3$ that the dioxo units should be ideal for carrying out the sequence of steps involved in selective oxidation. Using this idea that multiple dioxo sites are required, one might be able to develop strategies for promoter additives and for preparative techniques based on enhancing the probability of such sites. Perhaps the role of such sites could be tested using bidentate or multidentate ligands, e.g.,

that would bond only to specific configurations.

Ammoxidation

In mechanistic studies of ammoxidation by Bi-molybdates, Grasselli et al. (9) have suggested a critical role of bis-imido sites,

$$\underset{\overset{\displaystyle Mo}{\diagup \; \diagdown}}{HN \diagdown \quad \diagup NH}$$

Indeed, assuming an active site analogous to that in Figure 5 but with oxo groups replaced by imido groups leads to a scheme analogous to Figure 4, where spectator imido groups play the role of spectator oxo groups in steps (a), (b), and (c) of Figure 4.

At intermediate NH_3 pressures, oxo-imido species

$$\underset{\overset{\displaystyle Mo}{\diagup \; \diagdown}}{O \diagdown \quad \diagup NH}$$

are probably present on the surface. Because of the difference between spectator-oxo and spectator-imido effects, reaction at the N is greatly favored over reaction at the oxygen. Thus, for allyl trapping by N, ΔH = -30 kcal, whereas for allyl trapping by O, ΔH = -20 kcal [see (ii)a, (ii)b]. Thus, in thermodynamic equilibrium, only reaction at the N would be observed. Kinetic data of Grasselli et al. (10) have been interpreted to indicate that allylic N insertion is approximately three times *faster* than allylic O insertion, i.e., $\dfrac{K(\text{acrylonitrile})}{K(\text{acrolein})} \sim 3$. Comparing the theory with experiment suggests kinetic control with an activation energy of $\Delta H^{\ddagger} = 0$ for (ii)a and $\Delta H^{\ddagger} \approx 2$ kcal for (ii)b. We are currently in the process of examining the energetics for various possible catalytic sequences involved in ammoxidation.

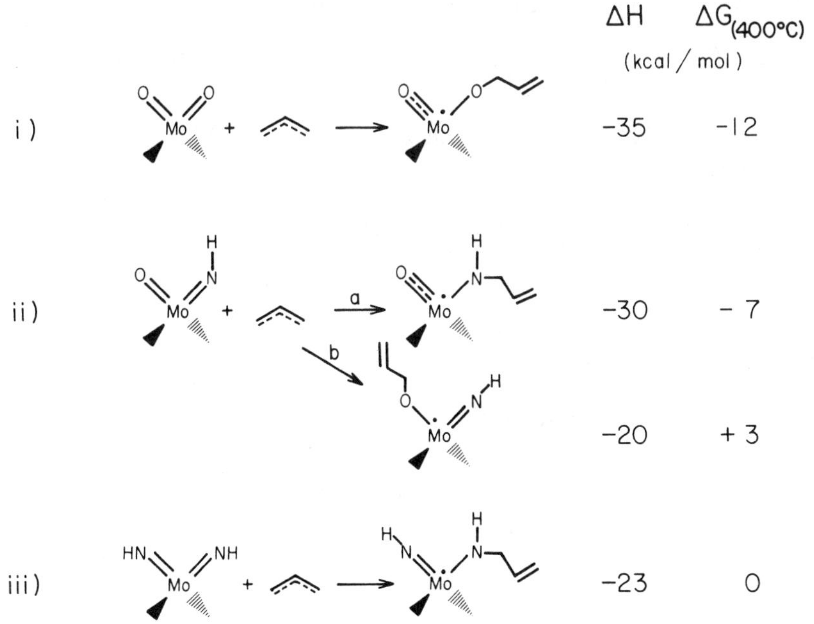

	ΔH	$\Delta G_{(400°C)}$
	(kcal / mol)	
i)	-35	-12
ii)	-30	-7
	-20	$+3$
iii)	-23	0

Acknowledgments

We would like to thank Bob Grasselli, Jim Burrington, and Keith Hall for spirited discussions of various aspects of chemistry on molybdates. We also gratefully acknowledge partial support of this work from the Department of Energy (under a contract with the Jet Propulsion Laboratory) and the Donors of the Petroleum Research Fund of the American Chemical Society. One of the authors (JNA) wishes to acknowledge support in the form of a fellowship from the Fannie and John Hertz Foundation. This chapter is Contribution No. 7101 from the Arthur Amos Noyes Laboratory of Chemical Physics.

Literature Cited

1. Kihlborg, L. *Arkiv Kemi* 1963, 21, 357-364.
2. Rappé, A. K.; Goddard III, W. A. *Nature* 1980, 285, 311; *J. Am. Chem. Soc.* 1980, 102, 5114; *ibid.* 1982, 104, 448.
3. Rappé, A. K.; Goddard III, W. A. *J. Am. Chem. Soc.* 1982, 104, 3287.
4. (a) Keulks, G. W.; Krenzke, L. S.; Noterman, T. M. *Adv. Catal.* 1978, 27, 183; (b) Brown, F. R.; Makovsky, L. E.; Rhee, K. H., *J. Catal.* 1977, 50, 162, 385; (c) Patterson, T. A.; Carver, J. C.; Leyden, D. E.; Hercules, D. M., *J. Phys. Chem.* 1976, 80, 1700; (d) Matsuura, I.; Schut, R.; Kirakawa, K. *J. Catal.* 1980, 63, 152; (e) Holm, V.; Clark, A. *ibid.* 305 (1968); (f) McCain, C. C.; Gough, G.; Godin, G. W. *Nature* 1963, 198, 989; (g) Wragg, R. D.; Ashmore, P. G.; Hockey, J. A. *J. Catal.* 1971, 22, 49; (h) Portefaix, J. L.; Figueras, F.; Forissier, M. *ibid.* 1980, 63, 307.

5.   Otsubo, T.; Miura, H.; Morikawa, Y.; Shirasaki, T. *J. Catal.* 1975, 36, 240.
6.   Sleight, A. W. In "Materials Science Series - Advanced Materials in Catalysis"; Burton, J. J., Ed.; Academic Press: New York, 1977; p. 181.
7.   Grzybowska, B; Haber, J.; Janas, J. *J. Catal.* 1977, 49, 150-153.
8.   Martin, W.; Lunsford, J. H. *J. Am. Chem. Soc.* 1981, 103, 3728-3732.
9.   Grasselli, R. K.; Burrington, J. D. *Advan. Catal.* 1981, 30, 133; Burrington, J. D.; Kartisek, C. T.; Grasselli, R. K. *J. Catal.* 1980, 63, 235; *ibid.* 1982, 75, 225.
10.   Burrington, J. D.; Kartisek, C.T.; Grasselli, R. K. *J. Catal.* 1984, 87, 363-380.
11.   van der Elzen, A. F.; Rieck, G. D. *Acta Cryst.* B 1972, 29, 2433.
12.   Our current estimates for σ-allyl are currently based on calculations for CH₃ with corrections to allyl based on Benson-type estimates (13). In addition, we have not yet estimated the stability of the π-allyl species. Most important, we have not calculated the transition states and energy barriers for any of the steps.
13.   Benson, S. W. "Thermochemical Kinetics"; Wiley: New York, 1976.

RECEIVED October 4, 1984

Thermodynamic and Structural Aspects of Interfacial Effects in Mild Oxidation Catalysts

PIERRE COURTINE

Université de Technologie de Compiègne, B.P. 233, 60206 Compiègne Cedex, France

This paper presents a comprehensive description of mild oxidation catalysts containing one or more oxides or oxysalts and analyzes the thermodynamic and structural influence of interfacial effects between the constituent solid phases on activity and selectivity ($\underline{1}$).

Early attempts to correlate catalytic performance with solid state properties ($\underline{2-11}$) emphasized the intrinsic properties of constituent transition metal ions, such as the presene of double metal-oxygen bonds, the nature of surface oxygen species, and the acidity of the active sites, as if activity and selectivity were a complex function of the chemical composition of the catalyst ($\underline{12-17}$).

The role of the solid phase was first developed from electronic theory ($\underline{10}$) and then from the concept of valence and stoichiometry control ($\underline{18}$, $\underline{19}$). Recent experimental results have shown that one thermodynamic phase, and not a simple gathering or "sea" of ions at the surface of the active solid, can be responsible for catalytic activity and selectivity.

The first real characterization of active phases has been made for the high temperature polymorph of $CoMoO_4$ (called (a) by us and later (b) by other authors) in the selective oxidation of butane to butadiene ($\underline{20,21}$), as well as for USB_3O_{10} ($\underline{22}$) and bismuth molybdates ($\underline{23,24}$) for oxidation, and ammoxidation of propylene. Additional examples include solid solutions such as $(Mo_xV_{1-x})_2O_5$ (with $0<x<0.3$) for benzene conversion to maleic anhydride ($\underline{25,26}$) and the solid solution up to 15% of Sb_2O_5 in the $SnO_2-Sb_2O_5$ system for propylene oxidation to acrolein ($\underline{14,27}$).

In cases where the oxidation state of the solid changes during the reaction, the active solid state phase present at steady state must be characterized. For example, the reduced vanadium phosphate phase forms during the oxidation of butene to maleic anhydride ($\underline{28,29}$). And for supported catalysts such as V_2O_5/TiO_2 anatase, reduction to $VO_{2.17}$ (i.e. V_6O_{13}) occurs during the oxidation of

0097–6156/85/0279–0037$06.00/0

o-xylene (15,30). Finally, it has been found that the addition of
several other phases to a known active phase such as bismuth
molybdate improves its performance in the ammoxidation of propylene
by producing a multicomponent molybdate (MCM) catalyst (31).

The question arises, why do bi- or multi-phasic catalysts
generally show better activity and selectivity than the active
phase alone? The aim of this paper is to answer this question by
exploring the role of interfacial effects. We shall examine first
how the thermodynamic and structural properties of one phase
influence its interactions, not only with the gaseous reactants,
but also with coexisting solid phases as a result of its bulk,
surface, and defect structure. We will also examine the conditions
necessary for these interactions and set up a structural
classification of the main components of mild oxidation catalysts.
This will lead finally to a discussion of the role of interfacial
effects in catalyst performance using some illustrative examples.

Thermodynamic and Structural Properties of Single Phase Catalysts

Interdependence of Properties

We have already noted (28,29) that a single active phase possesses
a set of interdependent properties, and that any correlation
between one property and catalytic performance cannot adequately
explain catalytic behavior. Scheme I illustrates this using V_2O_5 as
an example. However, the nature of the active site, which can
contain a normal lattice ion, a group of ions, or extended or point
defects, depends on its solid state environment. The structure of
the host matrix can greatly influence the chemical and physical
properties of the site.

Thermodynamic Properties

In most studies it has been stated that the hydrocarbon is oxidized
by lattice oxygen of the oxidized form of the catalyst, KO.
Subsequently, the reaction of the reduced form, K, with O_2
regenerates the initial state according to the well-known Mars and
van Krevelen mechanism (32):

$$A + KO \rightarrow AO + K \qquad\qquad r_{red} \qquad\qquad (1)$$

$$K + 1/2\ O_2 \rightarrow KO \qquad\qquad r_{ox} \qquad\qquad (2)$$

This mechanism, though not always applicable (33), describes
the importance of oxygen donation by the active phase in selective
oxidation catalyst (34).

Kinetic studies, structural considerations, isotopic oxygen
exchange experiments, and equilibrium adsorption measurements have
allowed some authors to establish correlations between selectivity
and the change of standard enthalpy or free energy of the redox
process. Although these correlations concern the bulk and not the

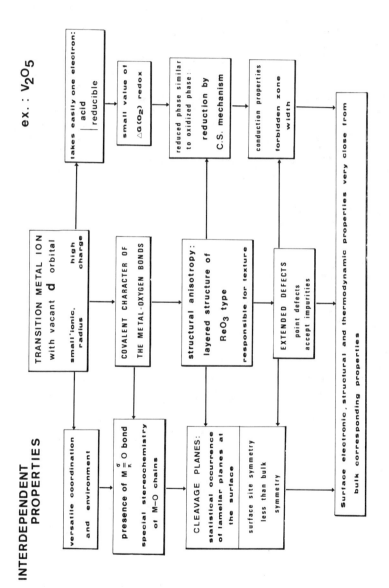

Scheme I. Relationship between interdependent properties and catalytic performance of V_2O_5.

surface of the catalyst ($\underline{12}$), it was found in certain cases that the surface structures bear some relations with the bulk structure ($\underline{28,29,35-41}$), particularly in the case of layered solids.

Thermodynamic aspects must be taken into acount, not only concerning the reaction itself, but also with regard to the solid state. Most of the mild oxidation catalysts exhibit either a low melting point (V_2O_5, MoO_3...), or first or second order polymorphic transitions near the reaction temperature, as in the case of $M^{2+}MoO_4$ (M^{2+} = Co^{2+}, Fe^{2+}, Ni^{2+}) ($\underline{20, 21}$), or the $\aleph$, $\aleph'$, $\aleph"$ modifications of Bi_2MoO_6 ($\underline{42}$). Their Tammann temperatures are low. As a consequence, these lattices are in a metastable state, near these temperatures. If two structures are closely related, as we shall see in the next section, the two solids will be able to form either coherent interfaces or solid solutions. In these cases, "hybrid crystals," in which microdomains of both phases coexist, can be formed according to UBBELOHDE's theory ($\underline{43}$), which considers strain energy ε and internal surface energy η ($\underline{44}$). The free enthalpy of a domain (1) in a matrix of structure (2) is given by:

$$G_1 = f(p, V, T, \varepsilon_{12}, \eta_{12}...)$$

and the classical phase rule must be modified such that:

$$v = c + 2 - \emptyset + \Sigma \pi$$

where $\Sigma \Pi$ represents the "history of the sample," including the additional parameters due to coherence ($\underline{44}$). In such systems, hysteresis is expected to occur since ϵ_{12} (strains due to microdomains of (1) in structure (2) is different from ϵ_{21} (microdomains of (2) in structure (1)). This has been effectively observed by Bordes ($\underline{28,29}$) in thermogravimetric measurements of the equilibrium between $\overline{VOPO_4}$ (α or β) and $(VO)_2P_2O_7$, the oxidized and reduced forms, respectively, of selective catalysts for butene oxidation.

Several conclusions can be drawn up from these observations:

i) On both sides of the coherent boundaries the active ions are in an excited state as compared to the normal ions in the host lattice.

ii) Whereas in a reduction process in which strain produced at such interfaces is compensated mainly by mechanical relaxation (e.g., C.S. planes), the lattice of the catalyst at the steady state prefers to release a part of its excess energy towards the reactants and recovers it when reoxidation occurs ($\underline{29}$).

iii) In supported, bi- or multi-component catalysts, the existence of coherent interfaces, even in a very small area, constitutes a necessary condition for any mass, energy, or electron transfer due to the redox mechanism to an adjacent phase ($\underline{45}$).

Structural Properties of Some Oxide Catalysts

Mild oxidation catalysts are generally composed of phases having roughly two types of structures: i) oxides with the ReO_3-like structural, and ii) oxy-salts having a transition metal ion coordinated to oxygen atoms in a molecular complex.

Simple Oxides Having ReO_3-like Structures

These oxides contain d^0 and d^1 transition metal ions with a high formal valence state which are linked to oxygens by strong covalent bonds. Their framework is depicted as an array of distorted tetrahedra, pentahedra, or octahedra as in the "Wadsley phases" belonging to the (V-O), (Mo-O), (V-Mo-O), (W-O) systems (44,46). In an idealized octahedral arrangement, one of the axial metal-oxygen bonds is short, and has substantial $\sigma-\pi$ bond character, and is usually adjacent to a relatively long, metal oxygen σ bond. This feature is also present in (V-P-O) system (28,29).

These structural features play important roles in several catalytic systems (Figure 1).

1) Anisotropy in the oxygen environment lowers the symmetry of crystals with layered structures such as V_2O_5, MoO_3 and even the oxysalts $\alpha-$, $\beta-$,$\delta-VOPO_4$ (28,29,47).

2) Since the morphology of the catalyst depends on structure, it can be either enhanced or modified by the method of preparation (28,47). Since in most cases, the specific area of mild oxidation catalysts is small (2-30 $m^2.g^{-1}$), the external aspects of the microcrystals (200 to 5,000 A) and main cleavage (hkl) planes are easily observed by scanning and transmission electron microscopy. Sheets are frequently encountered for ReO_3-like structures, and (010) MoO_3, (001) $\alpha-VOPO_4$, (010)$\delta-(VO)_2P_2O_7$ (47) are the more often observed planes.

The effect of these structural features on catalysis are two fold:

a) Specific surface patterns may be selective for different mild oxidation reaction as in the case of so-called "structure-sensitive reactions" (47-53). For example, (010) MoO_3 has been shown to catalyze the dehydrogenation of alcohols whereas (001) is selective for dehydration only.

b) The surface patterns can form coherent interfaces in supported catalysts (V_2O_5/TiO_2 anatase), and at steady state, between the oxidized and reduced form of the solid ($VOPO_4-(VO)_2P_2O_7$, $V_2O_5-V_6O_{13}$).

3) Another important property is the presence of extended defects in ReO_3 structures and the possibility of nucleation of crystallographic shear planes (C.S.) during reduction (44,46,54-56). By sharing edges of octahedra, C.S. planes facilitate the transport of oxygen through the lattice (57) during catalysis and influence the rate of reoxidation of the superficial sites as well as the rate of oxidation of the substrate (28,29,55,57). During such reactions, coherent intergrowth domains may be formed, as illustrated by the reduction of $H-Nb_2O_5$ (54) (Figure 2) and in the reduction of (α or β) $VOPO_4$ in $(VO)_2P_2O_7$ (Figure 3).

Oxysalts

The second important structural series of mild oxidation catalysts are the oxysalts which contain discrete metal-oxygen molecular complexes, $(MO)_n$. Two types of metal-oxygen bonds are usually present; strong metal oxygen covalent bonds such as (MoO_4^{2-}) or $(UO_2)^{2+}$, and ionic bonds such as Fe-O or Co-O. Considerable work

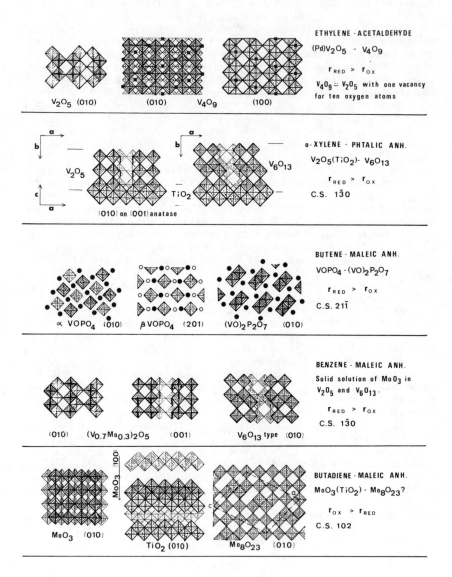

Figure 1.   ReO₃ matrices as mild oxidation catalysts.

has been done on the structural properties, preparation, reactivity
(58) and catalytic properties (20,23,24,31,58-60) of metal
molybdates.

Metal molybdates can be divided into three general groups:

a) Koechlinite structure. Previous workers have described
both the structure of Bi_2MoO_6, which contains successive layers of
(Bi_2O_2) pyramids and (MoO_2) corner-shared octahedra (61,62) and the
role of this structure in the mechanism of reduction and
reoxidation. The structural similarity between the (010) cleavage
planes of Bi_2MoO_6 and MoO_3 may account for some of their catalytic
properties since both bidimensional patterns are made of corner
sharing octahedra (ReO_3 structure). As will be discussed below, the
solid-solid reactions occuring between MoO_3 and Bi_2MoO_6 in
multicomponent catalysts would be facilitated by the existence of
coherent interfaces.

b) $A_xB_yO_z$ structures such as $B-CoMoO_4$, Te-Mo-O or U-Sb-O
systems. In the case of $\beta-CoMoO_4$, which is isostructural with
$MnMoO_4$, (63), the main cleavage plane is (110). This plane is
strickingly similar to the (010) plane of $(VO)_2P_2O_7$ (Figure 4).
The structure of (100) $\alpha-Te_2MoO_7$ is also related to this latter
compound and to the Bi_2MoO_6 structure (64). USb_3O_{10} contains
layers of U-Sb-O in which the oxygen atoms are selective in
ammoxidation of propylene while the other oxygen between the layers
are inactive (22,31).

c) Scheelite-derived structures. Examples of these structures are
$Fe_2(MoO_4)_3$ and $Bi_2(MoO_4)_3$, which are related to $CaWO_4$ (59) and
which can be extended to the molybdates $MMoO_4$ just mentioned as
shown in Figure 5.

All these compounds are thought to possess neither non-
stoichimetric reduced phases, nor extended defects, but rather
point defects, mainly cation and anion vacancies. The latter are
known to produce a considerable mobility in the lattice (58). In
these compounds, the defect structures readily account for the
rapid reoxidation of the bulk by rapid diffusion of oxygen and
electron transfer, as well as for the ability of the host matrix to
form coherent interfaces.

Discussion

Various interfacial effects take place whenever two solid phases
(or microdomains in one phase) come in contact and coherent phase
boundaries form as a result of small crystallographic misfits
between the phases. The role of these interfacial effects in
catalytic processes can be understood by examination of the
following specific cases of the initial relative rates of oxidation
r_{ox} and reduction r_{red} in the Mars and Van Krevelen mechanism.

$r_{red} \gg r_{ox}$

Under catalytic reaction conditions, the fresh catalyst can be
reduced by the reactants, and a reduced phase of the catalyst can
crystallize. This is the case for the pseudomorphic layered form
$\gamma(VO)_2P_2O_7$ obtained by the topotactic decomposition of

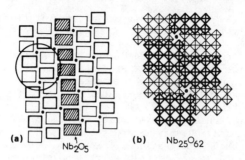

Figure 2. Example of coherent intergrowth with extended defects: (A) Lamellae of H-Nb$_2$O$_5$ (blocks 5x3 and 4x3) in Nb$_{25}$O$_{62}$ (4x3)$_2$,(B) Details of Nb$_{25}$O$_{62}$ (circle in a))

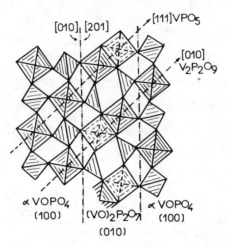

Figure 3. Coherent interface between slabs of (100) VOPO$_4$ and of (010) (VO)$_2$P$_2$O$_7$, along 100 and 201, respectively.

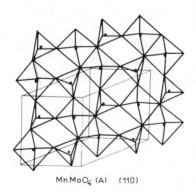

Figure 4. Pattern of the surface lattice plane common to MnMoO$_4$(a), CoMoO$_4$(a) and close to (010) (VO)$_2$P$_2$O$_7$.

$VOHPO_4, 0.5H_2O$ (47), which is active and selective in butane oxidation to maleic anhydride. The cleavage plane (010) of $(VO)_2P_2O_7$ (Figure 3) possesses vanadyl groups situated perpendicular from the surface as well as four different crystallographic vanadium sites (65,66). In this case, there is no need to consider interfacial effects since the active catalyst acts as a mixed-valence oxysalt (67), due to the neighboring vanadium ions in the formal mean valence states 3.7 to 4.4.

$$r_{red} \simeq r_{ox}$$

The situation becomes different in this instance since the evolution of the active phase corresponds to a biphasic system at steady state.

In the catalyst α, $\beta-VOPO_4-(VO)_2P_2O_7$ for butene to maleic anhydride, the reduced form is progressively nucleated by a C.S. mechanism during reduction of $\beta-VOPO_4$ (Figure 6) (28). This confers various morphologies to the crystals of $(VO)_2P_2O_7$, which are different than the layered "γ" form. Consequently, at steady state, microdomains of $VOPO_4$ are included in a large matrix of $(VO)_2P_2O_7$ and are in coherent contact. Streaks observed in TEM indicate disordered C.S. planes, and the hysteresis we obtained in microgravimetric equilibrium measurements show the existence of coherent microdomains under steady state conditions (28,29). As explained above, on both sides of coherent boundaries, the coordination of the vanadium atoms is more or less distorted and their electronic state more excited than in the normal matrix. Consequently, they are assumed to play the role of active and selective sites by a concerted mechanism (28) similar to that presented for USb_3O_{10} (22).

Another related example is the WO_3-based catalyst in oxidation of propylene to acrolein (57) for which the presence of C.S. planes was correlated with selectivity.

The well-studied catalyst V_2O_5 supported on TiO_2 anatase is more selective in the oxidation of o-xylene to pthalic anhydride (15,30,68-74) than V_2O_5 alone or supported on $\gamma-Al_2O_3$ (70). The maximum selectivity is generally reached at ⅃7-15 mole % of active phase although other compositions below this optimum composition have been shown effective (68). Two sets of results have led us to assume that the cleavage plane (010) of V_2O_5 can be anchored to the (001), (100) (53) and (011) planes of anatase (75) with a very small misfit:
a) Under N_2 at 600°C, V_2O_5 is reduced topotactically into successive suboxides (15b) by a C.S. mechanism starting from the coherent interface between V_2O_5 and TiO_2 lattices (53), whereas no reduction occurs for unsupported V_2O_5. Simultaneously anatase transforms into rutile with a very low activation energy (15 kcal/mole) (76) whereas in the absence of V_2O_5 this transformation occurs only above 850°C with a high activation energy (150 kcal/mole) (77).

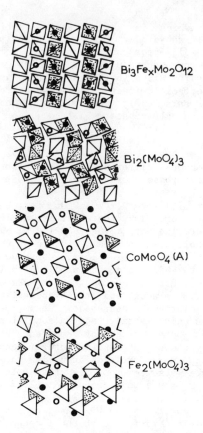

Figure 5. Structural relations between $Bi_3Fe_xMo_2O_{12}$, $Bi_2(MoO_4)_3$, $Fe_2(MoO_4)_3$ and $CoMoO_4$ (a): examples of oxysalts belonging to the second class of catalysts.

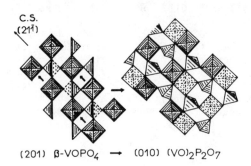

Figure 6. C.S. mechanism in the transformation of $VOPO_4$-$(VO)_2P_2O_7$.

This synergetic effect can be explained by the presence of coherent boundaries whose first consequence is to lower the potential barrier for electron transfer between the n-semiconductor TiO_2 and V_2O_5. The resulting elastic strains on both sides of the boundaries allow the reduction of V_2O_5 and the transmission of this cooperative transformation to anatase which becomes rutile. When this last transformation is complete, the interface vanishes and the reaction stops.

b) Under catalytic conditions, the same reduction of V_2O_5 occurs but begins at the surface while TiO_2 remains anatase since the temperature does not exceed 450°C. The coherent interface is not destroyed since the successive suboxides formed by C.S. mechanism have similar cleavage patterns to V_2O_5 (76). When the steady state is reached under o-xylene and air, the reduction stops near V_6O_{13} (30), corresponding to an optimal value of the misfit and of the strain factor (45).

A way to verify this interpretation was to expect the same phenomena if TiO_2 was replaced by another support having the same crystallographic configuration. We have shown indeed that the two experiments a) and b) could be repeated when V_2O_5 is supported on $AlNbO_4$ or $GaNbO_4$ (78,79) (Figure 7). The anatase to rutile transformation also occurs for MoO_3/TiO_2 and $CoMoO_4/TiO_2$ (79,80).

Other examples found in the literature can be interpreted in the same way. In the ammoxidation of 3-picoline (81) on V_2O_5/TiO_2 (rutile), the maximum activity and selectivity are obtained in nearly the same composition range as for oxidation of o-xylene. The catalyst was prepared by heating V_2O_5 and TiO_2 anatase at 1150°C with subsequent reduction by H_2 at 450°C for one hour. Under these conditions, anatase transforms into rutile and V_2O_5 melts. On cooling, epitaxial relationships are formed between (010) V_2O_5 and (110) of rutile. Since the electronic properties of rutile are very close to anatase, the same explanations apply.

Similar interfacial effects cannot be expected for $V_2O_5-Al_2O_3$ catalysts. The number of exposed active V=O species differ depending upon the nature of the support (70), and the development of the layered (010) cleavage plane of V_2O_5 is not enhanced on Al_2O_3 as it is by the presence of coherent boundaries with TiO_2 anatase (Figure 8).

Recent studies (68-71) have centered on vanadium oxide monolayers supported on anatase. EXAFS and XANES experiments have shown that tetrahedral vanadium species are anchored with an intrinsic disorder on the surface of anatase. Although it seems that short range matching exists with some crystalline faces of the support (which are not defined), the structural pattern of this monolayer was found to be different from the cleavage plane (010) of crystallized V_2O_5. This result could be expected by taking into account the large influence of the surface potential of anatase on the distortion of the vanadium environment, with the notion of cleavage plane becoming meaningless since it depends obviously on the existence of a bulk. The monolayer seems to be a borderline case as compared to our experiments where higher concentrations of crystalline V_2O_5 were used. In order to define more closely the configuration of the active sites in the interfacial zone, further

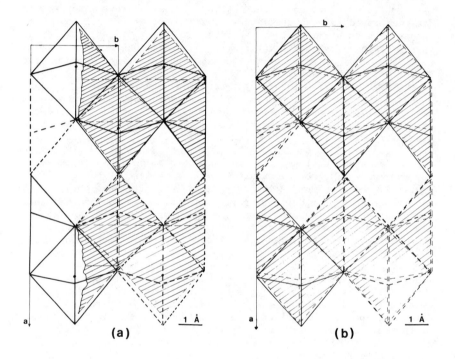

Figure 7. Crystallographic fit between (010) $AlNbO_4$ and a) (010)V_2O_5 and b) (010) V_6O_{13}. Reproduced with permission from Ref. 78. Copyright 1980, Academic Press, Inc.

work is needed on the preparation and morphology of V_2O_5-TiO_2 samples.

$$\frac{r_{red}}{} \ll r_{ox}$$

a) Single Catalytic Active Phases

Except when the rate limiting step is the diffusion of oxygen through the bulk at lower temperature (58,76,83,84), the catalytic phase is nearly fully oxidized. Molybdenum oxide catalyst for the dehydrogenation of alcohols (49,50), cobalt molybdate for dehydrogenation of butane (20,21) and bismuth molybdates belong to this category. The proposed mechanisms (58) are based on anionic and cationic defects which allow bulk migration of oxygen ions through the lattice and electronic transfer. Recently, conductivity measurements have been performed on a series of catalysts MoO_3, $CoMoO_4$, $Fe_2(MoO_4)_3$ and $Bi_2(MoO_4)_3$ (85, 86) for which a compensation effect is observed. Using a hopping model, this effect is due to a distribution of the activation energies corresponding to a distribution of the distances of jumps specific to each kind of lattice. We can, therefore, assume that the electron transfer proceeds in these phases by an activated bound small polaron mechanism.

There is no need in these cases to relate the selectivity to the presence of C.S. planes, which are difficult to conceive in these kinds of ionocovalent structures. Recent HREM experiments made on (Te-Mo-O) system, catalyst for ammoxidation of propylene, have not revealed the presence of crystallographic shear planes (87).

b) Supported Multicomponent Molybdates

The improvement of the catalytic performances of MCM catalysts can only be interpreted by means of interfacial effects between their components. XRD and TEM experiments performed on $M^{II}_{n-x}Fe_xBiMo_{12}O_y$ (M^{II} = Co, Ni, Mg ; $0 < x < 4$) (60) before and after reaction (58-60,88) have shown the presence of the pure

individual molybdates, which are incorporated by a special preparation. These phases can undergo various transformations depending on the gaseous environment. It has been shown that under reducing conditions, bismuth molybdates can exchange molybdenum oxide by the reactions (89-92)

$$Bi_2(MoO_4)_3 \rightleftharpoons Bi_2MoO_6 + 2MoO_2 + O_2$$
$$Bi_2Mo_2O_9 \rightarrow Bi_2MoO_6 + MoO_2 + 1/2O_2$$

whereas a disproportionation of $Bi_2Mo_2O_9$ is observed under air (90,93)

$$2\ Bi_2Mo_2O_9 \rightarrow Bi_2(MoO_4)_3 + Bi_2MoO_6$$

below 400°C and above 550°C where ४' is nucleated instead of ४. At low temperature the disproportionation looks like a spinodal

decomposition yielding modulated structures (<u>94,95</u>). $Fe_2(MoO_4)_3$
can be reduced also, in two steps (<u>96</u>):

$$Fe_2(MoO_4)_3 \rightleftharpoons 2 \ FeMoO_4 + MoO_3 + 1/2 \ O_2$$

$$MoO_3 \rightarrow MoO_2 + 1/2 \ O_2$$

According to the model presented (<u>60</u>), the catalyst particle should
contain an inner core consisting of a β-MMoO$_4$ structure, covered by
$Fe_2(MoO_4)_3$, with the outer shell of 100 Å thickness containing
Bi_2MoO_6 and $Bi_2(MoO_4)_3$. In used catalysts, all the results confirm
the presence of β-FeMoO$_4$.

Additional facts lead to the conjecture that coherent
interfaces are formed in such MCM catalysts :

(i) The isostructural β-MMoO$_4$ molybdates (M = Fe, Co, Ni, Mg)
can form extended solid solutions (<u>97</u>), or at least coherent
interfaces depending on the temperature (<u>86</u>).

(ii) Microgravimetric experiments performed at 750°C under
nitrogen show that the reduction of $Fe_2(MoO_4)_3$ alone leading to
β-FeMoO$_4$ is slower than when $Fe_2(MoO_4)_3$ is supported by β-MgMoO$_4$
(<u>85,86</u>). The small crystallographic misfit suggests the probable
existence of coherent boundaries (Fig. 9).

(iii) The same conclusions can be drawn up for the interfaces
between $Bi_2(MoO_4)_3$ and $Fe_2(MoO_4)_3$ which have very close structures.

Therefore, coherent boundaries should be present between each
shell of the catalyst particle, with the following consequences for
the catalysis namely, setting up of a metastable state of the
catalyst, and a reduction in the energy barrier for electron
transfer by the redox mechanism proposed for MCM catalysts (<u>98-
100</u>).

c) $CoMoO_4/MoO_3/TiO_2$ catalyst

Since β-CoMoO$_4$ is a catalyst for butane oxidation to
butadiene, and MoO_3/TiO_2 anatase for butadiene-maleic anhydride
(<u>28,101</u>), and since we know that (001) MoO_3 fits well with (110)
β-CoMoO$_4$, we have recently attempted to use multicomponent
$CoMoO_4/MoO_3,TiO_2$ in the direct oxidation of butane to maleic
anhydride (<u>80</u>). We have found that ·10 % by weight of CoMoO$_4$
supported by 90 % of (0.3 MoO_3 + 0.7 TiO_2) is effective in this
reaction, giving substantial yields of maleic anhydride (Figure
10). The butadiene formed does not desorb and reacts at the
surface of the catalyst according to the known mechanism for the
oxidation of butene to maleic anhydride. The notion of interfacial
effects has therefore been applied, with some success, to finding a
new catalyst.

Conclusion

To sum up we have shown that :

1. Mild oxidation catalysts are made up of phases, the selectivity
of which depends directly or indirectly on their thermodynamic and
structural properties.

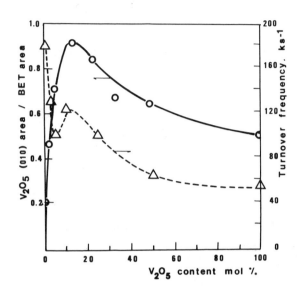

Figure 8. Promoting effect of TiO$_2$ anatase on the catalytic properties of V$_2$O$_5$. Reproduced from Ref. 70. Copyright 1979, 1980, 1981 American Chemical Society.

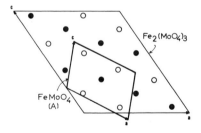

Figure 9. Crystallographic fit between (010) Fe$_2$(MoO$_4$)$_3$ (larger cell) and FeMoO$_4$(a) (smaller cell) (only the ion atoms are shown).

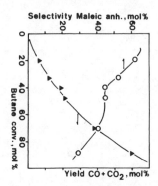

Figure 10. Catalytic results obtained with $CoMoO_4/MoO_3$, TiO_2 in the oxidation of butane to maleic anhydride (2% C_4H_{10}, = 1 sec.)

2. This allows the classification of these phases into two groups: ReO_3^- like structures and oxysalts such as molybdates. The first one consists of host matrices having extended defects; the second one, of point defects such as anionic and cationic vacancies.

3. Depending on the kinetic conditions of the catalytic reaction, they can act either as a single phase, or as a biphasic or multiphasic system.
    A single phase requires specific conditions to be highly active and selective. It has to possess a thermodynamically possible redox with regard to the reactants. It must be a mixed valence phase at the steady state. It must possess a particular morphology exhibiting statistically the selective lattice planes at the surface and must have optimal semiconductivity by "small polarons".
    The second case, corresponding to the first class, results from an evolution of the catalyst state to a biphasic system through nucleation of C.S. planes. Coherent interfaces must exist, near which active sites are in an excited state and in a metastable coordination as compared to the normal sites of the host matrices.

4. It is therefore not surprising that supported or multicomponent catalysts with complex formulae exhibit a better selectivity than the active phases alone. These phases can undergo solid-solid transformations near the reaction temperature, which are facilitated by the coherent interfaces. Despite the lack of quantitative knowledge on the values of interfacial strain factors for the free energy of the system, comparative studies could be performed using the present classification.
    This general rule about selectivity in mild oxidation catalysis, based on the very numerous experimental data gathered in this field, should be of substantial help for the characterization of selective catalysts and for the improvement of their preparation.

Acknowledgments

Thanks are due to Dr. E. Bordes for valuable discussions and practical assistance.

Literature Cited

1. Courtine, P. ACS Symposium, Las Vegas (1982)
2. Germain, J. E. Intr. Sci. Chem. Reports 1972, 6, 101.
3. Naito, S.; Kondo, T.; Ichikawa, M.;Tamaru, H. J. Phys. Chem. 1972, 76, 2184.
4. Whitney A. G.; Gay, I. D. J. Catal. 1972, 25, 123.
5. Shvets V. A.; Kasanskii, V. B. J. Catal. 1972, 25, 123.
6. Krylov, O. V. In "Catalysis by Non-Metals", Academic Press, 1970.
7. Krylov, O. V.; Parikkii, G. B.; Spiridonov, K. N. J. Catal. 1971, 23, 301.
8. Schuit, G. C. A.; Chim. Ind. 1969, 51, 1307.
9. Balandine A. A.,et al. Adv. Catal. 1969, 19, 1.
10. Wolkenstein, T. H. In "Theorie electronique de la catalyse", Masson, 1968.
11. Dowden, D. A. Catal. Rev. 1971, 5, 1.
12. Germain, J. E. J. Chim. Phys. 1973, 70, 1048.
13. Ai M.; Suzuki, S. Bull. Chem. Soc. Jap. 1974, 47, 3074.
14. Haber, J. Int. Chem. Eng. 1975, 15, 21.
15. a) Hucknall, D. J. In "Selective Oxidation of Hydrocarbons", Academic Press, 1974.
b) Cole, D. J.; Cullis, C. F.; Hucknall, D. J. J. C. S. Faraday I, 1976, 72, 2185.
c) Hauffe, K.; Raveling, H. Ber. Bunsenges. Phys. Chem. 1980, 84, 912
16. Trifiro, F.; Notarbatolo, S.; Pasquon, I. J. Catal. 1971, 22, 324.
17. Mitchell, P. Ch.; Trifiro, F. J. Chem. Soc. 1970, A 3183.
18. Stone, F. S. J. Sol. State Chem. 1975, 12, 271.
19. Cimino, A.; Indovina, V.; Pepe, F.; Stone, F. S. Gazz. Chim. Ital. 1973, 103, 935.
20. Daumas, J. C. Thesis, Paris, 1970.
21. Courtine, P.; Cord, P. P.; Pannetier, G; Daumas, J. C.; Montarnal, R. Bull. Soc. Chim. Fr. 1968, 4816.
22. Grasselli, R. K.; Suresh, D. D. J. Catal. 1972, 25, 273.
23. Batist, P. A.; der Kinderen, A. H. W .M.; Uwenburgh, Y. L.; Metz, F. A. M. G.; Schuit, G. C. A. J. Catal. 1968, 12, 45.
24. Bleyenberg, A. C. A. M.; Lippens, B. C.; Schuit, G. C. A. J. Catal. 1965, 4, 581.
25. Eick, F. A.; Kilhborg, L. Acta. Chem. Scand. 1968, 20.
26. Robb, F. Y.; Glausinger, W. S.; Courtine, P. J. Solid State Chem. 1979, 30, 171-181.
27. Wakabayashi, G. W.; Kamiya, G. W.; Ohta, N. Bull. Chem. Soc., Jap. 1967, 40, 2172.
28. Bordes, E. Thesis, Compiegne, France, 1979.
29. Bordes. E.; Courtine, P. J. Catal. 1979, 57, 236.
30. Van Den Bussche, F.; Jouy, M.; Courtine, P. 6e Colloque Franco-Polonais sur la Catalyse, Compiegne, 1977
31. a) Grasselli, R. K.; Callahan, J. U. S. Patent 3 414 631, 1968.
b) Grasselli, R. K.; Hardman, H. F. U. S. Patent 3 642 930, 1972.
c) Standard Oil Company of Ohio, Dutch Patent application 7 006 454, 1970.

32. Mars, P.; Van Keverlen, D. W. Chem. Eng. Sci. Suppl. 1954, 3, 41.
33. Boreskov, G. K. 2nd Jap. Sov. Catal. Sem., Tokyo, 1973.
34. Sachtler, W. M. H.; De Boer, N. H. Proc. 3rd Int. Cong. Catal. North Holland, Amsterdam, 1965, p. 252.
35. Skalkina, L. V.; Suzdalev, L. P.; Kolchim, J. K.; Margolis, L. Y. Kinet. Katal. 1969, 10, 378.
36. Matsuura, I. 6th Int. Cong. Catal., London B-21, 1976.
37. Moro-oka, Y.; Ozaki, A. J. Catal. 1966, 5, 116.
38. Moro-oka, Y.; Morikawa Y.; Ozaki, A. J. Catal. 1967, 7, 23.
39. Moro-oka, Y.; Ozaki, A. J. Catal. 1960, 12, 361.
40. Boreskov, G. K. Kinet. Catal. 1970, 11, 374.
41. Boreskov, G. K. Discuss. Farad. Soc. 1968, 285.
42. Gaucher, P.; Ernst V.; Courtine, P. J. Solid State Chem. 1983, 47, 47.
43. Ubbelohde, A. R. J. Chim. Phys. 1966, 62, 33.
44. Hyde, B. G.; Bursill, L. A. In "The Chemistry of Extended Defects in Non Metallic Solids"; Leroy-Eyring and O'Keeffe (ed.): North Holland, 1970.
45. Eon, J. G. Thesis, Compiegne, 1984.
46. Andersson, S. Bull. Soc. Chim. Fr. 1965, 1088.
47. a) Bordes, E.; Courtine, P.; Johnson, J. W. J. Sol. State Chem., 1984, 55, 270.
     b) Bordes, E; Raminosona, A.; Johnson, J. W.; Courtine, P. 10th Int. Symp. Reactivity of Solids, Preprints p. 503, Dijon, 1984.
48. Boudart, M. Adv. Catal. 1966, 26, 153.
49. Volta, J. C.; Desquesnes, W.; Moraweck, B.; Tatibouet, M. J. 7th Int. Congr. Catalysis, 1980, C4, Tokyo.
50. Kozlowski, R.; Ziolkowski, J.; Mokala, K; Haber, J. J. Solid State Chem. 1980, 35, 1.
51. Theobald, F. Thesis, Besancon, 1975.
52. Ziolkowski, J. J. Catal. 1983, 80, 263.
53. Vejux A.; Courtine, P. J. Solid State Chem. 1978, 23, 93.
54. Anderson, J. S. 7th Int. Symp. Reactivity of Solids, 1972, 1-21, Chapman and Hall.
55. O'Keefe, M. In "Fast Ion Transport in Solids"; North Holland, Amsterdam, 1973, p. 233.
56. Haber, J. Pure and Applied Chemistry, 1978, 50, 923.
57. a) De Rossi, S.; Iguchi, E.; Schiavello, M.; Tilley, R. J. D. Z. Phys. Chem. 1976, 103, 193.
     b) Barber, S.; Booth, J.; Pyke, D.R.; Reid, R.; Tilley, R. J. D. J. Catal. 1982, 77, 180.
58. Gates, B. C.; Katzer, J. R.; Schuit, G. C. In "Chemistry of Catalytic Processes", 1979.
59. a) Linn, W. J.; Sleight, A. W. Ann. N.Y. Acad. Sci. 1976, 272, 22.
     b) Sleight, A. W. In "Advanced Materials in Catalysis", Acad. N.Y., 1977.
60. Wolfs, M. W. J.; Batist, P. A. J. Catal. 32, 25.
61. Zeman, J. Heidelberger Beitr. Mineral. Petrogr. 1956, 5, 139.
62. Van Den Elzen, A. F.; Rieck, G. D.
     a) Acta Crystall. 1973, B29, 2433; B29, 2436
     b) Mat. Res. Bull. 1976, 10, 1163.

63. Abrahams, S. C.; Reddy, J. M. J. Chem. Phys. 1965, 43, 2533.
64. a) Bart, J. C. J.; Perini, N.; Giordano, N. Z. Anorg. Allgem. Chem. 1975, 412; 258.
     b) Bart. J. C. J.; Giordano, N. Gazz. Chim. Ital. 1979, 109, 73.
65. Middlemiss, N. Ph.D. Thesis, McMaster Univ., Hamilton, Canada, 1979
66. Gorbunova, Y.; Linde, S. A. Dokl. AKAD. Nauk. SSSR 1979, 245, 5864.
67. Bordes, E.; Courtine, P. Ass. Gen. Soc. Chim. Fr. 1982.
68. Bond, G. C.; Bruckman, K. Faraday Discuss. Chem. Soc. 1982, 72, 235
69. Bond, G. C.; Sarkany, A. J.; Parfitt, G. D. J. Catal. 1979, 57, 2195.
70. Inomata, M.; Miyamoto A.; Murakami, J. J. Phys. Chem. 1981, 85, 2372; J. Chem. Soc., Chem. Comm., 1979, 1010; ibid. 1980, 233.
71. Kozlowski,R.; Pittifer, R. F.; Thomas, J. M. J. Chem. Soc., Chem. Comm. 1983, 438
72. Bond, G. C.; Konig, P. J. Catal. 1982, 77, 309.
73. Vanhove, D.; Blanchard, M. Bull. Soc. Chim. Fr. 1971, 9, 3291.
74. Gasior, M.; Grzybowska, B. Bull. Acad. Pol. Sci. 1979, 27, 835.
75. Courtine, P. unpublished results.
76. Courtine, P.; Vejux, A. C. R. Acad. Sci. Paris 1978, 286 C, 135.
77. Shannon, R. D.; Pask, J. A. Amer. Mineral. 1964, 49, 1707 and J. Appl. Phys. 1964, 35, 3414.
78. Eon, J. G.; Courtine, P. J. Solid State Chem. 1980, 32, 67.
79. Eon, J. G.; Bordes, E.; Vejux, A.; Courtine, P. IXth Int. Symp. on Reactivity of Solids, Cracow, Pologne, 1980.
80. Jung, J. S.; Bordes, E.; Courtine, P. Int. Symp. on Adsorption and Catalysis, Brunel Univ., Uxbridge, U.K., 1984.
81. Anderson, A.; Lundin, J. J. Catal. 1980, 58, 383.; 65, 9.
82. Anderson, A. Thesis, Lund, Sweden, 1982.
83. Yoshida, S.: Iguchi, T.; Ishada, S.; Karama, K. Bull. Soc. Chim. Jap. 1972, 45, 376.
84. a) Keulks, G. W. J. Catal. 1970, 19, 232.
     b) Keulks, G. W.; Krenzke, L. D. 6th Int. Congr. Catalysis, B-20, London, 1976.
85. Gaucher, P. Thesis, Compiegne, 1984.
86. Gaucher, P.; Courtine, P. to be published.
87. Gai, P. L.; Boyes, E. D.; Bart, J. C. J. Phil. Mag. 1982, 45, 531.
88. Chaze, A. M.; Courtine, P. J. Chem. Res. (S), 1980, 96; (M), 1980, 954-970.
89. Trifiro, F.; Forzatti, P.; Villa, P. L. In "Preparation of Catalysts", Elsevier, 1976.
90. Grzybowska, B.; Haber, J.; Komorek, J. J. Catal. 1972, 25, 25-32.
91. Aykan, K. J. Catal. 1968, 12, 281-90.
92. Batist, Ph. A.; Van de Moesdijk, G. C. A.; Matsuura, I.; Schuit, G. C. A. J. Catal. 1971, 20, 40-57.

93. Egashira, M.; Matsuo, K.; Kagawa, S.; Seiyama, T. J. Catal.
     58, 409-18.
94. Schultz, A.; Stubican, V. S. Phil. Mag. 1968, 18 (155),
     929-37.
95. Kumar, J.; Ruckenstein, E.   J. Solid State Chem. 1980, 31,
     41-46.
96. Trifiro, F.; de Vecchi, V.; Pasquon, I. J. Catal. 1969, 15,
     8-16.
97. a) Courtine, P.; Daumas, J. C.   C. R. Acad. Sci. Paris 1969,
     268,
     b) Cord, P. P.; Courtine, P.; Pannetier, G.; Guillermet, J. Spe
     ctrochim. Acta 1972, 28A, 1601-13.
98. Lojacono, M.; Noterman, T.; Keulks, C.   J. Catal. 1975, 39,
     286; 40, 19.
99. Krylov, O. V. Kinet. Katal. 1981, 22, 15.
100. Wolfs, M. W. J.; Van Hoof, J. H. C.   In "Preparation of
     Catalysts"; Delmon, B.; Jocobs, P. A.; Poncelet, G., Eds.;
     Elsevier, Amsterdam, 1976.
101. Akimoto, M.; Echigoya, E.   Bull. Chem. Soc. Jap. 1075, 48,
     3518.

RECEIVED January 14, 1985

Structural and Thermodynamic Basis for Catalytic Behavior of Bismuth–Cerium Molybdate Selective Oxidation Catalysts

J. F. BRAZDIL[1], R. G. TELLER[1], R. K. GRASSELLI[1], and E. KOSTINER[2]

[1] Sohio Research Center, The Standard Oil Company (Ohio), Cleveland, OH 44128
[2] Department of Chemistry and Institute of Materials Science, University of Connecticut, Storrs, CT 06268

The $Bi_{2-x}Ce_xMo_3O_{12}$ two phase system has been examined for its activity in the catalytic oxidation of propylene to acrylonitrile. The two phases have been characterized as a solid solution of Bi in cerium molybdate and Ce in bismuth molybdate. Results of these oxidation studies have been correlated with structural results on the pure and doped end members. A plot of catalytic activity versus x shows three maxima which coincide with the maximum concentration of Ce in $Bi_2Mo_3O_{12}$, Bi in $Ce_2Mo_3O_{12}$ and equal concentrations of cerium in bismuth molybdate and bismuth in cerium molybdate. These results suggest that selective propylene ammoxidation occurs in a trifunctional matrix which contains metals that; activate propylene to form an allyl intermediate (Bi), insert oxygen into the allylic intermediate, (Mo) and contain a redox couple (Ce). Aspects of phase cooperation in a multiphase catalyst are also discussed.

Much of the understanding of the solid state mechanism of heterogeneous catalysis stems from fundamental studies of single phase model compounds (1–5). In many cases, the role of a metal component in a catalytic process has been discerned through its incorporation into solid solutions of relatively inert host matrices (6). In the case of the selective oxidation and ammoxidation of olefins to unsaturated aldehydes and nitriles, respectively (e.g. the ammoxidation of propylene to acrylonitrile), such studies have established several important tenets for the process. These include the need for the coexistence of key catalytic elements with the proper electronic structure, redox chemistry, and metal–oxygen bond strength (7). It is however well recognized that the most effective catalysts, be they mixed metals or mixed metal oxides, are usually multiphase in nature (8). Some progress has been made in understanding the source of the synergistic effects observed in these multiphase catalysts. For example, Sinfelt and coworkers (9) have been able to explain the

0097–6156/85/0279–0057$06.00/0

catalytic behavior of bimetallic, biphasic systems through the use of structural characterization tools such as EXAFS. In oxide catalysts for selective oxidation, rigorous studies of multiphase catalysts have been limited primarily to vanadium oxides (10-12). In this case, it has been shown that the coexistence of structurally related oxidized and reduced vanadium oxide phases is stabilized by the presence of structurally coherent phase boundaries. Recently, significant efforts have been made to develop a rational mechanism for the catalytic behavior of multicomponent bismuth molybdate based selective oxidation catalysts (13-15). However, the complexity of the catalyst systems investigated has prevented the development of a rigorous model for catalytic behavior.

Our recent work on the bismuth-cerium molybdate catalyst system has shown that it can serve as a tractable model for the study of the solid state mechanism of selective olefin oxidation by multicomponent molybdate catalysts. Although compositionally and structurally quite simple compared to other multiphase molybdate catalyst systems, bismuth-cerium molybdate catalysts are extremely effective for the selective ammoxidation of propylene to acrylonitrile (16). In particular, we have found that the addition of cerium to bismuth molybdate significantly enhances its catalytic activity for the selective ammoxidation of propylene to acrylonitrile. Maximum catalytic activity was observed for specific compositions in the single phase and two phase regions of the phase diagram (17). These characteristics of this catalyst system afford the opportunity to understand the physical basis for synergies in multiphase catalysts. In addition to this previously published work, we also include some of our most recent results on the bismuth-cerium molybdate system. As such, the present account represents a summary of our interpretations of the data on this system.

Experimental

Bismuth cerium molybdates were prepared by coprecipitation using aqueous solutions of $(NH_4)_6Mo_7O_{24}$, $(NH_4)_2Ce(NO_3)_6$, and $Bi(NO_3)_3 \cdot 5H_2O$. The catalysts were supported on SiO_2 (20% by weight) using an ammonium stabilized silica sol. Samples for diffraction analysis were unsupported. Samples were calcined in air at 290 and 425°C for three hours each followed by 16 hours at 500, 550, or 600°C. X-ray powder patterns were obtained using a Rigaku D/Max-IIA X-ray diffractometer using Cu K radiation.

Powder neutron diffraction data for $Bi_{1.8}Ce_{0.2}Mo_3O_{12}$ were collected at Brookhaven National Laboratory's High Flux Beam Reactor. Details of the experimental procedure and Rietveld refinement have been reported previously (18). Starting parameters for Ce doped and pure $Bi_2Mo_3O_{12}$ were taken from a single crystal x-ray diffraction study of $Bi_2Mo_3O_{12}$ by Van den Elzen and Rieck (19a).

Time-of-flight (TOF) Powder neutron diffraction data for $Bi_2Mo_3O_{12}$ were collected at Argonne National Laboratory's Intense Pulsed Neutron Source (IPNS) utilizing the Special Environment Powder Diffractometer (SEPD). Details of the instrument, data collection software, and Rietveld analysis software have been

previously published($\underline{20}$).   TOF   data   from the backscattering data
banks (2θ=150°)   were   used   in   the   analysis   and corresponded to
TOF(min) of 7500msec (d(min)=.99 Å) and TOF(max) of 23500msec
(d(max)=3.10 Å).   In addition to background ($\underline{5}$), halfwidth ($\underline{3}$),
absorbtion, extinction, lattice ($\underline{4}$), and scale parameters, all
positional and isotropic thermal parameters were varied in the
least-squaring process, for a total of 84 variables. The least-
squares process converged to R(p)=0.037, R(wp)=0.056, R(exp)=0.044.
Refinements were also carried out for TOF(min) values of 7000 and
8000 sec. There was no significant difference between the results
of those refinements and those presented here. Figure 1 displays
the neutron diffractogram of $Bi_2Mo_3O_{12}$.
     Single crystals of Bi doped cerium molybdate were prepared in
sealed tube experiments.   A   quantity   of material of composition
$BiCeMo_3O_{12}$ was packed into   a   welded   gold capsule pinched shut at
the open end, placed in a .quartz .tube .and sealed under vacuum
(10^{-4} mm Hg). The tube was held in a vertical position and kept at
950°C for one week.   Single   crystals corresponding to two phases
were isolated from these experiments.   Red crystals harvested
possessed cell parameters and composition (via EDX analysis) that
was consistent with Ce doped $Bi_2Mo_3O_{12}$. In a similar fashion, the
amber crystals that were also isolated from this procedure were
identified as having the $Ce_2Mo_3O_{12}$ structure with some Bi present.
EDX analysis of three amber crystals gave an average composition of
$Ce_{1.8}Bi_{0.13}Mo_3O_{11.9}$ (normalized to 3 Mo atoms, oxygen stoichiometry
calculated on the basis of charge balance assuming Ce^{3+}).
     Single crystal x-ray diffraction data were collected on one of
the amber crystals with a FACS-I automated Picker diffractometer
with Zr filtered Mo $K\alpha$ radiation. Lattice parameters (a=16.886(5),
b=11.839(3), c=15.797(5)Å, b=108.64(1) were determined by a least
squares fit of 24 reflections in the angular range 51<2θ<61°.
Details of the instrument, data collection, data reduction, and
analysis have been published ($\underline{21}$).   Starting parameters for the
full matrix least squares refinement were taken from $La_2Mo_3O_{12}$(19b)
with 90%Ce and 10%Bi occupation for each La site. Parameters
varied were: all positional parameters, anisotropic temperature
factors for metal atoms, isotropic temperature factors for the
oxygen atoms, scale and extinction factors and scattering factors
for the Ce/Bi sites.   Data in the angular range 20<60° were
investigated. Of the 4601 independent data 3578 were taken to be
observed on the basis of counting statistics. With 144 variables
(data to parameter ratio = 25) the refinement converged to R=0.085,
R_w=0.083. Refinement of the scattering factors indicated that the
Bi/Ce sites were not all equivalent, site 1 was found to be 86%Ce
14%Bi, site 2 92%Ce 8%Bi and site 3 73%Ce 27%Bi. On the basis of
these occupancies a $Ce_{1.7}Bi_{0.3}Mo_3O_{12}$ composition has been assigned
to this crystal.
     Propylene ammoxidation experiments were conducted using a
single pass, plug flow, tubular (3/8") stainless steel reactor
operated at atmospheric pressure at 420°C.   The gaseous feed
mixture consisted of propylene, ammonia, air and water in the
ratios 1/1.2/10.5/4. Liquid products were collected in an aqueous
scubber and analyzed for acylonitrile, acrylic acid and acetonitrile via gas chromatography, NH_3 and HCN via titration. Nonsoluble
gases were collected and analyzed via gas chromatography.

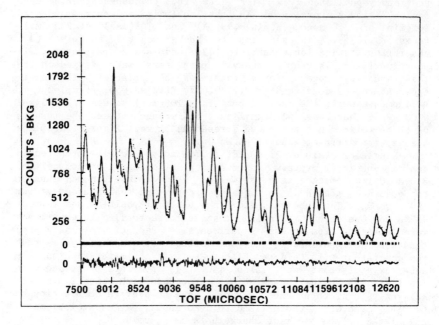

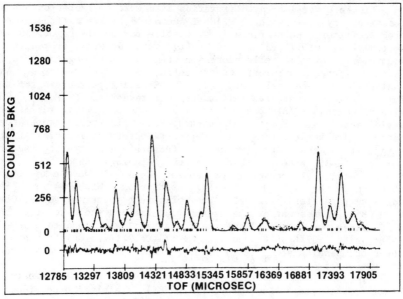

Figure 1.   Powder neutron time-of-flight diffraction data
for $Bi_2(MoO_4)_3$. Points represent the raw data, the solid
line the calculated profile, and the tick marks Bragg peak
positions. A difference plot is given below.

Results & Discussion

Scheelite-Type Structures

The scheelite structure-type is commonly adopted by compounds with the stoichiometry ABO_4 when the A cation is fairly large (>1.0 Å) and the B cation fairly small (on the order of 0.6 A). The prototype of this structure is the mineral scheelite, $CaWO_4$, which crystallizes(21) with four formula units per unit cell in the tetragonal space group $I4_1/a$ ($\underline{a}$ = 5.243, $\underline{c}$ = 11.376 Å). Each Ca^{2+} ion is surrounded by eight oxygen atoms from different tetrahedrally coordinated B ions, four at 2.44 Å and four at 2.48 Å. The nearly regular discrete WO_4^{2-} tetrahedra have four equal W-O distances (1.78 Å). The ideal structure is illustrated in Figure 2.

One characteristic of the scheelite structure-type is the number and extent of cationic oxidation states and defect (cation-deficient) structures that have been found. The single guide to the formation of scheelite-type structure seems to be the ability of A cations to be eight-coordinated (i.e., rather large) and B ions to attain tetrahedral coordination (note, however, that PO_4^{3-} or SiO_4^{4-} containing scheelites are unknown).

If we first consider the simpler cases of heterovalent substitution with this structure-type (22) we find, in addition to the 2+/6+ (A+/B+) valances typified by $CaWO_4$, the valance combinations 1+/7+ ($KRuO_4$), 3+/5+ ($GdMoO_4$), and 4+/4+ ($CeGeO_4$). Extending this by doubling (or tripling) the chemical formula we can obtain, by coupled substitution, mixed (or substituted)ions on the A site: (1+,3+)/6+ $[NaLa(MoO_4)_2]$,(1+,4+)/6+ $[Na_2Th(MoO_4)_3]$, and (2+,4+)/5+ $[PbTh(VO_4)_2]$.

Substitution on the "anion" site (the B site) has also been observed. Several examples are: 3+/(4+,6+) $[La_2(SiO_4)(WO_4)]$, 3+/(3+,6+) $[Bi_3(FeO_4)(MoO_4)_2]$, and 3+/(2+,6+) $[Bi_4(ZnO_4)(MoO_4)_3]$. If two or more ions occupy either the A or the B site, the possibility of having either an ordered or a disordered (random) structure will exist. For example, order on the A sites occurs in $KEu(MoO_4)$. The compound $Bi_3(FeO_4)(MoO_4)_2$ exists in both an ordered and disordered form(23), the ordered form resulting from the ordering of the FeO_4 and MoO_4 (B-ion) tetrahedra.

The crystal chemistry of scheelite-type structures is further complicated by defect stoichiometries with extensive vacancies at cation sites. In general, random cation vacancies are examples of point defects. However, at high concentrations, an ordering of vacancies can occur at which time the vacancy can no longer be considered as a point defect and a new periodic lattice (or unit cell) will have been generated.

As an example of randomly disordered vacancies, the coupled substitution of two Bi^{3+} ions and one vacancy for three Pb^{2+} ions in $PbMoO_4$ gives rise (24) to a series of solid solutions $Pb_{(1-3x)}Bi_{2x}\phi_x(MoO_4)_3$, where ϕ represents a cation vacancy. Depending on temperature, a fairly large range of random cation vacancies (up to 15%) has been observed.

At high concentration of cation vacancies (ϕ = 0.33), new ordered compositions are observed with unit cells corresponding to supercells of the scheelite structure. The general formula for

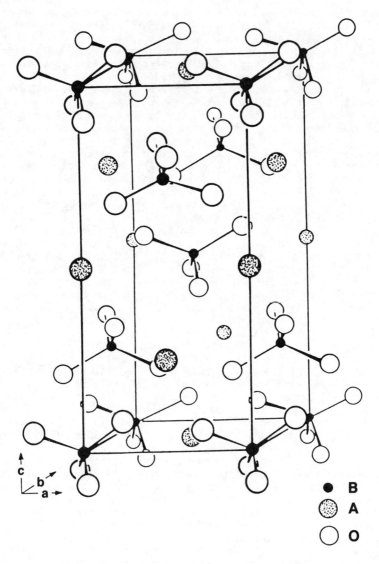

Figure 2. A representation of the scheelite $(CaWO_4)$ structure.
Reproduced by permission from Ann. N. Y. Acad. Sci., Vol. 272,
p. 23, authors: A. Sleight and W. Linn.

these cation deficient scheelites is $A_{0.67}\phi_{0.33}BO_4$, (= cation vacancy). An equivalent notation (multiplying through by 3) would be $A_2\phi(BO_4)_3$ or, more simply, $A_2(BO_4)_3$. Three different schemes for cation ordering in defect scheelite have been observed – the $Eu_2(WO_4)_3$ structure (25) (which is found for several rare earth molybdates and tungstates), the $La_2(MoO_4)_3$ structure (26) (found for larger rare earth molybdates only), and $Bi_2(MoO_4)_3$ (18), a unique structure. The cation ordering schemes for these three structures as derived from the ideal scheelite structure is shown in Figure 3, which is a projection of each of these structures down their respective monoclinic b-axes (for the defect compounds) with reference to the corresponding c-axis projection of $CaWO_4$.

It is evident from this schematic representation that these structures not only differ in the manner in which the vacancies order (and therefore in their unit cell metrics) but in distortions of the BO_4 polyedra. Diffraction studies have shown that these distortions, which are most pronounced in the structure of $Bi_2(MoO_4)_3$ (vida infra), also affect the B ion coordination.

Catalytic Activity and Phase Composition

All catalysts were tested for propylene ammoxidation activity.

$$C_3H_6 + NH_3 + 1.5\ O_2 \longrightarrow C_3H_3N + 3H_2O$$

The variation in activity for acrylonitrile formation as a function of composition for the $Bi_{2-x}Ce_xMo_3O_{12}$ system is shown in Figure 4. Maxima in activity, as measured by first order rate constant for propylene disappearance, occur at three compositions. A partial phase diagram for the system, constructed with powder x-ray diffraction is displayed in Figure 5.

The phase diagram consists of three regions, two of which are single phase regions. On the Bi-rich side, Ce is soluble in $Bi_2Mo_3O_{12}$. The maximum solubility of Ce in bismuth molybdate is approximately 10 mole percent. The Ce-rich side consists of a solid solution of Bi in $Ce_2Mo_3O_{12}$, which has the $La_2Mo_3O_{12}$ structure. The maximum solubility of Bi in this phase is approximately 50 mole percent. The intermediate region consists of a mixture of both phases. Comparison of the variation in catalytic activity with composition as presented in Figure 4 with the phase diagram reveals that maxima in catalytic activity occurs at the solubility limits of the two single phases. An understanding of the basis for this behavior requires a better definition of the solid state structural aspects of the catalyst system.

Analysis of the Single Phase Catalysts

In an attempt to identify any Bi/Ce ordering or site preference, diffraction data was collected on two end members of the series. A powder neutron diffraction data set was collected for a sample of $Bi_{1.8}Ce_{0.2}Mo_3O_{12}$ composition. Additionally, since neutron diffraction data had not been taken on the the end member of the series, $Bi_2Mo_3O_{12}$, and comparisons to this model were deemed

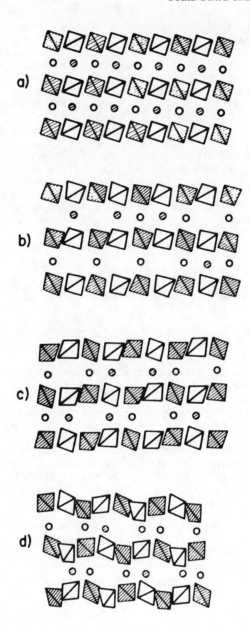

Figure 3. Ideal scheelite structure compared to ordered defect
structures. Projection is down the c axis with only 1/2 the unit
cell shown in this direction. Shaded circles and tetrahedra are
at the top of level, unshaded 1/4 of the way down the unit cell.
(a) $CaWO_4$. (b) $La_2(MoO_4)_3$. (c) $Eu_2(WO_4)_3$. (d) $Bi_2(MoO_4)_3$.
Reproduced by permission from Ann. N. Y. Acad. Sci., Vol. 272,
p. 23, authors: A. Sleight and W. Linn.

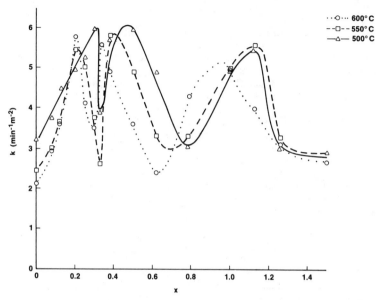

Figure 4. Catalytic activity as a function of composition and reaction temperature for the $Bi_{2-x}Ce_x(MoO_4)_3$ system. Catalytic activity is expressed as k, the first order rate constant for propylene disappearance. Reproduced from Ref. 17. Copyright 1983 American Chemical Society.

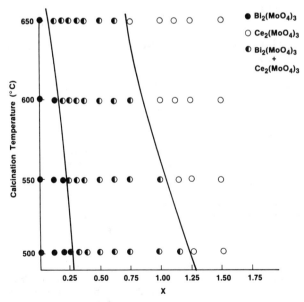

Figure 5. Partial phase diagram for the $Bi_{2-x}Ce_x(MoO_4)_3$ system. Reproduced with permission from Ref. 16. Copyright 1983, Academic Press, Inc.

significant, powder neutron diffraction data were collected on this
material as well.

The structure of $Bi_2Mo_3O_{12}$ (perhaps better written
$Bi_{2/3}\phi_{1/3}MoO_4$), is based on scheelite. As discussed above, there
is considerable distortion from the ideal scheelite structure, this
is probably due to the Bi lone pair of electrons (vida infra). As
a result of these distortions in the anion packing, the Mo atoms
are penta-coordinate (as opposed to tetrahedrally coordinated in
$CaWO_4$) and Mo_2O_8 "dimers" are observed, with Mo... Mo separations
of 3.4 Å and two bridging O atoms. A listing of Mo-O distances
derived from the neutron diffraction data for $Bi_2Mo_3O_{12}$ is given in
Table 1. Note that the O coordination about each of the three
crystallographically distinct Mo atoms is similar. Each Mo atom
contains one short (double) Mo-O bond, one singly bound O atom, two
oxygens with intermediate bond orders and one weakly bound oxygen
atom. The result is a very asymmetric oxygen coordination sphere
about each Mo atom.

Table I. A Comparison of Interatomic Mo-O Distances in
$Bi_2Mo_3O_{12}$ and $Bi_{1.8}Ce_{0.2}Mo_3O_{12}$

	$Bi_2Mo_3O_{12}$	$Bi_{1.8}Ce_{0.2}Mo_3O_{12}$
Mo(1)-O(4)[a]	1.70(1)	1.78(3)
-O(5)	1.78(1)	1.67(2)
-O(2)	1.80(2)	1.92(3)
-O(10)	1.89(1)	1.85(2)
-O(10)'	2.20(1)	2.27(2)
Mo(2)-O(1)	1.65(2)	1.63(3)
-O(9)	1.75(2)	1.68(3)
-O(6)	1.85(2)	1.98(4)
-O(3)	1.97(2)	1.90(3)
-O(8)	2.16(2)	2.23(3)
Mo(3)-O(12)	1.74(2)	1.72(2)
-O(11)	1.63(2)	1.88(3)
-O(8)	1.87(2)	1.93(3)
-O(7)	1.90(1)	1.83(3)
-O(6)	2.48(2)	2.34(3)

a) Bond distances are given in angstroms, with estimated
standard deviation of the last digit given in parenthesis.

Results (Table I) from refinement of the powder neutron diffraction
data for the Ce doped material, $Bi_{1.8}Ce_{0.2}Mo_3O_{12}$, indicate that
significant structural alterations have resulted from Ce incorporation into the solid. The three Mo atoms are no longer
chemically equivalent. The coordination about Mo(1) is unchanged
from that of the parent compound. A second Mo atom contains two
short (double) bonds to oxygen atoms. These molybdate dioxo
(dimolybdenyl) groups are believed to be an important feature for
selective oxidation catalysis (7). The third Mo coordination
sphere contains no Mo=O bond. Bond order calculations (27) about

this latter Mo site indicate the total Mo-O bond order to be 5. This apparent reduction of Mo from 6 to 5 is most likely accompanied by some Ce oxidation to 4 and/or oxygen removal from the solid. The distribution of the Ce dopant is not uniform. Of the three cation sites in the parent compound, the first two are fully occupied by Bi atoms, while the third is empty. This results in a sequence of cation occupation on the [010] face as shown below in A and B. Site a is 92% Bi, 4% Ce, site b is 88% Bi, 12% Ce, and site c, normally vacant, is 4% Ce occupied. Examination of the oxide environment about site c indicates that there is insufficient room for a Ce cation. Occupation of the site must therefore result in some local disorder, the most likely manifestation thereof being a vacancy in site a (the distance between sites a and c are too short to allow simultaneous occupation). Note also that site a is not 100% occupied. Consequently, about 96% of the time the distribution of cations (Bi/Ce) in $Bi_{1.8}Ce_{0.2}Mo_3O_{12}$ is as in A below (normally found in bismuth molybdate), and 4% of the time it is as in B.

A)   ...... M M □ M M □ M M □ ......
B)   ...... M M Ce □ M □ M M □ ......

M=Bi/Ce

The distribution of cations represented in B is reminiscent of that of $Ce_2Mo_3O_{12}$ on the [010] face.

A structural study of a Ce rich catalyst was also undertaken. A single crystal x-ray diffraction analysis of a crystal of composition $Ce_{1.7}Bi_{0.3}Mo_3O_{12}$ was performed, and some results (along with comparisons to isostructural $La_2Mo_3O_{12}$) are presented in Table II. The results of the experiment indicate that the Bi substitutes for Ce in the structure. Unlike $Bi_2Mo_3O_{12}$, the parent compound $Ce_2Mo_3O_{12}$ has a relatively undistorted scheelite-based structure with 1/3 cation vacancies. The difference between the bismuth and cerium molybdate structures lies in the ordering of these vacancies as indicated above. The comparison between $La_2Mo_3O_{12}$, and the Bi doped cerium molybdate is quite good. The Mo atoms are approximately tetrahedrally coordinated to four oxygen atoms, and the Ce/Bi atoms are eight coordinate. Because the gross distortions in bismuth molybdate are attributed to the Bi lone pairs, this relative lack of scheelite distortion in cerium molybdate is not unexpected. At significantly lower concentrations of Bi atoms lone pair - lone pair interactions are minimized.

As in the bismuth rich structure discussed above, the Bi atoms in Bi doped cerium molybdate are not randomly distributed on the Ce sites. The compositions of M(1), M(2), and M(3) (see Table II) are 86%Ce and 14%Bi, 92%Ce and 8%Bi, 73%Ce and 27%Bi, respectively. This compositional difference is reflected in the M(3)-O distances as well. Note that the difference between the maximum and minimum M(3)-O distances is larger for Bi doped cerium molybdate than lanthanum molybdate. Evidently, a 27% occupation of Bi (with the attendant lone pair) is sufficient for a small but noticable distortion in the average local O environment. Summarizing the structural results on Ce incorporation into bismuth

Table II. A Comparison of Mo-O Distances in $La_2Mo_3O_{12}$ and $Ce_{1.7}Bi_{0.3}Mo_3O_{12}$

	$La_2Mo_3O_{12}$	$Ce_{1.7}Bi_{0.3}Mo_3O_{12}$
Mo(1)-O(4)	1.73	1.73
-O(3)	1.75	1.74
-O(1)	1.78	1.82
-O(2)	1.82	1.85
Mo(2)-O(7)	1.75	1.74
-O(5)	1.76	1.77
-O(6)	1.76	1.78
-O(8)	1.82	1.82
Mo(3)-O(9)	1.73	1.75
-O(12)	1.75	1.74
-O(11)	1.79	1.82
-O(10)	1.81	1.79
Mo(4)-O(14)	1.76	1.74
-O(13)	1.76	1.75
-O(15)	1.77	1.77
-O(16)	1.81	1.82
Mo(5)-O(18)	1.75	1.76
-O(19)	1.81	1.80

	$La_2Mo_3O_{12}$		$Ce_{1.7}Bi_{0.3}Mo_3O_{12}$
La(1)-O(4)	2.48	M(1)-O(4)	2.45
-O(12)	2.48	-O(12)	2.46
-O(7)	2.50	-O(7)	2.51
-O(16)	2.51	-O(16)	2.47
-O(5)	2.52	-O(5)	2.45
-O(16)	2.52	-O(16)	2.48
-O(17)	2.57	-O(17)	2.53
-O(10)	2.57	-O(10)	2.53
La(2)-O(14)	2.46	M(2)O(14)	2.44
-O(9)	2.48	-O(9)	2.44
-O(3)	2.52	-O(3)	2.53
-O(6)	2.52	-O(6)	2.47
-O(2)	2.53	O(2)	2.47
-O(13)	2.53	-O(13)	2.52
-O(18)	2.54	-O(18)	2.52
-O(11)	2.57	-O(11)	2.50
La(3)-O(15)	2.44	M(3)-O(15)	2.38
-O(1)	2.46	-O(1)	2.40
-O(2)	2.49	-O(2)	2.40
-O(8)	2.53	-O(8)	2.54
-O(11)	2.55	-O(11)	2.48
-O(8)	2.55	-O(8)	2.54
-O(17)	2.55	-O(17)	2.53
-O(10)	2.59	-O(10)	2.62

A) The estimated standard deviation for each Mo-O distance is .01. Distances for $La_2Mo_3O_{12}$ are taken from Reference 26. All bond distances are given in angstroms.

molybdate and Bi incorporation into cerium molybdate one finds nonrandom dopant substitution. Apparently the symmetry of the different +3 cation sites, coupled with the different steric requirements of Ce^{3+} and Bi^{3+} ions, plays an important role in determining the occupations of the various sites. Additionally, in the case of bismuth molybdate, a radical alteration in the Mo-O coordination spheres results upon Ce incorporation. Coupled with this distortion there is an apparent reduction of one Mo atom, and the creation of a dimolybdenyl group on another Mo site.

In comparing the two structures, one fails to find a structure type or distortion (such as Mo...Mo interactions or grossly distorted metal environments) common to both materials in the structures described here. Yet, some commonality is expected to exist based on the close correspondence of the catalytic data. There are several possible explanations for this: a) considerably more distortion is found in more Bi rich cerium molybdate compounds (e.g. $Bi_{0.9}Ce_{1.1}Mo_3O_{12}$) and a structure analysis of this material is needed, b) local environments within the two structures described here are similar, but the average structures that result from diffraction experiments do not reveal this similarity or c) incorporation of another metal more greatly effects the surface of each material not the bulk, and catalytic behavior is a reflection of surface structure.

Increased propylene ammoxidation activity of each phase upon alterion doping is due to the juxtaposition of all necessary elements for oxidation catalysis in a single phase. The requirements of a good oxidation catalyst are a) activation of the substrate molecule, b) oxidation activity (oxygen inserting) and c) facile redox capabilities to ease electron conduction and site reconstruction. For reasons discussed extensively in the literature (7), we assign these roles to Bi, Mo, and Ce ion sites respectively in the catalysts described here. The solid solution formation observed in these materials enables all of these functions to be represented in one phase and on one surface of the catalyst.

Analysis of the Multiphase Catalyst

The maximum in catalytic activity observed for the multiphase region of the phase diagram necessarily arises from interactions between the separate phases. The bismuth rich and cerium rich solid solutions can readily form coherent interfaces at the phase boundaries due to the structural similarities between the two phases which can permit epitaxial nucleation and growth. A good lattice match exists between the [010] faces of the compounds, this match is displayed in Figure 6. We have also shown that regions of an [010] face of a Ce doped bismuth molybdate crystal resembles cerium molybdate compositionally. This means that the interface between the two compounds need not have sharp composition gradients. It is structurally possible for the Bi-rich phase to possess a metal stiochiometry at the surface that matches that of the Ce-rich phase.

In order to assess the nature of this interfacial region, the thermodynamic treatment of Cahn and Hilliard (28) was employed (17) to derive the following free energy function:

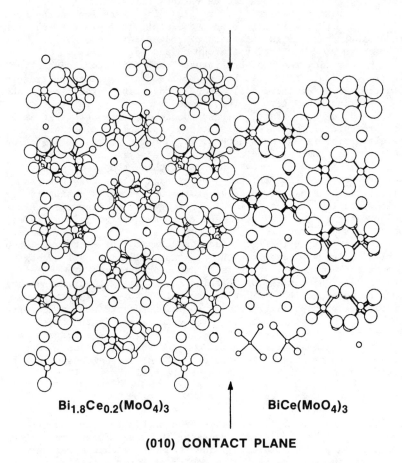

$Bi_{1.8}Ce_{0.2}(MoO_4)_3$ $BiCe(MoO_4)_3$

(010) CONTACT PLANE

Figure 6. Crystallographic match between the [010] faces of the $Bi_{1.8}Ce_{0.2}(MoO_4)_3$ and $BiCe(MoO_4)_3$ solid solutions.

$$\Delta f = RT \ X_{Bi} \ln \ [a_{Bi}/a^e_{Bi}] \ + X_{Ce} \ln \ [a_{Ce}/a^e_{Ce}]$$

where a_{Bi} and a_{Ce} are the activities of bismuth and cerium, respectively, in the interfacial region and a^e_{Bi} and a^e_{Ce} are the activities of bismuth and cerium, respectively, in the equilibrium solid solution. The function can be regarded as the free energy of an equilibrium mixture of the two solid solution phases.

The free energy function Δf was calculated as a function of composition in the two phase region of the $Bi_{2-x}Ce_x(MoO_4)_3$ system (Figure 7). Comparison of Figures 5 and 7 reveals that energy of an equilibrium mixture of the phases is minimized at the composition which gives maximum catalytic activity. It is apparent that at compositions where Δf is a minimum, the difference in the chemical potentials of the components in the interfacial region and the equilibrium solid solutions is minimized. Thus, an interfacial region which is chemically similar to the saturated solid solutions appears optimum for maximum catalytic efficiency.

Physically, the relationship between catalytic activity and Δf can be understood from a study of single phase bismuth cerium molybdate solid solutions. The results show that maximum activity is achieved when there exists a maximum number and optimal distribution of all the key catalytic components; bismuth, molybdenum and cerium in the solid. Therefore, it reasonably follows that the low catalytic activity observed for the two phase compositions where $\Delta f \neq \Delta f(min)$ results from the presence of interfacial regions in the catalysts where the compositional uniformity deviates significantly from the equilibrium distribution of bismuth and cerium cations present in the solid solutions. These compositions may contain areas in the interfacial region which are more bismuth-rich or cerium-rich than the saturated solid solutions. Conversely, at $\Delta f(min)$, the catalyst is similar to an ideal mixture of the two optimal solid solutions. The compositional homogeneity of the interfacial region approaches that of the saturated solid solutions. Therefore, the catalytic behavior of compositions at $\Delta f(min)$ is similar to that of the saturated solid solutions.

Summary and Conclusions

Several conclusions can be drawn about the solid state mechanism of selective olefin ammoxidation by both single phase and multiphase oxide catalyst. Firstly, optimum catalytic performance is achieved when there is maximum interaction between key catalytic components in a solid oxide matrix. Maximum interaction occurs in a single phase saturated solid solutions since these contain the maximum number and dispersion of the catalytically important co-existing elements. Secondly, these key catalytic components for selective olefin ammoxidation are identified as: an α-H abstracting element (Bi), an olefin chemisorption and nitrogen insertion element (Mo) and a multivalent redox couple ($Ce^{3+} \rightleftharpoons Ce^{4+}$). The multivalent redox couple enhances oxygen ion, electron and anion vacancy transport in the solid which enhances catalytic activity by increasing the reconstruction/reoxidation rate of the active Bi and Mo containing sites. This results in an effective increase of the

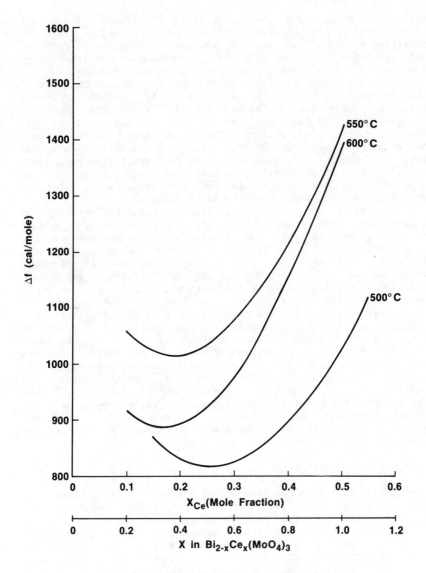

Figure 7. Free energy function Δf for $Bi_{2-x}Ce_x(MoO_4)_3$ as a
function of composition.
Reproduced from Ref. 17. Copyright 1983 American Chemical
Society.

number of active sites available at the surface at any given time. Thirdly, we have found that the chemical and structural nature of the interfacial region between co-existing phases in $Bi_{2-x}Ce_x(MoO_4)_3$ catalysts has a profound effect on catalytic behavior. Thermodynamic calculations show that compositions which give maximum catalytic activity also give minima in the free energy of mixing of the two phases relative to the saturated solid solutions. This can be explained on the basis that at the free energy minimum the chemical and compositional similarity between the interfacial region and the equilibrium solid solutions is greatest.

The simultaneous existence of coherent phase boundaries between the separate phases of multicomponent catalyst is also an important criteron for maximum catalytic activity. In bismuth-cerium molybdates, close structural similarity between the two saturated solid solutions permits mutual epitaxial growth which produces a "pseudo" single phase catalyst. As a consequence, oxygen ion, anion vacancy and electron transport between the phases can readily occur. In addition, oxygen ion and electron transfer between the individual phases will be facilitated when the compositional nonuniformity of the region at the interface is minimized. A compositionally non-uniform region in which the redox couple (cerium in this case) is not properly distributed will exhibit diminished oxygen ion mobility across the coherent phase boundaries and the overall catalytic activity for selective olefin oxidation will be less than optimum.

Acknowledgments

We gratefully acknowledge M. H. Rapposch for collecting the x-ray data and assisting in the structure solution, and L. C. Glaeser for catalyst preparation and testing. The authors thank the US department of energy for supporting IPNS at Argonne as a national users facility, and the Standard Oil Company (Sohio) for permission to publish this work.

Literature Cited

1. Grasselli, R. K.; Suresh, D. D. *J. Catal.*, 1972, 25, 273.
2. Bart, J. C. J.; Giordano, N. *J. Catal.*, 1980, 64, 356
3. Brazdil, J. F.; Suresh, D. D.; Grasselli, R.K. *J. Catal.*, 1980, 66, 347.
4. Aso, I.; Furukawa, S.; Yamazoe, N.; Seiyama, T. 1980, *J. Catal.*, 64, 29.
5. Aso, I.; Amamoto, T.; Yamazoe, N.; Seiyama, T. *Chem. Lett.*,1980, 365.
6. Stone, F.S. *J. Solid State Chem.* 1975, 12, 271.
7. Grasselli, R. K.; Burrington, J. D.; Brazdil, J. F. *Farad. Disc.*,1982, 72, 203 and references therein.
8. Schuit, G. C. A.; Gates, B. C. *Chem. Tech.* 1983, November 693.
9. Sinfelt, J. H.; Via, G. H.; Lytle, F. W. *J. Chem. Phys.*, 1980, 72, 4832.

10. Vejux, A.; Courtine, P.   J. Solid State Chem.,1978,
     23, 93.
11. Bordes, E.; Courtine, P.   J. Catal.,1979, 57, 236
12. Eon, J. G.; Courtine, P.   J. Solid State Chem. 1980,
     32, 67.
13. Carson, D.; Coudurier, G.; Forissier, M.; Vedrine,
     J. C. J. Chem. Soc., Farad Trans. 1, 1983, 79, 1921
     1983.
14. Ueda, W.; Moro-oka, Y.; Ikawa, T. J. Chem Soc., Farad,
     Trans I.,1982, 78, 495.
15. Chaze, A. M.; Courtine, P.   J. Chem. Research, 1983,
     96.
16. Brazdil, J. F.; Grasselli, R. K., J. Catal. 1983, 79,
     104.
17. Brazdil, J. F.; Glaeser, L. C.; Grasselli, R. K.
     J. Phys. Chem. 1983, 87, 5485.
18. Teller, R. G.; Brazdil, J. F.; Grasselli, R. K.;
     Thomas, R.; Corliss, L.; Hastings J. J. Solid State
     Chem., 1984, 52,313.
19a. Van den Elzen, A. F.; Rieck, G. D. Acta. 1973, Cryst.,
     B29,2433.
19b. Zalkin, A.; Templeton, D. H. J. Chem. Phys. 1964,40,
     501.
20. Von Dreele, R. B.; Jorgensen, J. D.; Windsor, C. G.
     J. Appl. Cryst., 1982, 15, 581 and Jorgensen, J. D.;
     Faber, J.   ICANS-II, proc. VIth Int. Collab. Adv.
     Neutron Sources, Argonne Natl. Lab., June 28-July 2,
     1982, 1983 (ANL-82-80.
21. Rapposch, M. H.; Anderson, J. B.; Kostiner, E. Inorg.
     Chem., 1980, 19, 3531.
22. Sleight, A. W.   In "Advanced Materials in Catalysis";
     Ed. Burton and Garten, Academic Press, NY 1977,181.
23. Jeitschko, W.; Sleight, A. W.; McClellan, W. R.;
     Weiher, J. F.   Acta Cryst. 1976, B32, 1163.
24. Sleight, A. W.; Aykan, K.; Rogers, D. B. J. Solid
     State Chem., 1975, 13, 231.
25. Templeton, D. H.; Zalkin, A.   Acta Cryst. 1963, 16,
     762.
26. Jeitschko, W. Acta Cryst., 1973, B29, 2074.
27. Brown, I. D.; Wu, K. K. Acta. Cryst., 1976, B32, 1957.
28. Cahn, J. W.; Hilliard, J. E. J. Chem. Phys., 1958, 28,
     258.

RECEIVED February 26, 1985

5

Structure and Activity of Promoted Uranium-Antimony Oxide Catalysts

R. A. INNES, A. J. PERROTTA, and H. E. SWIFT

Gulf Research & Development Company, Pittsburgh, PA 15230

At one time the preferred catalyst for propylene ammoxidation was a uranium-antimony oxide composition whose active phase was USb_3O_{10}. We have found that the partial substitution of certain tetravalent metals for the pentavalent antimony in this phase greatly increases catalytic activity. Catalysts with the empirical formula USb_2MO_{9-10}, where M=Ti, Zr, or Sn, were 6, 11, and 13 times as active as the old catalyst, while exhibiting as good or better selectivity to acrylonitrile. The high activity of the modified catalysts is attributed to the generation of oxygen vacancies in the USb_3O_{10} lattice. The stability of these catalysts is enhanced by the addition of small amounts of molybdenum or vanadium which prevent decomposition of the active phase by acting as a catalyst for reoxidation.

Acrylonitrile is manufactured by passing propylene, ammonia, and air over a mixed-oxide catalyst at 400-500°C. The process is also a major source of acetonitrile and hydrogen cyanide which are obtained as the result of side reactions. Catalysts used in this process are generally mixed oxides of bismuth or antimony with other multivalent metals such as molybdenum, iron, uranium, and tin. At one time, the preferred catalyst for propylene ammoxidation was a uranium-antimony oxide composition (1-4). This catalyst contained excess Sb_2O_4 and a silica binder in combination with the catalytically active phase USb_3O_{10} (3,4). Both uranium and antimony in the active phase assume the +5 oxidation state.

We have found that the partial substitution of certain tetravalent metals for pentavalent antimony greatly increases catalytic activity. For example, catalysts with the empirical formula USb_2MO_{9-10}, where M=Ti, Zr, or Sn, were respectively 6, 11, and 13 times as active as the original uranium-antimony oxide catalyst, while exhibiting as good or better acrylonitrile

0097-6156/85/0279-0075$06.00/0

selectivity. This paper will discuss how the replacement of antimony by a tetravalent metal affects the crystalline phase distribution, how the resulting differences in both composition and structure relate to the catalytic properties, and how catalyst stability is enhanced by the addition of small amounts of molybdenum or vanadium.

Experimental Methods

Unless otherwise noted, catalysts were prepared by coprecipitating the hydrous oxides of uranium, antimony, and a tetravalent metal from a hydrocholoric acid solution of their salts by the addition of ammonium hydroxide. The precipitates were washed, oven dried, then calcined at 910°C overnight or at 930°C for two hours to form crystalline phases. Attrition resistant catalysts, containing 50% by weight silica binder, were prepared by slurrying the washed precipitate with silica-sol prior to drying. In some cases, small amounts of molybdenum or vanadium were added by impregnating the oven dried material with ammonium paramolybdate or ammonium metavanadate solution. The details of these preparations may be found elsewhere (5-8).

The crystalline phases present in each catalyst were determined from X-ray powder diffraction patterns obtained with Cu-Kα radiation and a nickel filter. The region scanned was $2\theta = 10°$ to 70°. Infrared transmission spectra from 650 to 4000 cm^{-1} were obtained using a grating infrared spectrophotometer, a demountable cell with sodium chloride windows, and catalyst samples prepared as paraffin oil mulls. Magnetic susceptibilty measurements were made using the Faraday method and an apparatus which has been described elsewhere (9). This apparatus was designed for low temperature studies so our experiments were limited to the 4-105°K range. The magnetic field strength was 20,369 oersted.

A standard microactivity test was used to determine the effect of substituting tetravalent metals for antimony. A 0.5 cm^3 sample of 20-40 mesh catalyst was weighed and charged to a 0.48 cm I.D. tubular stainless steel reactor. The catalyst was heated to 450°C in air flowing at 32.5 cm^3 (STP) min^{-1}. The reaction was then carried out in cyclic fashion. Ammonia and propylene were added to the air stream at rates of 3.0 and 2.5 cm^3 (STP) min^{-1}, respectively. The furnace temperature was adjusted so that the reaction temperature was 475°C, as measured by a sheathed thermocouple located within the catalyst bed. After 15 minutes on stream, the product stream was sampled and analyzed by gas chromatography. After another 15 minutes on stream, the propylene and ammonia flows were shut off and the catalyst was regenerated by allowing the air flow to continue. Propylene and ammonia flows were then resumed to begin the next cycle. This procedure was repeated for 5 or 6 cycles and the results averaged. Product analyses were made with a gas chromatograph equipped with a thermal conductivity detector, a 6' x 1/4" column packed with 5 Å molecular sieves and a 15' x 1/8" column packed with 6' of Porapak T followed by 9' of Porapak QS. Oxygen–argon, nitrogen, and carbon monoxide were analyzed on the molecular sieve column, while carbon dioxide and heavier products were determined on the Porapak column. A microreactor holding 5.0 cm^3 of catalyst was used to determine the optiumum acrylonitrile yield and study the effect of silica binder and molybdenum and vanadium addition.

Altering the Active Phase

Small increases in activity may be obtained by adding a variety of multivalent metal oxides to the optimum uranium–antimony oxide catalyst (10,11). Heretofore, antimony and uranium have been emloyed in at least a four to one atomic ratio to ensure that USb_3O_{10} is the only uranium-containing phase formed. The presence of excess antimony prevents the formation of $USbO_5$ which is a less selective catalyst. Our approach was basically different. Instead of adding small amounts of promoters to the optimum uranium-antimony oxide composition, we attempted to alter the active phase (USb_3O_{10}) through crystallization in a hypothetical binary system USb_3O_{10}–USb_2TiO_{10}. Ti^{+4} has an ionic crystal radius of 0.68 Å, which is close to the 0.62 Å ionic radius of Sb^{+5} (12) making it a logical candidate to replace Sb in the USb_3O_{10} structure.

Grasselli and co-workers (3,4) have determined the crystal structure of the USb_3O_{10} phase by analogy to the single crystal work of Chevalier and Gasperin (13) on UNb_3O_{10}. In another paper (14), Chevalier and Gasperin report that compounds of the type $U(Nb_{1-x},Ti_x)Nb_2O_{10}$ have the same structure. Based on this work, they proposed that the uranium in UNb_3O_{10} is hexavalent and that one atom of niobium is tetravalent. Thus, to compensate for the replacement of Sb^{+5} with Ti^{+4}, it was expected that uranium would be converted from the +5 to the +6 oxidation state and that this would have a profound effect on the catalytic properties.

Effect on Crystalline Phase Distribution

A series of catalysts was prepared to study the effect of substituting titanium, zirconium, or tin for antimony in the USb_3O_{10} lattice. The crystalline phases present in these materials were determined by X-ray powder diffraction. To provide a basis for comparison with the prior art, catalyst 1 listed in Table I was prepared following the published recipe (2). This catalyst represents the old uranium–antimony oxide catalyst without any silica binder. The crystalline phases detected in catalyst 1 were USb_3O_{10} and Sb_2O_4 as expected (3,4).

Table I. Crystalline Phases Detected By X-Ray Powder Diffraction

Catalyst	Atomic Ratio			Crystalline Phases*
	U	Sb	Ti	
1	1.0	4.6	–	I, Sb_2O_4
2	1.0	3.0	–	I, sm-II, sm-Sb_2O_4
3	1.0	2.4	0.6	I
4	1.0	2.0	1.0	I
5	1.0	1.5	1.5	I, II, sm-TiO_2
6	1.0	1.0	2.0	I, II, TiO_2
7	1.0	0.5	2.5	I, II, TiO_2, UTiO5
8	1.0	–	3.0	UTiO$_5$, TiO_2
9	1.0	4.0	0.9	I, II, Sb_2O_5, sm-TiO_2 sm-Sb_2O_4
10	1.0	4.0	0.9	I, Sb_2O_5, sm-TiO_2

*   I = USb_3O_{10} type phase, II = $USbO_5$ type phase
     sm = small amount

A series of catalysts was then prepared having the empirical formula $USb_{3-x}Ti_xO_y$, where x ranged from 0 to 3. Titanium appeared to substitute for antimony in the USb_3O_{10} lattice up to x=1, since only a single crystalline phase closely resembling USb_3O_{10} was obtained. No peaks were seen corresponding to TiO_2 or $USbO_5$. The X-ray diffraction patterns for the x=0.6 and x=1.0 compositions are compared in Table II with that of USb_3O_{10} prepared by our method. These patterns were almost identical to published patterns for USb_3O_{10} except that the 004 reflection (3) shifted from 3.83 to 3.90 Å as x increased from 0 to 1. Attempts to increase titanium substitution beyond x=1.0 resulted in the formation of TiO_2 and $USbO_5$ at the x=1.5 level, and eventually UTiO$_5$ at higher levels.

Table II. X-Ray Powder Diffraction Patterns

Ref. (2,3)		Catalyst 2		Catalyst 3		Catalyst 4		Catalyst 17	
USb_3O_{10}		USb_3O_{10}		$USb_{2.4}Ti_{0.6}O_y$		USb_2TiO_y		USb_2ZrO_y	
d(Å)	I/Io	d(Å)	I/Io	d(Å)	I/Io	d(Å)	I/Io	d(Å)	I/Io
3.85	66	3.83	44	3.87	52	3.90	55	3.92	67
3.18	100	3.16	100	3.18	100	3.18	100	3.22	100
2.45	61	2.44	53	2.45	72	2.45	71	2.47	73
1.92	12	1.91	10	1.94	15	1.96	12	1.95	16
1.83	29	1.83	20	1.83	40	1.83	36	1.84	40
1.66	30	1.66	35	1.65	68	1.67	65	1.67	75
1.65	33	1.64	31						
1.59	14	1.58	14	1.59	15	1.59	17	1.60	46
1.47	19	1.46	17	1.46	22	1.47	21	1.48	23
1.33	18	1.33	19	1.34	20	1.34	20	1.34	21

Tabulation of peaks exceeding I/Io>10

To determine whether titanium substitution would occur in the presence of excess antimony, catalyst 9 (Table I) was prepared as described in a Distillers patent (10), while catalyst 10 was prepared by our standard coprecipitation method. Small amounts of TiO_2 could be detected in both catalysts. Catalyst 9, which was calcined at a lower temperature than catalyst 10, contained $USbO_5$ in addition to TiO_2. The shift in d-spacing for the 004 reflection noted with the titanium-substituted phases was not seen for catalysts 9 and 10. Thus, the presence of excess antimony appeared to inhibit titanium substitution. These compositions were well above (on the excess antimony side) the binary join expected to facilitate titanium substitution for antimony.

Zirconium, with a larger ionic radius (0.79 Å), did not substitute as easily as titanium in the USb_3O_{10} lattice. In only one case, x=1.0, was a pure $USb_{3-x}Zr_xO_y$ phase obtained. The other catalysts contained $USbO_5$, Sb_2O_4, Sb_2O_5, and ZrO_2 type phases. The X-ray diffraction pattern of USb_2ZrO_y is compared in Table II with the unsubstituted and titanium-substituted phases. As with the titanium catalyst the d-spacing for the 004 reflection was increased.

Tin substitution was also attempted, but the x-ray diffraction patterns gave no indication of substitution. The peaks corresponding to SnO_2 increased in direct proportion to the amount of stannic chloride used in their preparation, and $USbO_5$ and USb_3O_{10} were present in amounts consistent with the U/Sb ratio. Table III shows the X-ray diffraction pattern obtained for the composition USb_2SnO_y.

Table III. X-Ray Diffraction Pattern For USb_2SnO_y

d(Å)	I/Io	USb_3O_{10}	$USbO_5$	SnO_2
			Crystalline Phases	
3.92	48		x	
3.86	60	x		
3.36	38			x
3.25	80		x	
3.20	100	x		
2.64	23			x
2.50	33		x	
2.46	50	x		
2.37	8			x
1.97	6		x	
1.93	8	x		
1.87	12		x	
1.84	19	x		
1.77	19			x
1.69	22		x	
1.68	22	x		
1.66	27	x		
1.59	9			

Effect on Catalytic Properties

Figures 1 and 2 show how the catalytic activity and selectivity for the production of acrylonitrile varied with titanium level for the $USb_{3-x}Ti_xO_y$ series. For purposes of comparison we assumed first order kinetics and computed the relative activity per gram using the formula:

$$\text{Relative Activity} = \frac{\ln(1-X)^{-1}}{(0.4045)\ (\text{grams of catalyst})}$$

where X is the fraction of propylene converted. A relative activity of 1.0 corresponds to the activity of the old uranium-antimony oxide catalyst (Catalyst 1). Selectivity was defined on a carbon weight basis.

Substituting titanium for antimony in the USb_3O_{10} phase dramatically increased catalytic activity. The relative activity for the $USb_{3-x}Ti_xO_y$ series peaked at x=1.5. The best acrylonitrile selectivity was obtained at x=0.6 and x=1.0. Reduced activity and selectivity at higher titanium levels corresponded to $USbO_5$ and $UTiO_5$ formation. The USb_2TiO_y catalyst seemed to offer the best combination of activity and selectivity. Under optimum conditions (Table IV) it yielded 83-84 mol% acrylonitrile per pass compared to 78% for the old uranium-antimony oxide catalyst (1,2,4) which required six times the contact time to obtain comparable conversions.

Replacing antimony with zirconium increased catalytic activity 11-fold. Figures 3 and 4 show that activity peaked at x=1.0. The USb_2ZrO_y catalyst was less selective than the corresponding titanium-substituted catalyst but compared favorably to the old uranium-antimony oxide catalyst.

Table IV. Optimum Acrylonitrile Yield With USb_2TiO_y Catalyst

Reaction Temperature, °C	475	475
Contact Time, s	0.65	0.72
C_3H_6/Air/NH_3	1.0/11/1.1	1.0/10/1.1
% C_3H_6 Conversion	97.6	98.9
% O_2 Conv	85.9	93.9
% Selectivities		
$CO + CO_2$	11.8	13.7
HCN	0.3	0.4
Acetonitrile	1.3	1.4
Acrylonitrile	86.5	84.1
Other	0.1	0.4
% Yield Per Pass	84.4	83.2

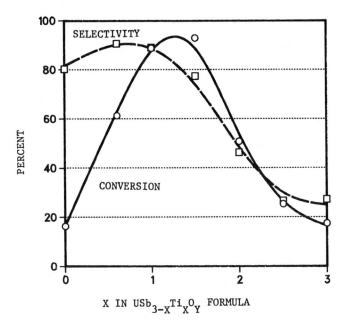

Figure 1.   Effect of Ti substitution for Sb.

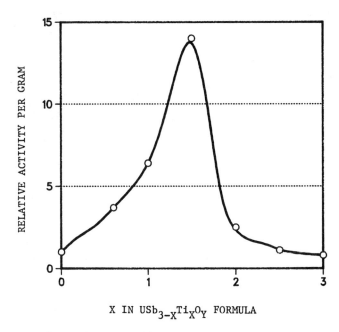

Figure 2.   Relative activity of $USb_{3-x}Ti_xO_y$ catalysts.

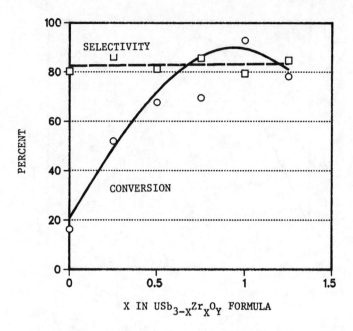

Figure 3.   Effect of Zr substitution for Sb.

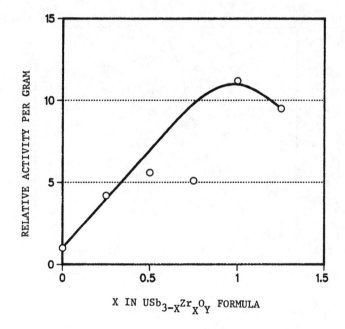

Figure 4.   Relative activity of $USb_{3-x}Zr_xO_y$ catalysts.

Although the X-ray diffraction patterns indicate that very little tin was incorporated in the USb_3O_{10} phase, tin had the greatest effect on catalyst activity. A catalyst consisting of equimolar amounts of $USbO_5$ and USb_3O_{10} without any promoters is less active and less selective for the production of acrylonitrile than than USb_3O_{10} (2). Also, SnO_2 by itself is a poor catalyst. Yet, intimate mixing of these phases produced highly active catalysts. The x=1.0 and x=1.25 compositions had relative activities of 13.0 and 13.9, while exhibiting good selectivity for acyrlonitrile production (Figures 5 and 6).

Table V shows the effect of Ti, Zr, and Sn addition when excess antimony was present. Although each increased catalyst activity, the effect was much smaller than for the $USb_{3-x}M_xO_y$ compositions. Titanium addition about doubled the relative activity compared to the standard uranium–antimony oxide catalyst, while Zr and Sn addition had a smaller effect. The poor selectivity of the Distillers-type catalyst, No. 9, is attributed to the presence of $USbO_5$.

Table V. Promoting Effect In Presence Of Excess Antimony

Catalyst No.	Catalyst Composition	% C_3H_6 Conv.	% AN Sel.	Rel. Act.
1	$USb_{4.6}Ox$	16.0	82.1	1.0
10	$USb_{4.0}Ti_{0.9}O_y$	45.7	84.5	2.3
9	$USb_{4.0}Ti_{0.9}O_y$	33.7	58.4	1.8
17	$USb_{4.6}Zr_{1.0}O_y$	32.2	82.6	1.8
35	$USb_{4.6}Sn_{1.0}O_y$	19.8	80.3	1.2

Oxidation State of Uranium

Figure 7 shows a portion of the infrared transmission spectrum for the $USb_{3-x}Ti_xO_y$ catalysts. Infrared bands at 925 and 865 cm^{-1} in USb_3O_{10} have been attributed to the $U-O_{III}$ stretch (3), while a band at 715 cm^{-1} is believed to be an Sb–O stretch. As x was varied from 0 to 1.5, the infrared adsorption bands at 925 and 865 cm^{-1} shifted to 950 and 895 cm^{-1}, while the band at 740 cm^{-1} shifted in the opposite direction to 715 cm^{-1}. As x was increased above 1.5, these bands disappeared.

The effective molar paramagnetic moment of USb_2TiO_y was less than that of the standard uranium–antimony oxide composition (Figure 8) but still significant. As temperature was increased from 4 to 105°K, the effective magnetic moment of the old uranium–antimony oxide catalyst increased to a value corresponding to one unpaired electron which is consistent with U^{+5}. At low temperatures the effective magnetic moment of USb_2TiO_y was significantly lower than for the old uranium–antimony catalyst, but as the temperature was

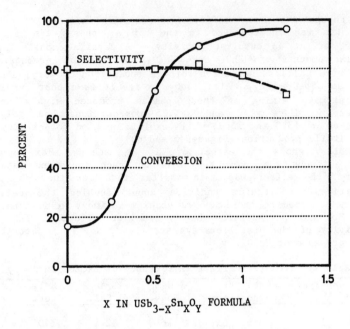

Figure 5.   Effect of Sn substitution for Sb.

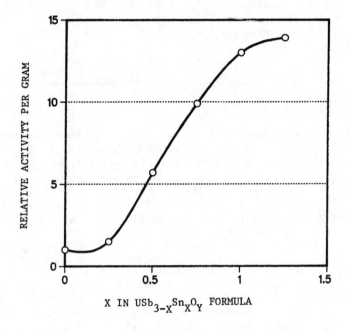

Figure 6.   Relative activity of $USb_{3-x}Sn_xO_y$ catalysts.

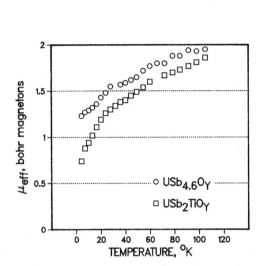

Figure 7. Infrared spectra of $USb_{3-x}Ti_xO_y$ catalysts.

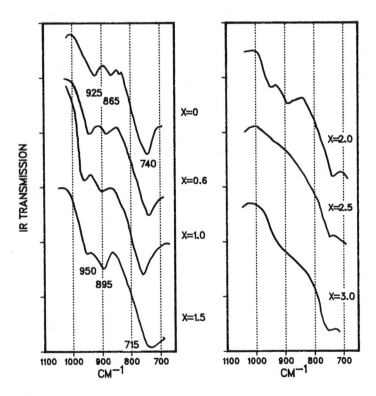

Figure 8. Effective magnetic moment of uranium in original and titanium substituted catalysts.

increased to 105°K it also approached values corresponding to one unpaired electron.

While titanium substituted for antimony and this had a dramatic effect on catalytic activity as expected, there is a question as to how much of the uranium was converted from the +5 to the +6 oxidation state. The shifts in the infrared bands indicate a shortening of the $U-O_{III}$ bond distance and a lengthening of the Sb-O bond distance which is consistent with an increase in hexavalent character, but the magnetic measurements show that a substantial portion of the uranium remained in +5 state. If the valence of uranium is not changed, then the replacement of Sb^{+5} by Ti^{+4} must generate oxygen vacancies in the USb_3O_{10} lattice. It is these sites that may be responsible for the high activity of the promoted catalysts.

Increasing Catalyst Stability

The USb_2TiO_{9-10} catalyst was made attrition resistant by adding an equal weight of silica binder prior to calcination. X-ray diffraction patterns show that the same USb_2TiO_{9-10} phase was formed as in the unsupported catalyst. Acrylonitrile selectivity was unaffected by the addition of silica and the activity per gram of active phase was as good or better than with the unsupported catalyst.

Like the original uranium-antimony oxide catalyst, the titanium substituted catalysts were able to operate only a short time without regeneration. Otherewise, the catalyst became over-reduced, the USb_3O_{10} type phase decomposed, and selectivity suffered. The addition of small amounts of molybdenum or vanadium prevented over-reduction enabling the catalyst to operate without regeneration.

Replacing 0.10 atom of uranium with molybdenum stabilized the catalyst with only a small effect on activity and acrylonitrile selectivity (Table VI). Further replacement of uranium by molybdenum markedly reduced catalyst activity, so that contact time had to be increased to maintain a high conversion. The addition of molybdenum resulted in a greater production of by-product HCN and correspondingly less carbon oxides (Table VII). A similar effect was obtained with vanadium. However, the vanadium seemed to increase activity as well as stabilize the catalyst.

Conclusion

The catalytic activity of the uranium-antimony oxide catalyst for propylene ammoxidation has been increased an order of magnitude by modifying the catalytically active phase rather than by adding various promoters to the optimum uranium-antimony oxide composition. This modification was accomplished by substituting titanium, zirconium, or tin for antimony in compositions with the empirical formula $USb_{3-x}M_xO_y$. Titanium and zirconium replaced

Table VI. Molybdenum Or Vanadium Addition To Improve Catalyst Stability

X Value Mo	V	Contact Time, s	% C_3H_6 Conversion	Acrylonitrile Selectivity	Hours Stable without Regeneration
-	-	1.7	98	86	0.5
0.025		1.7	98	86	0.5
0.50		1.7	99	81	0.75
0.10		1.7	96	85	>150.
0.15		3.2	92	83	>150.
0.20		6.5	96	84	>150.
-	0.10	1.1	98	84	>150.
0.05	0.05	1.1	99	83	>150.

Catalyst composition 50 wt% $U_{0.9}Sb_2Ti(Mo,V)_xO_{9-10}$/50 wt% SiO_2

Contact-time = 1.7 s

Temperature = 475°C

C_2H_6/air/NH_3 = 1.0/11/1.1

Table VII. Effect Of Molybdenum Addition On Selectivities

Catalyst Composition	50% USb_2TiO_{9-10} 50% SiO_2	50% $U_{0.9}Sb_2TiMo_{0.1}O_{9-10}$ 50% SiO_2
% C H Conversion	98.2	96.3
Selectivities:		
CO + CO_2	12.0	7.1
HCN	0.4	6.7
Acetonitrile	1.6	1.7
Acrylonitrile	86.0	84.5

Contact-time = 1.7 s

Temperature = 475°C

C_3H_6/air/NH_3 = 1.0/11/1.1

antimony in the USb_3O_{10} lattice with only small changes in the X-ray diffraction pattern. Tin was not incorporated into the USb_3O_{10} lattice but still had a strong effect on catalytic activity. Since a significant amount of uranium remains in the +5 oxidation state, we believe that the replacement of Sb^{+5} with Ti^{+4} and Zr^{+4} generates oxygen vacancies in the crystal lattice which enhance catalytic activity. The stability of these catalysts is increased by the addition of small amounts of molybdenum or vanadium which may catalyze reoxidation of the catalyst preventing over-reduction.

Acknowledgments. We thank Professor W. E. Wallace of the University of Pittsburgh and his students for making the magnetic susceptibility measurements.

Literature Cited

1. Callahan, J. L.; Gertisser, B. U.S. Patent 3,308,151, issued to The Standard Oil Co. (Ohio), (Aug. 3, 1965).
2. Grasselli, R. K.; Callahan, J. L. J. Catal. 1969, 14, 93-103.
3. Grasselli, R. K.; Suresh, D. D.; Knox K. J. Catal. 1970, 18, 356.
4. Grasselli, R. K.; Suresh, Dev D. J. Catal., 1972, 25, 273-291.
5. Innes, R. A.; Perrotta, A. J. U.S. Patatent 4,040,983, issued to Gulf Research & Development Company (Aug. 9, 1977).
6. Innes, R. A.; Perrotta, A. J. U.S. Patatent 4,045,373, issued to Gulf Research & Development Company (Aug. 30, 1977).
7. Innes, R. A.; Kehl, W. L. U.S. Patent 4,222,899, issued to Gulf Research & Development Co. (Sept. 16, 1980).
8. Innes, R. A.; Kehl, W. L., U.S. Patent 4,296,046, issued to Gulf Research & Development Co. (Oct. 20, 1980).
9. Butera, R. A.; Craig R. S.; Cherry, L. V. Rev. Sci. Instr. 1961, 32, 708-711.
10. Ball, W. J.; Barclay, J. L.; Boheman, J.; Gassen, E. J.; Wood, B. Brit. Patent 1,007,929, issued to Distillers Company Limited, (Oct. 22, 1965).
11. Callahan, J. L.; Grasselli, R. K.; Knipple, W. R. U.S. Patent 3,328,315, issued to The Standard Oil Company (Ohio), (June 27, 1967).
12. "Lange's Handbook of Chemistry"; Twelfth Edition, Dean, J. A., Ed.; Section 3, pp. 120-123, McGraw-Hill Inc., New York, N.Y. (1979).
13. Chevalier, M.; Gasperin, M. C. R. Acad. Sci. 1968, C 267, 481.
14. Chevalier, M.; Gasperin, M. C. R. Acad. Sci. 1969, C 268, 1426.

RECEIVED October 4, 1984

Phase Relationships in the Cerium–Molybdenum–Tellurium Oxide System

J. C. J. BART[1], N. GIORDANO[1], and P. FORZATTI[2]

[1]Instituto di Chimica Industriale, Universita di Messina, Messina, Italy
[2]Dipartimento di Chimica Industriale ed Ingegneria Chimica del Politecnico, Piazza Leonardo da Vinci 32, 20133 Milano, Italy

The complex solid state relations of the cerium–molybdenum–tellurium oxide system were studied to determine the boundaries of single phase regions and phase distributions of a typical multicomponent ammoxidation catalyst. Between 400° and 600°C in air the $(Ce,Mo,Te)O$ system contains the following phases: CeO_2, MoO_3, TeO_2, $\alpha-Te_2MoO_7$ and $\beta-Te_2MoO_7$, $\beta-Ce_2Mo_3O_{13}$, $Ce_2Mo_3O_{12.25}$, $\alpha-Ce_2Mo_4O_{15}$ and $\beta-Ce_2Mo_4O_{15}$, $CeTe_2O_6$, $Ce_2(TeO_4)_3$, a solid solution $(Ce,Te)O_2$, $Ce_6Mo_{10}Te_4O_{47}$, $Ce_2Mo_2Te_2O_{13}$, $Ce_4Mo_{11}Te_{10}O_{59}$, $Ce_2Mo_2Te_4O_{17}$, $Ce_{10}Mo_{12}Te_{14}O_{79}$, fields of primary crystallization of each of these compounds in the $(Ce,Mo,Te)O$ system are indicated. A typical active $(Ce,Mo,Te)O$ ammoxidation catalyst is composed of the binary phase $\beta-Ce_2Mo_3O_{13}$ and/or $\alpha-Ce_2Mo_4O_{15}$ and the ternary oxide $Ce_4Mo_{11}Te_{10}O_{59}$ (eventually together with $Ce_6Mo_{10}Te_4O_{47}$ and MoO_3).

The structure of a highly active cerium–molybdenum–tellurium acrylonitrile catalyst([1]) has previously been described in terms of binary $(Ce,Mo)O$ and ternary $(Ce,Mo,Te)O$ phases([2]). It was concluded that none of the constituent oxides (CeO_2, MoO_3 and TeO_2)

0097–6156/85/0279–0089$06.00/0

or compounds of the binary (Te,Mo)O or (Te,Ce)O systems are present as the active phases in the ammoxidation catalyst.

We have recently identified and characterized several new ternary oxides (Ce,Mo,Te)O(3), after studying over 100 different compositions calcined in air at temperatures between 400° and 600°C. Combined with the knowledge of the (Te,Mo)O, (Ce,Mo)O, and (Ce,Te)O chemistry, which was developed in the last decade, it is now possible to describe the complex solid-state relations of the ternary (Ce,Mo,Te)O system. The results culminated in the identification of the active phase composition of a typical (Ce,Mo,Te)O acrylonitrile catalyst.

Experimental

Preparative methods and samples used for this study were those of previous work(3). Samples were subjected to x-ray diffraction (CuKα radiation) and spectra were interpreted with reference to the following support materials (taken from the binary oxide systems): Te_2MoO_7(4), $\alpha-Ce_2Mo_3O_{13}$ and $\beta-Ce_2Mo_3O_{13}$(5), $\alpha-Ce_2Mo_4O_{15}$ and $\beta-Ce_2Mo_4O_{15}$(5)$Ce_2Mo_3O_{12.25}$(5), $Ce_2(MoO_4)_3$(6), $CeTe_2O_6$(7), $Ce_2(TeO_4)_3$(7), (Ce,Te)O$_2$(7) together with TeO_2 (ASTM 11-693), CeO_2 (ASTM 4-593) and MoO_3 (ASTM 9-209). Ternary oxides (Ce,Mo,Te)O were identified on the basis of previous work(3). The relative amounts of the phases formed were estimated by comparison of the heights of the characteristic peaks in non-overlapping positions. As especially TeO_2- and MoO_3-rich samples are often non-crystalline after heating above 550°C, such preparations were calcined additionally at 500°C for 8 hours followed by slow cooling in order to enhance crystallinity.

Results and Discussion

Figure 1 shows the distribution of the three constituent oxides (CeO_2, MoO_3 and TeO_2) in the various ternary phase compositions at temperatures from 400° to 600°C. It is clearly seen that CeO_2 exhibits the lowest overall reactivity. Noteworthy is the

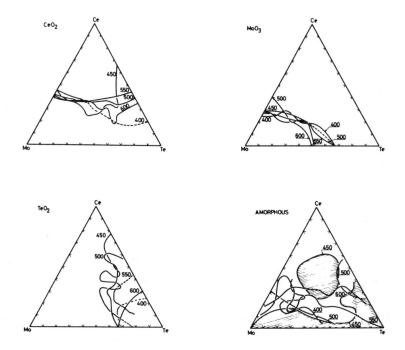

Figure 1. Single-phase boundaries for the component oxides of the (Ce,Mo,Te)O system between 400° and 600° C and regions of formation of non-crystalline reaction products.

considerable extension of the compatibility range for TeO_2 between 400° and 450°C, which correlates with the decomposition process of H_6TeO_6. Also of interest is the affinity of TeO_2 for the other components (in particular CeO_2) above 500°C.

The presence of $-Te_2MoO_7$ (Figure 2) in the CeO_2-poor area of the phase triangle is expected based on the known behavior of the TeO_2-MoO_3 system($\underline{4},\underline{8}$). Noteworthy is the presence of considerable non-crystalline material in this part of the diagram, especially at 550° and 600°C, which relates partly to the formation of the $\beta-Te_2MoO_7$ glass (Figure 1). The glass-forming tendency of Te_2MoO_7 ($\underline{4},\underline{9}$) leads to an underestimate of the extension of the compatibility range of the compound.

Among the products of the (Ce,Mo)O system, only yellow-green $\beta-Ce_2Mo_3O_{13}$ is found below 450°C; the compound occupies an extensive phase field up to 500°C. Above 500°C the phase distribution is more complex with formation of $\alpha-Ce_2Mo_4O_{15}$ and $Ce_2Mo_3O_{12.25}$ at increasingly higher temperatures. $Ce_2Mo_3O_{12.25}$ is abundant especially at 550° and 600°C (Figure 2). At 600°C, $\alpha-Ce_2Mo_4O_{15}$ shows a more restricted phase field in comparison to $Ce_2Mo_3O_{12.25}$. Brown-red $\beta-Ce_2Mo_4O_{15}$ is formed at about 600°C as a minor product under our reaction conditions. The observed formation sequence of the cerium-molybdenum oxide phases agrees with the phase relations found in the (Ce,Mo)O system($\underline{2}$), even though the binary compounds are formed at lower temperatures in the ternary system. In fact, whereas $\beta-Ce_2Mo_3O_{13}$ is detected at 500°-550°C in (Ce,Mo)O, this compound forms already at 400°C in the (Ce,Mo,Te)O system. Similarly, in the binary system in air $\alpha-Ce_2Mo_4O_{15}$ and $Ce_2Mo_3O_{12.25}$ are formed at 550° and 650°C, respectively, as opposed to 450°C in the ternary system. $\beta-Ce_2Mo_4O_{15}$ forms by polymorphic transformation of $\alpha-Ce_2Mo_4O_{15}$ at about 650°C in air in the binary system and at a slightly lower temperature in this study. The absences of $\gamma-Ce_2Mo_3O_{13}$ and the scheelite $Ce_2(MoO_4)_3$ are not surprising and are in accordance with previous data($\underline{2}$).

With regard to the phases of the (Te,Ce)O subsystem (Figure 3), we notice that the solid solutions $\alpha-$ and $\beta-(Ce,Te)O_2$

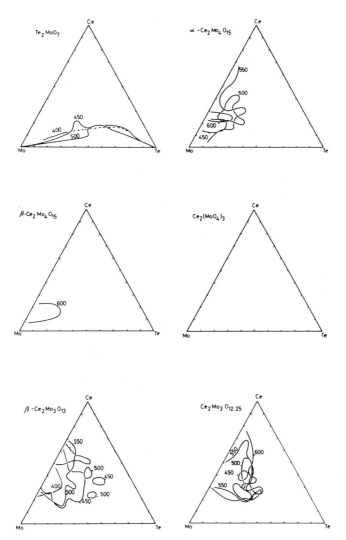

Figure 2. Single-phase boundaries for (Te,Mo)O and (Ce,Mo)O phases in the (Ce,Mo,Te)O system between 400° and 600°C.

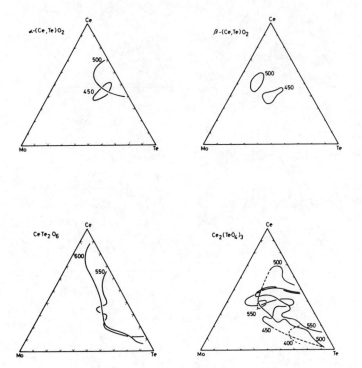

Figure 3. Single-phase boundaries for (Ce,Te)O phases in the
(Ce,Mo,Te)O system between 400° and 600°C.

are both formed in rather restricted compositional ranges between 450° and 500°C. These solid solutions are not stable at higher temperatures($\underline{7}$,$\underline{10}$). The Te(VI)– containing scheelite $Ce_2(TeO_4)_3$ is formed at a temperature where H_6TeO_6 decomposes to TeO_2 (above 400°C). The stability range (up to 550°C) of this widely distributed phase is slightly more restricted than in the binary (Te,Ce)O system. Formation of $CeTe_2O_6$ at 550°C conforms with our knowledge of the (Te,Ce)O system($\underline{7}$). This phase also occupies a broad compatibility range. The sequence of phase formation with increasing temperature, namely from $(Ce,Te)O_2$ to $Ce_2(TeO_4)_3$ and $CeTe_2O_6$, is in good agreement with the binary system($\underline{7}$).

As may be seen from Figure 4, the molybdenum– and/or tellurium–rich ternary compounds $Ce_4Mo_{11}Te_{10}O_{59}$, $Ce_2Mo_2Te_4O_{17}$, and $Ce_6Mo_{10}Te_4O_{47}$ occupy important and extensive composition ranges within the (Ce,Mo,Te)O phase triangle. The central portion of this diagram is taken up by $Ce_6Mo_8Te_6O_{45}$, $Ce_2Mo_2Te_2O_{13}$, and $Ce_{10}Mo_{12}Te_{14}O_{79}$, the latter two with broad compatibility ranges. Also, some minor components are formed in the solid state equilibria of the (Ce,Mo,Te)O system, namely $Ce_2Mo_3Te_2O_{16}$ and $(Ce_4Mo_{13}Te_3O_{51})$.

Under our experimental conditions, four of the aforementioned ternary compounds are formed already at 400°C. Amongst these is $(Ce_4Mo_{13}Te_3O_{51})$, which occupies a small compatibility area and is not observed at higher temperatures. The molybdenum–rich compound $Ce_6Mo_{10}Te_4O_{47}$ is stable over a broad temperature range (400°–550° C), and is extensively present in the 400°–450°C range. Also, $Ce_2Mo_2Te_2O_{13}$ is stable in the 400°–550°C range. $Ce_4Mo_{11}Te_{10}O_{59}$ is the only ternary compound which is stable over the full temperature range investigated; its phase field is most extensive in the 450°–500°C temperature interval.

Compounds $Ce_2Mo_2Te_4O_{17}$, $Ce_{10}Mo_{12}Te_{14}O_{79}$ and $Ce_6Mo_8Te_6O_{45}$ are all formed at about 450°C and are stable up to at least 600°C. $Ce_2Mo_2Te_4O_{17}$ exhibits the most extensive compatibility range of all ternary compounds (in particular above 500°C); $Ce_{10}Mo_{12}Te_{14}O_{79}$ finds its major extension at 600°C. Finally, the minor phase $Ce_2Mo_3Te_2O_{16}$ is formed at 500°C and is stable up to over 600°C.

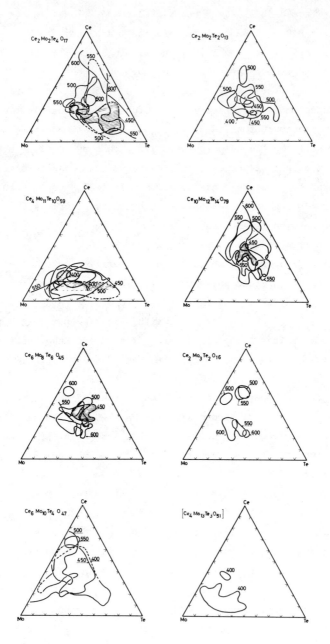

Figure 4. Single-phase boundaries for (Ce,Mo,Te)O phases in
the temperature range between 400° and 600°C.

As may be seen from Figure 1, x–ray amorphous or micro-crystalline material was detected in fairly extensive areas in the phase triangle at various temperatures. In particular, many cerium–poor samples calcined at 600°C show a vitreous or highly sintered aspect and are dark in colour. However, only in a few cases completely amorphous samples were observed but the regions of glass formation obviously strongly depend on the cooling rate; this aspect was not further investigated. It appears that considerable amounts of (Te,Mo)O glass–like material (of the β–Te_2MoO_7 type) are formed at 550° and 600°C. We have noticed that the x–ray scattering maximum I_{max} of the amorphous fraction in the samples calcined at 600°C varies from d=3.34 Å at the molybdenum–rich side to 3.19 Å at the tellurium–rich side (cfr. 3.33 Å in β–Te_2MoO_7 and 3.29 Å in $Ce_4Mo_{11}Te_{10}O_{59}$). At 550°C the tellurium– and molybdenum–rich amorphous materials exhibit maxima in correspondence to d=3.34 Å but with increasing cerium concentration this value drops to ca. 3.24 Å. Also, the non–crystalline molybdenum–rich fraction at 500°C conforms to glassy (Te,Mo)O (d_{max}=3.33 Å), but the molybdenum–poor microcrystalline part at the same temperature (Figure 1) is different (d_{max}=3.15 Å). At lower temperatures the presence of non–crystalline material is probably partly due to incomplete decomposition and interaction of the starting products. The extensive amorphous phase formation in the central portion of the phase diagram at 450°C may be due to CeO_2 or $Ce_6Mo_{10}Te_4O_{47}$ (d_{max}=3.15 Å); in the cerium–poor amorphous fraction d_{max} varies from ca. 3.36 to 3.27 Å at the Mo– and Te–rich sides, respectively. At 400°C these values are 3.30 and 3.19 Å, respectively. At this temperature, some completely amorphous samples were found at the Ce:Mo:Te=(15–20):45:(35–40) ratios.

Despite the great complexity of the system (with 20 different phases) all x–ray powder spectra were interpreted. As may easily be ascertained from the figures, the proposed phase distributions properly account for the presence of each of the cations over the full compositional range with the exception of the phase ranges with less than about 5 at% of one of the components. This is reasonable and corresponds to the sensitivity limit of the x–ray

method. Only in the 400°C series, extensive areas in the cerium-
and tellurium-rich ranges of the phase triangle do not account for
the presence of Te or Mo.   This is, however, a consequence of less
complete data at this temperature.   Based on the reported thermo-
analytic data for the new ternary phases(3), it appears that below
the TeO_2-$Ce_2(MoO_4)_3$ binary juncture the liquid surface descends
steeply; compositions rich in MoO_3 and TeO_2 melt at considerably
lower temperatures.

   In some reactant mixtures as many as six crystalline phases
were detected, e.g. as in the case of the composition
Ce:Mo:Te=7:8:5 after calcination at 550°C for 8 hours.   The
presence of more phases than those permitted under thermodynamic
equilibrium is a consequence of the incompleteness of the reactions
between the components under our experimental conditions.   It is
also noticed that various areas of the phase diagram are highly
sensitive to the preparative conditions; e.g. the composition
Ce:Mo:Te=5:8:7 is composed of $Ce_2Mo_2Te_4O_{17}$, $Ce_4Mo_{11}Te_{10}O_{59}$ and
β-$Ce_2Mo_3O_{13}$ after calcination at 550°C for 8 hours but consists of
$Ce_6Mo_8Te_6O_{45}$, $Ce_2Mo_3Te_2O_{16}$ and an amorphous fraction after
calcination at 500°C and 550°C each for 8 hours.   Another typical
example is given in Table I, but various other such cases were also
encountered.   Without variations in the reaction param-eters, the
results are normally perfectly reproducible.   No reduced phases
(such as reduced molybdenum oxides and $TeMo_5O_{16}$) were ever
observed, as indeed expected.

Conclusions

On the basis of the solid-state relationships of the (Ce,Mo,Te)O
system, it is now possible to derive the phase distribution of a
typical unsupported (Ce,Mo,Te)O acrylonitrile catalyst with the
composition of Ref.(11), as indicated in Table I.   The results
agree with previous conclusions with regard to the role of (Te,Mo)O
and (Te,Ce)O oxides in this system and the most likely composition
of the active (Ce,Mo)O phases (β-$Ce_2Mo_3O_{13}$ and α-$Ce_2Mo_4O_{15}$) (2).
XPS results (Ce(III) rather than Ce(IV) in the catalyst) favour the

Table I. Phase distribution in the unsupported (Ce,Mo,Te)O catalyst.

Activation conditions		Phase distribution[b]						
T (°C)	t (h)	MoO_3	$-Ce_2Mo_3O_{13}$	$Ce_2Mo_3O_{12.25}$	$-Ce_2Mo_4O_{15}$	$Ce_4Mo_{11}Te_{10}O_{59}$	$Ce_6Mo_{10}Te_4O_{47}$	amorphous
400	8	s	-	-	-	+	+	-
450	8	s	+	-	-	+	-	-
500	a+8	s	+	-	+	+	-	-
550	a+8	-	-	-	+	-	-	+ (c)
600	a+8	-	-	+	+	-	-	+ (c)
600	8	-	-	+	-	-	-	+ (c)

a   As preceeding line.

b   +, present; -, absent; s, small amount.

c   Amorphous fraction with d_{max}=3.33 Å.

presence of $\alpha-Ce_2Mo_4O_{15}$ and lead to a revision of the proposed oxygen content of the catalyst(1). The absence of $Ce_2(MoO_4)_3$ as a component in the phase diagram is of significance with respect to the catalyst composition and activity. In fact, it is well known that this phase is essentially inactive in selective ammoxidation of propylene(12). The compatibility range of MoO_3 in the (Ce,Mo,Te)O system casts doubt on the presence of this compound as a significant component of the catalyst. This conclusion agrees with previously presented results(2) and our current study (Table I).

The new feature, derived from this work, is the identification of the active ternary phase(s), namely $Ce_4Mo_{11}Te_{10}O_{59}$ and/or $Ce_6Mo_{10}Te_4O_{47}$ with the latter not being stable much above 500°C. It is noticed that according to the phase distribution of samples with the stoichiometry of the active catalyst, no tellurium-containing phase is detected by x-ray diffraction above 500°C (Table I). However, the maximum of the amorphous fraction (d_{max}=3.33 Å) points to $Ce_4Mo_{11}Te_{10}O_{59}$ (d_{max}=3.29 Å), a compound which is easily obtained in non-crystalline form above 500°C(3). The previously reported unknown ternary oxide(2) has now been identified as $Ce_4Mo_{11}Te_{10}O_{59}$. Therefore, it is likely that the active phase of the (Ce,Mo,Te)O ammoxidation catalyst consists of an $\alpha-Ce_2Mo_4O_{15}$-rich mixture containing $Ce_4Mo_{11}Te_{10}O_{59}$.

In evaluating our results and conclusions in relation to the industrial catalyst(1), an account should also be taken of the effect of the silica support, which has not been considered here and yet is likely to play a role. In fact, as shown previously(7,10), the phase distribution of the (Te,Ce)O system differs from that of the (Te,Ce)O/SiO$_2$ system at the same activation temperature due to the dilution and interaction effect which affects the respective formation rates and stability ranges of the various phases. In fact, from x-ray diffraction data of a fresh SiO$_2$-supported active ternary phase, the formation of some CeO_2 is inferred, contrary to the results of the unsupported system. For these reasons, additional spectroscopic and catalytic activity studies are necessary to confirm our suggestions with regard to the nature of the active phases contained in the industrial $(Ce,Mo,Te)O/SiO_2$ catalyst.

Acknowledgments

One of us (P.F.) acknowledges support from the Italian Ministry of Education.

Literature Cited

1. Caporali, G.; Ferlazzo, N.; Giordano, N. German Patent 1.618.685, Nov. 23, 1972.
2. Bart, J. C. J.; Giordano, N. I.&E.C. Prod. Res. Dev., 1984, 23, 56
3. Bart, J. C. J.; Forzatti, P.; Garbassi, F.; Cariati, F., Proc. Third Intl. Symp. Ind. Uses Selenium & Tellurium, Stockholm, 1984.
4. Bart, J. C. J.; Petrini, G.; Giordano, N. Z. Anorg. Allg. Chem., 1975, 412, 258.
5. Castellan, A.; Bart, J. C. J.; Bossi, A.; Perissinoto, P.; Giordano, N. Z. Anorg. Allg. Chem., 1976, 422, 155.
6. Bart, J. C. J.; Giordano, N. J. Less Common Metals, 1975, 40, 257.
7. Bart, J. C. J.; Giordano, N.; Gianoglio, C. Z. Anorg. Allg. Chem., 1981, 481, 153.
8. Petrini, G.; Bart, J. C. J. Z. Anorg. Allg. Chem., 1981, 474, 229.
9. Dimitriev, Y.; Bart J. C. J.; Dimitrov, V.; Arnaudov, M. Z. Anorg. Allg. Chem., 1981, 479, 229.
10. Bart, J. C. J.; Giordano, N. J. Catal., 1982, 75, 134.
11. Hucknall, D. J. "Selective Oxidation of Hydrocarbons"; Acad. Press: London, 1974, p. 56.
12. Brazdil, J. F.; Grasselli, R. K. J. Catal., 1983, 79, 104.

RECEIVED March 20, 1985

Molybdate and Tungstate Catalysts for Methanol Oxidation

C. J. MACHIELS, U. CHOWDHRY, W. T. A. HARRISON, and A. W. SLEIGHT

Central Research & Development Department, Experimental Station, E. I. du Pont de Nemours and Company, Wilmington, DE 19898

Pure ferric, chromium and aluminum molybdates as well as complete series of solid solutions of iron-chromium and iron-aluminum molybdates were synthesized using a solution technique to ensure obtaining pure, single phase, homogeneous powders. The surface area of these molybdates varied from 5 to 15 m^2/g and homogeneity was confirmed using scanning transmission electron microscopy. The selective oxidation of methanol to formaldehyde was studied over these pure and mixed molybdates. No significant differences were observed between the phases in specific activity, selectivity, and kinetic parameters.

The most selective catalysts for the oxidation of methanol to formaldehyde are molybdates. In many commercial processes, a mixture of ferric molybdate and molybdenum trioxide is used. Ferric molybdate has often been reported to be the major catalytically active phase with the excess molybdenum trioxide added to improve the physical properties of the catalyst and to maintain an adequate molybdenum concentration under reactor conditions(1,2). In some cases, a synergistic effect is claimed, with maximum catalytic activity for a mixture with an Fe/Mo ratio of 1.7(3). A defect solid solution was also proposed(55). Aging of a commercial catalyst has been studied using a variety of analytical techniques(4) and it was concluded that deactivation can largely be accounted for by loss of molybdenum from the catalyst surface.

In this laboratory, the mechanism of methanol oxidation over molybdate and tungstate catalysts has been studied using a variety of techniques. Steady state and pulse reactor studies using labeled reactants have established that methanol conversion to formaldehyde is a redox reaction with lattice oxygen being involved(6). A kinetic isotope effect has recently been reported(5), and it shows that the rate limiting step in the reaction sequence is removal of a hydrogen from the methyl group. Fourier transform infrared studies have shown that a methoxy,

0097–6156/85/0279–0103$06.00/0

CH_3O, group is a surface intermediate during the reaction(53), and temperature programmed reaction studies have elucidated the nature of surface reactions and have allowed estimation of the number of catalytically active sites on powder surfaces(8). In this paper, we will discuss results of the oxidation of methanol over a series of molybdates including solid solutions of ferric, chromium and aluminum molybdates and also over a new ferric tungstate phase. The mixed molybdates of iron/chromium, iron/aluminum and chromium/aluminum were made for the first time in pure well-characterized forms. Results are compared with our earlier work over commercial mixtures of ferric molybdate and molybdenum trioxide and a number of pure molybdates(6).

There is a voluminous body of literature in patents(9-19), papers(20-27) and reports(28-34) on the preparation and catalytic properties of the methanol oxidation catalyst, often without detailed reference to the chemical composition of the products. Indeed, early investigations(35-37) suggested that no compound was formed in the reaction between Fe_2O_3 and MoO_3. However, Kozmanov et al.(38-44) have since prepared the pure iron (III) molybdate, $Fe_2(MoO_4)_3$, by solid state reaction of a stoichiometric mixture of iron and molybdenum oxides at 700°C. Jager(40) reported the formation of a bright green compound, while Nassau et al.(41), in their detailed study of trivalent molybdates, reported a tan compound prepared with careful annealing of the oxides at 600°C. X-ray powder measurements by Nassau(41) and Trunov and Kovba(43) suggested that $Fe_2(MoO_4)_3$ crystallized in an orthohombic space group, while an early single crystal study by Klevtsov and co-workers(45-47) reported the crystal symmetry to be monoclinic. Despite contradictory reports(48), centric monoclinic is now the accepted structure of the room temperature phase of $Fe_2(MoO_4)_3$ as confirmed by Chen(49) in this laboratory. The space group is $P2_1/a$ with a=15.707, b=9.231, c=18.204Å, and β=125.25°. The structure consists of a rather open 3-dimensional network of corner sharing FeO_6 octahedra and MoO_4 tetrahedra, with four crystallographically different iron sites which lead to some novel low temperature magnetic properties(50-51).

Sleight and Brixner(52) have shown the presence of a ferroelastic phase-transition at 499°C between a low-temperature monoclinic and high-temperature orthohombic phase on the basis of DSC measurements.

Many studies have been made of solution phase preparations of the ferric molybdate system. Aruanno and Wanke(54) used a preparation from ferric chloride and ammonium heptamolybdate solutions, following the work of Shelton et al.(11). This has been developed by Italian workers under Pernicone(55) into the Montedison process(56-60); they made an interesting statistical study of the precipitation stage(61). Pernicone suggests that excess MoO_3 may be incorporated into the $Fe_2(MoO_4)_3$ lattice(55) modifying the catalytic properties; however, not all workers are in agreement with this proposal(6,62).

Boreskov et al.(63,64) used $Fe(NO_3)_3$ and $(NH_4)_6Mo_7O_{24}$ solutions to prepare a precipitate and a number of other workers have followed this procedure in detail. In particular, Trifiro and co-workers(65) have explored this route. Their method involves a novel aging procedure of recovering the solid in the mother liquor at 100°C for a few hours(66,67) to control the Fe:Mo ratio in the final product.

Another solution method prepares an amorphous hydrated molybdate from Na_2MoO_4 and $Fe(NO_3)_3$ solutions(55), following the method of Kerr et al.(68) to produce a product with a precisely defined iron to molybdenum ratio. Ferric molybdate has been prepared from gels involving ferric nitrate and ammonium heptamolybdate/hydrogen peroxide solutions by Tsigdinos and Swanson(69). All of these solution methods involve stages of drying and calcining the initial precipitate to achieve the final product; full details are given in the respective references. Single crystals of $Fe_2(MoO_4)_3$ may be prepared hydrothermally following the work of Klevtsov(70) and Marshall(71).

Aluminum molybdate, $Al_2(MoO_4)_3$ was first prepared by Doyle and Forbes(72) quickly followed by other workers (41,73). DSC measurements by Sleight(52) indicate the structure to be monoclinic below 200°C; the structure is isomorphous with chromium molybdate(41).

There have been few studies on mixed compounds of iron, chromium and aluminum molybdates. Abidova and co-workers(74-75) made a somewhat inconclusive study on the mixed Fe/Al/Mo oxide system. Based on DSC and reflectance measurements on the 2-component systems, they concluded that the 3-component mixture would be a complex multiphase system. Another study(76) used co-precipitation of iron and chromium nitrates and ammonium heptamolybdate. However, their Mössbauer effect data suggested inhomogeneity in the final product.

One other notable method has been used in the preparation of mixed transition metal molybdates, amongst many other oxide systems. This novel method(77) involves preparation of the mixed metal oxides via an amorphous precursor such as a citrate salt of the appropriate metals, and then thermal decomposition of the complex to yield the resulting mixed oxides. The experimental procedures are described in four French patents(78-81), giving details of many different preparations including a proposed MoO_3 rich, chromium doped iron molybdate, prepared as a possible selective oxidation catalyst.

Catalyst Preparation

For the iron/aluminum series, preparations from mixtures of the oxides Fe_2O_3, Al_2O_3 and MoO_3 or from the nitrates $Fe(NO_3)_3$, $Al(NO_3)_3$ and the ammonium molybdate failed as did preparations from mixtures of the end-member molybdates, $Fe_2(MoO_4)_3$ and $Al_2(MoO_4)_3$. All the products had very poor homogeneity as determined by semi-quantitative analytical electron microscopy; similar results were experienced over the entire range of composition for x=0-2 in $Fe_{2-x}Al_x(MoO_4)_3$. Pelleted preparations fired at 700°C for several weeks showed no improvement in homogeneity with time. Samples fired at up to 1000°C lost MoO_3 as indicated by the presence of Fe_2O_3 x-ray lines in Guinier photographs, but still without noticeable improvement in product homogeneity. A composition range no better than ± 20 percent in x was the best obtained.

Precipitation methods from solution of the trivalent metal nitrates and ammonium heptamolybdate were also attempted. The method was successful for ferric molybdate following the aging method of Trifiro et al.(66) to produce a precipitate of the correct stoichiometry. It was notable that after this aging process, but before annealing, the dried precipitate showed some crystallinity, but atomic absorption measurements showed an excess of oxygen, presumably indicating most of the precipitate was hydrated. Annealing in air at 400°C just above the temperature at which the last water of hydration is lost for 24 hours was found to produce a good $Fe_2(MoO_4)_3$ x-ray pattern without any observable contamination by MoO_3 or Fe_2O_3. Similar results were experienced for mixed molybdates doped with small amounts of chromium and aluminum but those preparations containing more than 25% of dopant showed MoO_3 contamination when examined by x-ray diffraction and the method failed to produce the desired result for chromium or aluminum. It appears that the iron molybdate is preferentially formed at low pH values (typically 1.5 in these cases) while the chromium or aluminum ion remains in solution. Attempts to modify the pH of the solution by the addition of NH_4OH had little effect on the resulting products which were still heavily contaminated with MoO_3. Even for those samples doped with small amounts of Al or Cr, x-ray microanalysis showed poor product homogeneity as with the solid state preparations.

By far the most successful method of preparation was via an amorphous organic precursor to the required mixed molybdates, following the method of Delmon et al.(77). The method has proved successful for the pure iron, chromium and aluminum molybdates, and also for the mixed phases. A detailed outline of the method taking the example of $FeCr(MoO_4)_3$ is as follows:

i) 11.44g (2.83×10^{-2} mol) of ferric nitrate hydrate, $Fe(NO_3)_3 \cdot 9H_2O$ was dissolved in 100cc of distilled water at room temperature resulting in a yellow solution.

ii) 11.34g (2.83×10^{-2} mol) of chromium nitrate hydrate, $Cr(NO_3)_3 \cdot 9H_2O$ was dissolved in 100cc of distilled water at room temperature resulting in a dark blue solution.

iii) The iron and chromium solutions were mixed resulting in a blue solution, pH ca. 1.5 to which 20g (9.5×10^{-2} mol) of citric acid monohydrate $C_6H_8O_7 \cdot H_2O$ was added, and stirred to dissolve.

iv) 15g (1.21×10^{-2} mol) of ammonium heptamolybdate hydrate, $(NH_4)_6Mo_7O_{24} \cdot 4H_2O$ was dissolved in 200cc of pure water resulting in a clear solution.

v) The molybdate solution was added to the nitrate solution; no precipitate formation was observed and the result was a blue solution.

vi) This solution was dried on a steambath overnight, to a blue-green glass and this was transferred to a vacuum oven at 70°C for 1 hour to complete the drying of the precursor. The resulting glass is very hygroscopic gaining a sticky appearance in only a few minutes.

vii) The precursor was ground to a green powder. X-ray diffraction indicated that it was totally amorphous.

viii) The precursor was calcined at 400°C for 24 hours in air.

X-ray diffraction indicated a crystalline molybdate pattern without contamination; the product was a yellowish powder. The procedure is discussed in detail by Delmon et al.(82-84). The crucial step appears to be the rapid dehydration of the starting solution before any of the components can crystallize out of solution separately. Delmon(85) suggests that a rotary vacuum evaporation would be an effective method of drying the precursor. The actual structure of the precursor is not well defined, but appears to require at least one equivalent of citrate ion per mol of metal ion(83), as presumably the citrate complexes all the metal species in solution. The resulting powder patterns, after annealing, indicated no contamination. Delmon(83) suggests that any multifunctional acid containing at least one carboxyl and one hydroxyl function may be effective. Experiments with tartaric acid on the iron/chromium system produced results similar to citric acid; a calcination temperature of $500^{\circ}C$ was necessary before crystallization occurred.

The preparation of the new ferric tungstate phase has been described previously(7). It is schematically shown in Figure 1.

Catalyst Characterization

A continuous range of solid solution, such as the series $Fe_{2-x}Cr_x(MoO_4)_3$ provides a good opportunity for the quantitative, comparison of two analytical techniques - "classical" atomic absorption analysis and x-ray microanalysis. X-ray microanalysis of thin samples using scanning transmission electron microscopy has become an effective quantitative technique in the last few years(86), as opposed to the well-known electron microprobe analyses of bulk specimens(87).

All the work described below was carried out on a Vacuum Generators HB501 instrument with an accelerating voltage of 100kV, and at a typical magnification of 1 million. Powdered samples were dispersed onto carbon coated 3mm copper grids from a suspension in water, which led to a satisfactory dispersion over the grid. Particles analyzed measured no more than 1000Å in size whenever possible, to minimize absorption effects, and those particles lying near the center of grid-squares were selected to minimize the intensity of the background CuK_α emission peaks due to the grid. The chromium, iron and molybdenum K_α lines were used in the analysis, their average energies being 5.43, 6.43 and 17.50 KeV respectively. The MoL_α line was not selected for the quantitative analysis due to the high Bremsstrahlung background at low energy, and hence the difficulty in estimating an accurate background subtraction. At low energy, there are also absorption effects due to the beryllium detector window, and for this reason, the relatively feeble AlK_α peak may give unreliable quantitative results when only a small quantity of Al is present. Thus, in this study, the Fe/Al and Cr/Al molybdates were not examined by x-ray microanalysis. For each sample investigated, at least 30 crystallites were examined, and the resulting x-ray emission spectra were analyzed using standard Kevex software; background subtractions were made automatically, and peak intensity ratios were calculated. For each sample, a histogram of the Fe:Mo and

Cr:Mo ratios was plotted to ensure that the material was homogeneous and of single phase(86). There was a remarkable lack of impurity phases; in particular, MoO_3 was notable by its absence. Finally, Cliff-Lorimer K_{xy} values were determined for the pure iron and chromium molybdates, and the Fe:Mo and Cr:Mo ratios across the series were obtained. The x-ray microanalysis results are plotted in Figure 2.

Atomic absorption measurements were carried out by Galbraith Laboratories, Inc. These are plotted in Figure 3. In general, both methods are in excellent agreement with the predicted results. The x-ray results show that a solid solution exists across the whole range of composition of x in $Fe_{2-x}Cr_x(MoO_4)_3$, rather than any mixture of phases. Independent comparison of both sets of results with the theoretical values for each compound shows the x-ray results to be the closest to the predicted values, with a typical accuracy of $\pm$ 1 per cent compared to an estimated $\pm$ 3 per cent for the atomic absorption measurements.

The particle size of some of the mixed molybdates produced by the citric acid technique were determined using a Micromeritics "Sedigraph" instrument. The average particle size is quite large - in each case, the median particle size is about 20μm. Scanning electron micrographs of gold-coated samples suggest that this is a good approximation, many of these larger particles being agglomerates.

Surface areas were recorded for the whole series of each solid solution by the standard N_2 B.E.T. method. The results are listed in Table I. Especially notable are the relatively high surface area of those compounds rich in aluminum and $Al_2(MoO_4)_3$ itself. Such values are considerably higher than by previously attempted methods.

Table I
Surface Areas of Molybdates

	m^2/g
$Fe_2(MoO_4)_3$	7
$Al_2(MoO_4)_3$	14
$Cr_2(MoO_4)_3$	7
$Fe_{2-x}Al_x(MoO_4)_3$	?-17
$Fe_{2-x}Cr_x(MoO_4)_3$	4-13
$Cr_{2-x}Al_x(MoO_4)_3$	8-16

x=0.0 - 0.2 - 0.4.....2.0

All these molybdates are isostructural with ferric molybdate with an open 3-dimensional network of MoO_6 octahedra and MoO_4 tetrahedra. A ferroelastic transition exists from the low temperature monoclinic form to the high temperature orthorhombic form. The transition temperature varies from 200 C for pure aluminum molybdate to 385 C for pure chromium molybdate and 500 C for pure ferric molybdate. For the mixed molybdates, the transition temperature was found to be a linear function of composition as is illustrated in Figure 4 for the mixed iron-aluminum molybdates.

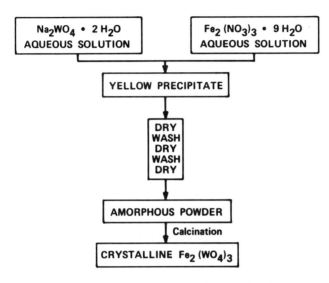

Figure 1.   Preparation Steps of Ferric Tungstate Phase,

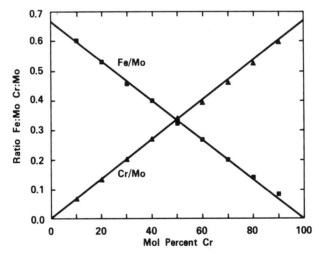

Figure 2.   X-ray Microanalysis for Fe/Cr Molybdate
Samples.

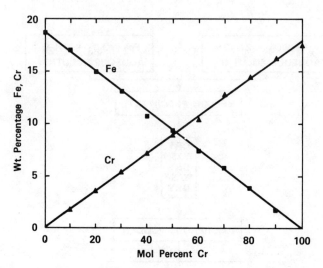

Figure 3.   Atomic Absorption for Fe/Cr Molybdate
Samples.

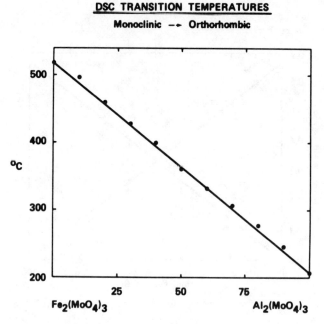

Figure 4.   Phase Transition Temperature for Fe/Al
Molybdate Samples.

Methanol Oxidation

The three pure molybdates, the ferric tungstate, and the mixed
molybdates with a 1/1 cation ratio were tested as catalysts for
methanol oxidation in a continuous flow reactor with external
recycle. The equipment and technique were described previously(6);
differential rate and selectivity data were obtained. The mixed
chromium-aluminum sample had very poor mechanical properties; no
recycle could be used as a result of excessive pressure drop over
the catalyst bed. The ferric tungstate sample showed behavior
quite different from that of the molybdate, results are shown in
detail elsewhere(7). The rate of reaction of methanol to
dimethylether was the same over both the tungstate and the
molybdate phases, but the reaction rate to formaldehyde was twenty
times larger over the molybdate than over the tungstate. As a
result, the product distribution was different for the tungstate
with dimethylether being the main product. Product distributions
for the three pure molybdates and the mixed molybdates were all
similar to those obtained previously for the commercial methanol
oxidation catalyst and various other molybdate phases(6). Figures
5 and 6 illustrate this product distribution for the mixed iron-
aluminum phase, selectivities are plotted over a range of methanol
conversion of 20-90%. Selectivity to formaldehyde can be increased
to over 90% by running at higher temperature, in a single pass
configuration and by adding water to the feed.

The kinetics of the reaction were determined by varying the
partial pressures of oxygen, water, and methanol as well as the
temperature. Other partial pressures were kept nearly constant;
nitrogen was the diluent. Kinetic observations also were similar
as previously reported(6) as is illustrated in Figures 7, 8 and 9
for different phases. The methanol reaction rate was nearly
independent of the oxygen partial pressure, except at very low
oxygen pressures in the reactor in which case the catalyst begins
to be reduced. It was shown previously(6) that a reduced catalyst
is much less active. The reaction rate has a positive
dependence on methanol partial pressure, but the reaction is
inhibited by the addition of water. Water does however increase
selectivity to formaldehyde at the expense of dimethoxymethane,
methylformate and dimethylether.

The kinetic data are fitted well by a power rate expression,
parameters are shown in Table II. They are in the expected ranges
with apparent activation energies ranging from 18 to 20 kcal/mol.
In order to compare the activity of the various phases, turnover
numbers were calculated at the following conditions: 250 C,
150 torr oxygen and 40 torr methanol partial pressure. These
turnover numbers are expressed as molecules of methanol reacting
per surface molybdenum atom and per second; they are listed in
Table III. Clearly there is little difference in activity between
the pure and mixed phases studied here and their activity is about
equal to that of a commercial mixture of ferric molybdate and
molybdenum trioxide. Pure molybdenum trioxide is less active by
about a factor 3, but we have shown by TPD studies that the
predominant (010) phase of molybdenum trioxide does not chemisorb
methanol(8). In comparison, various bismuth molybdate phases that

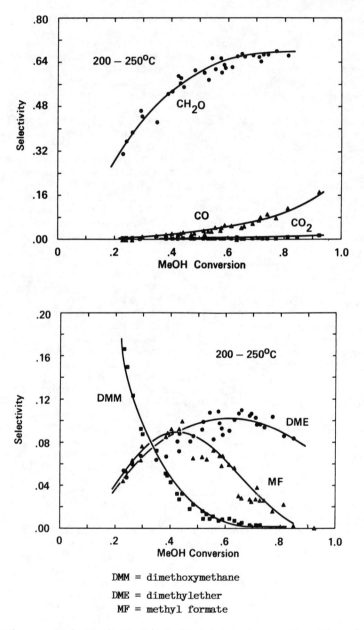

DMM = dimethoxymethane
DME = dimethylether
MF = methyl formate

Figure 5&6. Product Distribution versus Fractional
Methanol Conversion for FeAl(MoO$_4$)$_3$.

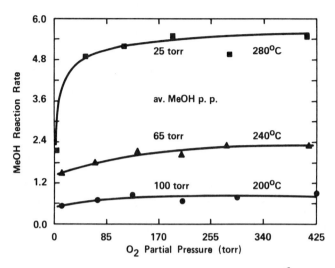

Figure 7. Reaction Rate of Methanol (10^{-6} moles/sec, g. catalyst) versus Oxygen Partial Pressure for $Fe_2(MoO_4)_3$.

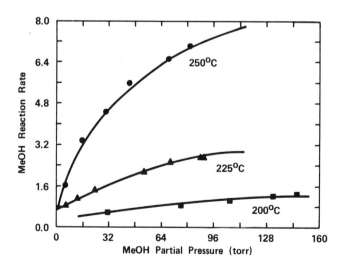

Figure 8. Reaction Rate of Methanol versus Methanol Partial Pressure for $FeAl(MoO_4)_3$.

are well known selective oxidation catalysts for other reactions
are much less active for methanol oxidation.

Table II. Power Rate Law Parameters

	E_a kcal/mol	Apparent Reaction Orders [O_2]	[MeOH]	[H_2O]
$Fe_2(MoO_4)_3$	19.0	0.17	0.58	
$Al_2(MoO_4)_3$	18.4	0.26	0.40	
$Cr_2(MoO_4)_3$	17.9	0.20	0.47	
$FeAl(MoO_4)_3$	20.3	0.16	0.49	-0.58
$FeCr(MoO_4)_3$	19.2	0.15	0.47	-0.55

Table III. Turnover Numbers

Molecules of MeOH Reacted Per Second, Per Surface
Molybdenum Atom at: 250°C
                                150 torr O_2
                                40 torr MeOH

$Fe_2(MoO_4)_3$	0.05
$Al_2(MoO_4)_3$	0.03
$Cr_2(MoO_4)_3$	0.08
$FeAl(MoO_4)_3$	0.06
$FeCr(MoO_4)_3$	0.06
MoO_3	0.02
$MoO_3/Fe_2(MoO_4)_3$	0.06
Bi_2MoO_6	0.003
$Bi_2Mo_2O_9$	0.009
$Bi_2(MoO_4)_3$	0.007
$Bi_3(MoO_4)_2(FeO_4)$	0.01

        Figure 10 shows a schematic representation of the reaction
mechanism over molybdenum trioxide. There is competitive
dissociative adsorption of methanol and water on molybdenum sites.
The slow step in the sequence is the breaking of a carbon-hydrogen
bond in the methyl group of the surface methoxy(5). Dimethylether
when fed over the catalyst with water does not react either to
methanol or other products up to 300 C. This implies that the
ether does not adsorb dissociatively as methoxy groups.
Dimethoxymethane and methyl formate when fed over the catalyst with
water react quantitatively at temperatures as low as 150 C.
Dimethoxymethane gives methanol and formaldehyde in a 2/1 ratio
with this ratio decreasing at higher temperature. Methylformate
gives methanol and CO in a 1/1 ratio with the methanol reacting
further at temperatures above 200 C.

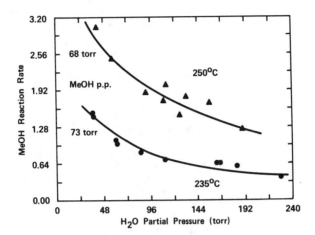

Figure 9.   Reaction Rate of Methanol versus Water
Partial Pressure for $FeCr(MoO_4)_3$.

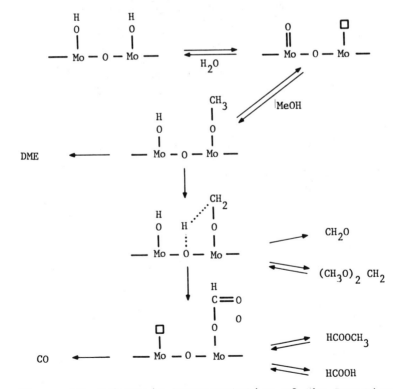

Figure 10. Schematic Representation of the Reaction
Mechanism.

Acknowledgment

The authors are thankful for the assistance of C. E. Lyman in
analyzing the catalysts by scanning transmission electron
microscopy.

Literature Cited

1. "Catalyse de Contact", Le Page, ed., Editions Technip, p. 385
    (1978).
2. Allessandrini, G., Cairati, L., Forzatti, P., Villa, P. and
    Trifiro, F., J. Less-Common Met., 54, 373 (1977).
3. Popov, B., Osipova, K., Malakhov, V. and Kolchin, A., Kinet.
    Catal. (Engl. Transl.), 12, 1464 (1971).
4. Burriesci, N., Garbassi, F., Petrera, M., Petrini, G. and
    Pernicone, N., "Catalyst Deactivation", B. Delmon, G. Froment,
    eds., Elsevier, p. 115 (1980).
5. Machiels, C. J. and Sleight, A. W., J. Catal., 76, 238 (1982).
6. Machiels, C. J. and Sleight, A. W., Proceedings of the 4th
    Intrntl. Conf. on the Chemistry and Uses of Molybdenum,
    Golden, Colorado, p. 411 (1982).
7. W. T. A. Harrison, U. Chowdhry, C. J. Machiels, A. W. Sleight,
    A. K. Cheetham submitted to J. Solid State Chem.
8. F. Ohuchi, U. Chowdhry, Proceedings NATAS Meeting,
    Williamsburg, VA, Sept. 1983.
9. Anon., British Patent 589, 292, 2 Aug. 1944.
10. F. J. Shelton et al., United States Patent 2,812,309, 22
    Aug. 1954.
11. F. J. Shelton et al., United States Patent 2,849,492, 1
    Sept. 1957.
12. F. J. Shelton et al., United States Patent 2,849,493, 5
    Sept. 1957.
13. V. Langebeck, G. Poblat, and G. G. Reif, U.S.S.R. Patent
    116,517, 19 Jan. 1959.
14. Anon., Italian Patent 589,718, 1959.
15. Anon., Austrian Patent 217,444, 10 Oct. 1961.
16. Anon., Austrian Patent 218,539, 11 Dec. 1961.
17. Anon., French Patent 1,310,499, 18 Apr. 1961.
18. Anon., French Patent 1,310,500, 18 Apr. 1961.
19. G. K. Boreskov et al. U.S.S.R. Patent 158,649, 15 Dec. 1964.
20. G. Fagherazzi and N. Pernicone, J. Catal., 1970, 16, 321.
21. G. Alessandrini, L. Cairati, P. Forzatii, P. L. Villa and
    F. Trifiro, J. Less-Comm. Met., 1977, 54, 373.
22. F. Trifiro, V. de Vecchi and I. Pasquon, J. Catal., 1966,
    15, 8.
23. N. Pernicone, J. Less-Comm. Met., 1974, 36, 289.
24. P. L. Villa, A. Sazbo, F. Trifiro and M. Carbucicchio, J.
    Catal., 1977, 47, 122.
25. F. Figueras, C. Pralus, M. Perrin and A. J. Renouprez,
    C. R. Acad. Sci., Paris, Ser. C., 1976, 282, 373.
26. P. Courty, H. Ajot and B. Delmon, C. R. Acad. Sci., Paris,
    Ser. C., 1973, 276, 1147.
27. M. Carbucicchio and F. Trifiro, J. Catal., 1976, 45, 77.

7. MACHIELS ET AL. *Molybdate and Tungstate Catalysts* 117

28. P. O. Warner and H. F. Barry, Molybdenum Catalyst Biography (1950-1964). Climax Molybdenum Company, Greenwich, CT.
29. E. N. Lovey and D. K. Means, Molybdenum Catalyst Biography (1964-1967) Supplement 1. Climax Molybdenum Company, Greenwich, CT.
30. R. Rudolph, Molybdenum Catalyst Biography (1967-1969) Supplement 2. Climax Molybdenum Company, Greenwich, CT.
31. G. A. Tsigdinos, Molybdenum Catalyst Biography (1975-1972) Supplement 3. Climax Molybdenum Company, Greenwich, CT.
32. F. C. Wilhelm, Molybdenum Catalyst Biograph (1973-1976) Supplement 4. Climax Molybdenum Company, Greenwich, CT.
33. W. W. Swanson, Molybdenum Catalyst Biography (1975-1976) Supplement 5. Climax Molybdenum Company, Greenwich, CT.
34. G. A. Tsigdinos, Molybdenum Catalyst Biography (1977-1978) Supplement 6. Climax Molybdenum Company, Greenwich, CT.
35. G. Tamman and F. Westerchold, Zeit. Anorg. Allg. Chem., 1925, 149, 21.
36. A. N. Zelikman and L. V. Delyaevskaya, Zh. Prikl. Khim., 1954, 27, 1155; (Trans.) J. Appl. Chem. (USSR), 1954, 27, 1091.
37. A. N. Zelikman, Zh. Inorg. Khim., 1956, 1, 2778; (Trans.) J. Inorgan. Chem. (USSR) 1956, 1,000.
38. Yu. D. Kozmanov, Zh. Fiz. Khim. 1957, 31, 1861.
39. Yu. D. Kozmanov and T. A. Ugol'nikova, Zh. Inorg. Khim., 1958, 3, 1267: (Trans.) J. Inorg. Chem. (USSR) 1958, 3-V 284.
40. W. Jager, A. Ramel and K. Beker, Arch Eisenhut., 1959, 30, 435.
41. K. Nassau, H. J. Levenstein and G. M. Loiacono, J. Phys. Chem. Sol., 1965, 26, 1805.
42. K. Nassau, J. W. Shiever and E. T. Keve, J. Sol. St. Chem., 1971, 3, 411.
43. V. K. Trunov and L. M. Kovba, Izv. Akad. Nauk SSSR Neorg. Mat., 1966, 2, 151; (Trans.) Onorg. Mat. (USSR) 1966, 2, 127.
44. A. Marcu et al., Rev. Chim. (Bucharest) 1970, 24, 405.
45. P. V. Klevtsova, R. F. Klevtsova, L. M. Kefeli and L. M. Plyasova, Izv. Akad. Nauk SSR Neorg. Mat., 1965, 1, 918; (Trans.) Inorg. Mat. (USSR), 1965, 1, 843.
46. L. M. Plyasova, S. V. Borisov and N. V. Belov, Kristallografria, 1967, 12, 33; (Trans.) Sov. Phys. Crystallog., 1967, 12, 25.
47. L. M. Plyasova, R. F. Klevtsova, S. V. Borisov and L. M. Kefeli, Doklad. Akad. Nauk SSR, 1966, 11, 189.
48. M. H. Rapposch, E. Korstiner and J. B. Anderson, Inorg. Chem., 1980, 19, 3531.
49. H. Chen, Mater. Res. Bull., 1979, 14, 1583.
50. P. D. Battle, A. K. Cheetham, G. J. Long and G. Longworth, Inorg. Chem., 1982, 21, 4223.
51. Z. Jirak, R. Salmon, L. Fouirness, F. Menil and F. Hagenmuller, Inorg. Chem., 1982, 21, 4128.
52. A. W. Sleight and L. H. Brixner, J. Sol. St. Chem., 1973, 7, 172.
53. R. P. Groff, J. Catal., 1984, 86, 215-218 (1984).
54. S. Aruanno and S. Wanke, Can. J. Chem. Eng. 1975, 53, 301.
55. N. Pernicone, Proc. Climax First International Conference on the Chemistry and Uses of Molybdenum, Reading, England, 1973, p. 155 and references therein.

56. Anon., Chem. Proc. Eng. 1970, 4, 100.
57. Anon., Italian Patent 599,419, 1959.
58. Anon., British Patent 909,376, 1962.
59. Anon., German Patent 1,144,252, 1963.
60. Anon., United States Patent 3,152,997, 1967.
61. F. Trania and N. Pernicone, Chem. Ind: (Milan), 1970, 52, 1.
62. V. Massarotti, G. Flor and A. Marini, J. Appl. Crystallogr.,
     1981, 14, 64.
63. G. D. Kolovertnov, G. B. Boreskov, V. A. Dzis'ko, B. I.
     Popov, D. V. Tavasova and G. G. Belugian, Kinet. Katal.,
     1965, 6, 1052; (Trans.) Kinet. Catal. (USSR), 1965, 6, 950.
64. G. K. Boreskov, G. D. Kolovertnov, L. M. Kefeli, L. M.
     Plyasova, L. G. Karachia, V. N. Mastikhin, B. I. Popov, V.
     A. Dzis'ko and D. V. Taracova, Kinet. Katal, 1966, 7, 144;
     (Trans.) Kinet. Catal. (USSR), 1966, 7, 125.
65. G. Alessandriani, L. Gairati, F. Forzatti, P. L. Villa and
     F. Trifiro, Proc. Second Climax International Conference on
     the Chemistry and Uses of Molybdenum, Oxford, England, 1976,
     p. 186, and references therein.
66. F. Trifiro, P. Forzatti, and P. L. Villa; in, Preparation of
     Catalysts, Elsevier, Amsterdam, 1976, p. 143.
67. H. Voge and C. R. Adams, Adv. Catal., 1967, 17, 151.
68. P. F. Kerr, A. W. Thomas and A. M. Langer, Amer. Mineral
     1963, 48, 14.
69. G. A. Tsigdinos and W. W. Swanson, Ind. Eng. Chem. Prod.
     Res. Div. 1978, 17, 210.
70. P. V. Klevtsov, Kristallografia, 1965, 10, 445; (Trans.)
     Sov. Phys. Crystallog., 1965, 10, 370.
71. D. J. Marshall, J. Mat. Sci., 1967, 2, 294.
72. W. R. Doyle and F. Forbes, J. Inorg. Nucl. Chem., 1965, 27,
     1271.
73. V. K. Trunov, V. V. Lutsenko and L. M. Kovaba, Izv. Vysh.
     Ucheb. Zabed. SSR, Khim, Khim. Tekhnol. 1967, 10, 375.
74. V. N. Vorobev, G. Sh. Talipov and M. F. Abidova, Zh. Obsch.
     Khim., 1973, 43, 450: (Trans.) J. Gen. Chem. (USSR), 1973,
     43, 452.
75. U. B. Khamikov, G. Sh. Talipov, K. A. Samigov, N. Rakhmatulaev
     and M. F. Abidova, Zh. Obsch. Khim., 1972, 42, 259; (Trans.)
     J. Gen. Chem. (USSR), 1972, 42, 248.
76. B. I. Popov, L. L. Sedova, G. N. Kustova, L. M. Plyasova,
     Yu. V. Maksimov and A. I. Matveev, React. Kin. Catal. Lett.,
     1976, 5:1, 43.
77. C. Marcilly and B. Delmon, C. R. Acad. Sci., Paris, Ser. C,
     1969, 268, 1975.
78. P. Courty, B. Delmon, C. Marcilly and A. Sugier, French
     Patent, 2,045,612, 9 Jun. 1969.
79. P. Courty, H. Ajot and B. Delmon, French Patent, 2,031,818,
     7 Feb. 1969.
80. Anon., French Patent, 1,604,707, 2 Jul. 1968.
81. P. Courty, H. Ajot and B. Delmon, French Patent, 1,600,128,
     30 Dec. 1968.

82. P. Courty and B. Delmon, C. R. Acad. Sci., Paris, Ser. C.,
    1969, 268, 1874.
83. P. Courty, H. Ajot, C. Marcilly and B. Delmon, Powder
    Technol., 1973, 7, 21.
84. C. Marcilly, P. Courty and B. Delmon, J. Amer. Ceram. Soc.,
    1970, 53, 56.
85. B. Delmon, The Catholic University, Louvain, Belgium, Private
    Communication, 1982.
86. A. K. Cheetham and A. J. Skarnulis, Anal. Chem., 1981, 53,
    1060.
87. S. J. B. Reed, Electron Microprobe Analysis, Cambridge
    University Press, England, 1975.

RECEIVED December 17, 1984

Characterization of Vanadium Oxide Catalysts in Relation to Activities and Selectivities for Oxidation and Ammoxidation of Alkylpyridines

ARNE ANDERSSON and S. LARS T. ANDERSSON

Department of Chemical Technology, Chemical Center, Lund Institute of Technology,
P.O. Box 124, S-221 00 Lund, Sweden

ESCA, XRD, IR, SEM and ESR were used to characterize
the composition and structure of V-Ti-O catalysts, in
both precursors and activated forms. Precursors
consist at low V/Ti ratios of non-stoichiometric
rutile containing Ti^{3+} and V^{4+} and with VO^{2+} on the
surface. With increasing V/Ti ratios V is dissolved up
to 6 atom % and V^{4+} clusters are formed in the rutile.
Excess vanadium forms non-stoichiometric V_2O_5 crystals
on the surface of the 25 μm rutile particles, which at
higher concentrations are completely embedded. Catalysts activated by reduction additionally contain
non-stoichiometric V_6O_{13} and V_2O_4 in amounts increasing with decreasing V/Ti ratios. Isolated V^{4+} ions in
the vanadium oxides also increase in concentration.
The catalytic performance of these catalysts in
oxidation and ammoxidation of some alkylpyridines is
discussed.

The V-Ti-O system has been extensively studied in connection with
catalytic oxidation and ammoxidation reactions of aromatic hydrocarbons. Two principally different types of catalysts can be dis-
tinguished. One type of catalyst is prepared by impregnation,
precipitation or mixing of the vanadium and titanium phases followed
by calcination in air below the melting point of V_2O_5 (1-4). The
simultaneous reduction of V_2O_5 and transformation of TiO_2 (anatase)
into rutile when heating below the V_2O_5 melting point has been
demonstrated to be due to topotactic reactions (5). The formation of
lower vanadium oxides can be of importance, because it has been
found that reduced phases determine the activity and selectivity of
catalysts (6,7).
    Another type of V-Ti-O catalyst is prepared by mixing V_2O_5 and
TiO_2 (anatase) phases, followed by heating the mixture above the
melting point of V_2O_5 (8,9). Clauws and Vennik (10) have found a
defect, associated with oxygen vacancies, by studying the optical
absorption of V_2O_5 crystals. The same defect was found in TiO_2-pro-
moted V_2O_5 crystals (11), but the intensity was greatly enhanced.

0097-6156/85/0279-0121$06.25/0
© 1985 American Chemical Society

This shows the importance of the incorporation of Ti^{4+} in creating
oxygen vacancies.
    The catalysts dealt with in this presentation belong to the
second type of catalysts described above. They have been characterized by means of XRD, ESCA, ESR, SEM and IR methods.
    It has been described in patents that ammoxidation catalysts
can be activated by treatment with ammonia and/or hydrogen (12) or
with carbon monoxide (13). Therefore, both precursors and H_2 reduced
catalysts will be considered in this presentation. It will be shown
that the performance of the catalysts are related to their characteristics. The adsorbed state of reactants will also be discussed.

Methods

Activity measurements. The measurements were performed at atmospheric pressure in a glass reactor. A thermocouple was positioned in
the center of the reactor. In the ammoxidation of 3-picoline, the
inlet reaction mixture was admitted at a rate of 32 liters/hr and
contained 232-254 moles of air, 13-14 moles of ammonia, and 56-62
moles of water vapor for each mole of 3-picoline. The reaction was
usually performed in the temperature interval 300-400°C. In the
oxidation of MEP (2-methyl-5-ethylpyridine) the molar ratios of
O_2/MEP and steam/MEP were 75 and 175 respectively, and the space
velocity was 7000 h^{-1}. In ammoxidation studies prereduced catalysts
were used and the measurements extrapolated in time to give data at
the start of the reaction. In oxidation studies unreduced catalysts
were used and the data were obtained at the steady state.

XRD. X-Ray diffraction analyses were carried out on catalysts by a
Philips X-ray diffraction instrument using a PW 1310/01/01 generator
and Cu Kα radiation.

IR. The infrared spectra were recorded on a Perkin-Elmer 580B
spectrophotometer connected to a data station from the same manufac-
turer. The KBr disc method was used. The spectra were stored on
disks and transferred to a Tektronix 4051 computer for evaluation.

ESCA. ESCA measurements were performed on an AEI ES 200B electron
spectrometer equipped with an Al-anode (1486.6 eV). The full width
at half maximum (FWHM) of the Au $4f_{7/2}$ line was 1.8 eV. Sample
charging was corrected for with the O 1s line at 529.6 eV, which has
been shown to be a suitable method in this system (14).
    For the quantitative analysis calibrated sensitivity factors,
obtained from pure oxides, were used. These were O 1s = 1, Ti
$2p_{3/2}$ = 1.37 and V $2p_{3/2}$ = 2.17.

ESR. A Varian E-3 spectrometer was used for the ESR studies. In the
quantitative measurements a calibrated V_2O_5/TiO_2 sample was run
between each catalyst. The error in these relative measurements was
less than 10 %. For the calibrated V_2O_5/TiO_2 sample the spin concentration was determined to within 30 % accuracy by calibration
against a $CuSO_4 \cdot 5H_2O$ single crystal. This measurement, and measurements for some of the samples, were performed on a Varian E9
equipped with a dual cavity. The g-values were measured within
±0.002.

<u>SEM</u>. Scanning electron microscopic investigations were performed
with a Jeol JSM-U3 or an ISI-100A instrument.

<u>Catalyst preparation</u>. The catalysts were prepared by heating V_2O_5
and TiO_2 (anatase) powders in a quartz crucible in a high temperature oven for 3 hrs at 1150-1250°C. The fused catalysts were divided
into small particles, and the 0.71-1.41 mm fraction was used in the
activity measurements. Activated catalysts were prepared by reduction of the precursors in 1 atm of hydrogen for 1 hr at 450°C.

Results and Discussion

Catalyst Precursor

<u>XRD</u>. X-Ray diffraction patterns of the precursors were composed of
the patterns of V_2O_5 and TiO_2 (rutile), except for the 0.5 and 1.0
mole % V_2O_5 catalysts for which only TiO_2 lines were observed (<u>15</u>).
Precursors with more than 50 mole % V_2O_5 also contained very small
amounts of TiO_2 (anatase). Scanning of the lines in the back-reflection region, 2θ = 115-142 degrees, showed that there was a small
shift of the TiO_2 lines. The lattice constants of the rutile phase
of TiO_2 were calculated. Cohen's least-squares method of eliminating
errors was used (<u>16</u>). The results are given in Table I. The unit
cell dimensions of the catalyst rutile phase has changed mainly in
the <u>a</u> direction. The length of the unit cell in the <u>c</u> direction was
practically the same as that of pure rutile. Bond and coworkers have
obtained the same result (<u>4</u>). The contraction of the unit cell can
be due to incorporation of V^{4+} in the TiO_2 phase. No changes of the
lattice parameters of the vanadium pentoxide phase could be detected, although it has been reported that Ti^{4+} can be dissolved in
V_2O_5 (<u>11</u>). This might be due to detectability problems. V_2O_5 does
not have any lines in the back-reflection region, where shifts are
most easily seen.

Table I. Lattice constants of TiO_2 (rutile)

Phase	Cell dimensions		Ref.
	<u>a</u> (Å)	<u>c</u> (Å)	
TiO_2, pure	4.594	2.959	(<u>17</u>)
VO_2, pure	4.530	2.869	(<u>18</u>)
TiO_2, catalyst	4.583	2.958	this work

<u>ESCA</u>. In the ESCA measurements on the fused V_2O_5/TiO_2 catalysts the
O 1s, V $2p_{3/2}$ and Ti $2p_{3/2}$ core lines were observed. The binding
energies (B.E.) indicate the presence of V_2O_5 and TiO_2 for all
samples and the values were 529.6, 516.6 and 457.9 eV, respectively
(See Table II). For the powder mixtures (un-fused) the values were
529.6, 516.6 and 458.5 eV, respectively. Thus there is a difference
in the Ti $2p_{3/2}$ B.E. which can be explained by the formation of
rutile during heating of the samples in addition to the doping of
the rutile phase with V^{4+}.

Table II.   Binding energies and half widths (eV) of the V $2p_{3/2}$ and
Ti $2p_{3/2}$ core lines

Sample	V $2p_{3/2}$	Ti $2p_{3/2}$
TiO_2, anatase		458.5
		(1.7)
TiO_2, rutile		458.2
		(1.9)
V_2O_5	516.6	
	(1.6)	
V_2O_5/TiO_2, catalyst	516.6	457.9
	(1.6)	(1.7)
Same after H_2SO_4+NH_3 treatment	516.1	458.1
	(2.8)	(1.9)
V_2O_5/TiO_2 (70/30), after $1100^{\circ}C$ in vacuum for 15 hr.	516.2	457.9
	(3.2)	(2.2)

In an attempt to study this effect a VO_2/TiO_2 (70/30) sample which
appears to be a solid solution from XRD data (only rutile lines
appear in spectra) was measured. The same Ti $2p_{3/2}$ B.E. as for the
catalysts was obtained. To measure the V^{4+} in the rutile phase some
samples were treated with sulphuric acid followed by ammonia to
dissolve the vanadium oxide phase. The ESCA analysis of the 10
mole % V_2O_5 catalyst treated in this manner showed the presence of 6
atom % V with a B.E. of 516.1 eV, obtained after subtraction of the
O 1s($K\alpha_{3,4}$) line from spectra. No vanadium oxides are detectable by
XRD on this sample. Thus, approximately 6 atom % V seems to be
dissolved in the rutile phase of the catalysts. The composition
$V_{0.04}Ti_{0.96}O_2$ has been suggested in the literature (4). That it is
probably present as V^{4+} is indicated by the lower V $2p_{3/2}$ B.E. and
in correspondence with the VO_2/TiO_2 sample.
      It is interesting to note that sample charging phenomena occur
for samples with low vanadium loading. For powder mixtures these are
observed at less than 50 atom % V and for fused samples at less than
6 atom % V. Pure V_2O_5 gives almost no charging whereas pure TiO_2
gives a charging of 4-5 V. With enough V_2O_5 so that there is contact
between V_2O_5 particles there should be a low charging effect. At low
V_2O_5 loadings, however, there is a considerably worsened contact
between the V_2O_5 particles throughout the bulk of the sample, and as
for pure TiO_2 a large charging effect arises. What is very interesting is that for the fused samples this occurs at a much lower
vanadium content. This indicates the excellent coverage of the V_2O_5
on the TiO_2 particles. It is further noteworthy that the 50 atom % V
powder mixture and the 6 atom % V fused sample contain approximately
the same atom % V as analysed by ESCA, keeping in mind the high
surface sensitivity.
      In Figure 1 the V content of the various V_2O_5/TiO_2 samples as
revealed by ESCA is plotted against the nominal V-content. Points
for V_2O_5/SnO_2 catalysts are also included and it is evident that
these behave similarly. The line for the powder mixtures falls close
to the theoretical line (dashed line). An almost perfect match could
be obtained by changing the sensitivity factors. However, since this
deviation might be due to particle size effects, sensitivity factors
from pure oxides were used instead. For the fused samples a much

higher atom % of V is obtained by ESCA than was expected. Evidently
the V_2O_5 phase completely covers the rutile particles. This was also
indicated by the charging phenomena discussed above.

In conclusion from the ESCA results, the fused V_2O_5/TiO_2
catalysts consist of a V^{4+} doped TiO_2 (rutile) phase embedded in the
V_2O_5 phase.

ESR. In Figure 2 the ESR spectra at R.T. for some freshly prepared
V_2O_5/TiO_2 catalysts are shown. Even after reduction in H_2 at $450°C$
for 1 hour pure TiO_2 showed only very weak bands from Ti^{3+} which
therefore does not interfere with the quantitative measurements of
V^{4+} discussed below. For 0.05 mole % V_2O_5 a weak band showing some
hyperfine splitting was observed. This was interpreted as VO^{2+} ions
on the rutile surface in accord with the assignments of some
slightly different spectra (19,20). These seem to broaden and
disappear on increasing the V_2O_5 content and a broad single band due
to V^{4+} in V_2O_5 appears. Thus, almost no fine structure is seen for
10 mole % V_2O_5, and is completely absent for the samples 30-100 mole
% V_2O_5. The g-value for the broad resonance was 1.972. The line
width increases from 100 to 150 Gauss with increasing V_2O_5 content
from 10 to 100 %. The 10 % sample showed no signal after the removal
of the vanadium oxides. This fact strongly indicates that the V^{4+}
ions in the rutile phase are not seen at R.T. due to strong spin-
lattice coupling, and thus do not contribute to the quantitative
measurements of V^{4+}.

In Figure 3 the ESR spectra at 77K are shown. The reduced TiO_2
sample shows a spectra assigned to Ti^{3+} (21). The intensity of the
Ti^{3+} signal is completely negligible compared to the intensity of
the other samples at 77K. The increase in intensity with the decrease in temperature (relative intensity 77K/R.T.) is very large
for all samples except for V_2O_5. Here a 3.6 fold increase is observed, which corresponds to the normal change in spin population
with decreased temperature. Estimates from instrument gain settings
point to an approximately 10^2-10^3 fold increase for the fused
V_2O_5/TiO_2 samples. The intensity at 77K for the pure V_2O_5 sample is
negligible - a few per cent at the most - compared with that of the
TiO_2 containing samples. The most significant effect is observed for
the H_2SO_4 and NH_3 treated samples which show no signal at all at
R.T. but one of similar magnitude as for all V_2O_5/TiO_2 samples at
77K. Evidently the low temperature signal is almost entirely due to
the rutile phase. It is possible that the temperature effect may be
due to a spin-lattice relaxation effect whereby substitutional V^{4+}
would not be observable at R.T. (22). It is known that V-V bonds are
formed in the V_xO_2-TiO_2 system, apparently by pairing of randomly
distributed V^{4+} ions at low concentrations of V (23). These would
undoubtedly give temperature dependent spin-spin relaxation effects.
The calculations of spin concentrations from measurements at 77K do,
however, give absurdly high values. It is therefore suggested that a
large part of the increase in signal intensity is due to a ferromagnetic phase transition. At higher concentrations it seems very
likely that clusters of V^{4+} or VO_2 islands are formed in the rutile
phase. Anatase doped with up to 2 atom % V was suggested to contain
a large proportion of VO_2 islands (24). Pure VO_2 does not give any
signals in the ESR spectra. It is well known that temperature
dependent phase transitions do occur in VO_2 and that these are

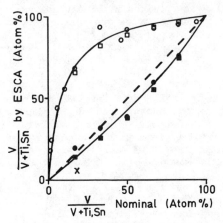

Figure 1. Atom % vanadium by ESCA versus nominal data for mixed V_2O_5 and TiO_2 or SnO_2 powders, fused and unfused. V_2O_5/TiO_2: ● unfused, o fused, x 10 mole % V_2O_5 treated with conc. H_2SO_4 and conc. NH_3 (aq.). V_2O_5/SnO_2: ■ unfused, □ fused.

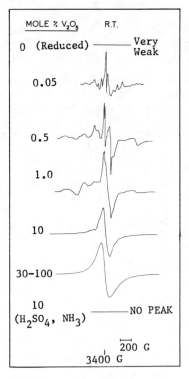

Figure 2. ESR spectra at room temperature (R.T.) for some V_2O_5/TiO_2 fused powder mixtures.

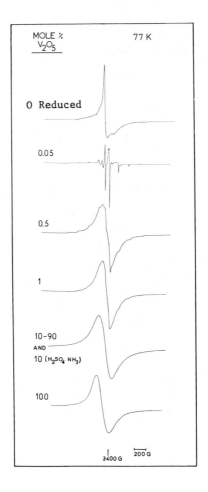

Figure 3. ESR spectra at 77K for some V_2O_5/TiO_2 fused powder
mixtures.
Reproduced with permission from Ref. 32. Copyright 1982 Academic
Press, Inc.

affected by contaminants. For example, $V_{0.96}Ti_{0.04}O_2$ gives a large increase in magnetic susceptibility at low temperatures (25). It is tentatively suggested that in our catalysts in the rutile phase Ti^{4+} doped VO_2 clusters give ferromagnetic species at lower temperatures.

Concerning the curve forms it is seen that for the 0.05 mole % V_2O_5 sample there is a hyperfine splitting indicative of a substitutional solution of V^{4+} in rutile (22,26). It is quite clear that on increasing the V_2O_5 content from 0.05 to 0.5 mole %, the ESR spectrum is strongly broadened, probably due to large spin-spin relaxation effects for V^{4+} in the rutile phase. Any traces of hyperfine splitting have almost disappeared at 1 mole % V_2O_5. All samples from 10 to 90 mole % V_2O_5 showed identical signals, which was also the case for the 10 % H_2SO_4 and NH_3 treated sample. The g-value for the broad resonance was 1.934 for the 30 mole % V_2O_5 sample and increased continuously to 1.957 for the 90 mole % V_2O_5 sample which differs from the values at R.T. Simultaneously, the linewidth increased from 180 to 210 Gauss.

SEM. The precursors with 50 to 10 mole % V_2O_5 treated H_2SO_4 and NH_3 were studied. Figures 4(a) and 4(b) show the TiO_2 particle size in these samples, which is about 25μm in both cases. Figure 4(c) shows sintered TiO_2. By comparing this with Figures 4(a) and 4(b), it can be concluded that the TiO_2 particles become larger when sintered in the presence of V_2O_5. The reason for this can be that V^{4+} is incorporated into the TiO_2 lattice, which leads to a decrease of the melting point of TiO_2. This decrease is then reflected in a greater lattice movability. The fact that the TiO_2 particle size is the same in precursors with both 50 and 10 mole % V_2O_5 indicates that the V^{4+} content in the TiO_2 phase is the same in both cases. A rutile phase $V_{0.06}Ti_{0.94}O_2$ was suggested from the ESCA data. V^{4+} ions are well known to be incorporated in TiO_2 (22,26). Figure 4(b) shows that the TiO_2 particles are agglomerated when the vanadium oxide content is low. The agglomeration can be caused by a vanadium oxide monolayer on the TiO_2 particles. According to the ESR study this monolayer can expose $(V=O)^{2+}$ units. When the vanadium oxide content is low, the monolayer binds the TiO_2 particles together. It has been reported that a monolayer catalyst can be obtained by treatment with an ammoniacal solution (27). During this treatment, V_2O_5 is dissolved and the monolayer of vanadium oxide remains on the carrier.

Figure 5 shows the fused 10 mole % V_2O_5 precursor both freshly prepared, after reduction with H_2 at $450°C$ and after treatment with H_2SO_4 and NH_3, respectively. From the fact that needle-like crystals appear on the TiO_2 surface after reduction, it can be concluded that the vanadium oxide phase in the precursors covers the TiO_2 particles. This was also the conclusion drawn from quantitative ESCA data.

Activated Catalysts

ESCA. Activation of the catalyst occurs in the initial period at the start up of the reactor. Alternatively, the catalyst may be pre-reduced to the same state as is present in the steady state of the reaction. ESCA studies of activated catalysts reveal changes in the O 1s and V $2p_{3/2}$ core lines, although the effects apparently are not great. In Figure 6 spectra of fresh, reduced and once used

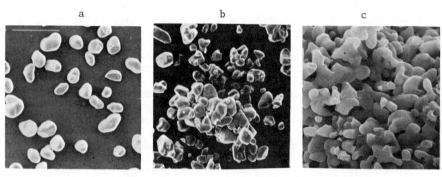

Figure 4.   SEM micrographs of V_2O_5/TiO_2 fused powder mixtures.
a) 50 mole % V_2O_5, 200x,   b) 10 mole % V_2O_5, 200x,   c) TiO_2,
3000x.

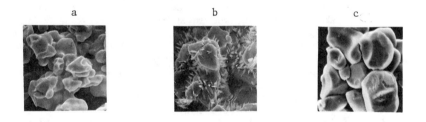

Figure 5.   SEM micrographs of 10 mole % V_2O_5 catalyst. 1000x.
a) Freshly prepared.   b) After H_2-reduction at 450°C for 1 hour.
c) After treatment with conc. H_2SO_4 and conc. NH_3 (aq.).

catalyst are shown. The significant differences in the spectra are the increased intensity of the high B.E. side of the O 1s core line and the asymmetric broadening and shift towards lower B.E. of the V $2p_{3/2}$ core line for both used and reduced catalysts. Evidently the catalysts are reduced and adsorbed oxygen species are formed. One interesting feature is the significantly larger amount of adsorbed oxygen species on the used catalysts than on the H_2 reduced catalysts. The latter may contain various adsorbed oxygen species, such as -OH and H_2O, which are formed in the reduction. For the used catalysts, however, one has to consider adsorbed oxygen containing intermediates or accumulated products.

It is relatively difficult to resolve such spectra as presented in Figure 6. This is mainly due to not knowing which components should be present, the slope of the baseline and the presence of the O 1s $K\alpha_{3,4}$ satellite lines. However, such an attempt is shown in Figure 6. This is an unoptimised simulation of the spectra with 8 Gaussian components and a base line due to inelastic scattering. The position of the two V $2p_{3/2}$ components corresponds to V_2O_5 and V_2O_4 (14). The V^{4+}/V_{tot} ratio calculated from these is 0.55. The first step in the refinement of the curve resolution would perhaps be to include a third V $2p_{3/2}$ peak for V_6O_{13} and possibly a fourth O 1s peak for some adsorbed species. This very tedious procedure is not applicable in larger scale quantitative evaluations. A much simpler method of measuring the degree of reduction is to measure the decrease in peak B.E. and increase in FWHM. In Figure 7 the V $2p_{3/2}$ B.E. and FWHM are shown for a 50 mole % V_2O_5 catalyst reduced in hydrogen at $450^{\circ}C$ for various periods of time. It can be seen that after 1 hour the B.E. has decreased at around 516.3 eV and the half width has increased to around 2.7 eV, both values in good agreement with data for V_6O_{13} (14).

In Figure 8 a similar plot is shown for catalysts with differing mole % V_2O_5, all reduced for 1 hour in hydrogen at $450^{\circ}C$. Here there is quite a clear trend of increasing degree of reduction with decreasing V_2O_5 content of the original sample. It appears that from roughly 100-70 mole % V_2O_5 the surface contains non-stoichiometric V_2O_5 and V_6O_{13}, while from 70-40 mole % V_2O_5 it seems as if V_6O_{13} is predominant. Finally, from 40 to 10 mole % V_2O_5 the data indicates the additional presence of V_2O_4. In conclusion, original samples with compositions around 50 mole % V_2O_5 contain after reduction predominantly V_6O_{13} and V_2O_5 in the surface layer.

XRD. In ground catalyst samples, with a composition between 30 and 100 mole % V_2O_5, lines corresponding to V_2O_5, V_6O_{13}, V_2O_4 and TiO_2 (rutile) could be identified. The number of discernible lines of vanadium oxides diminshed with increasing amount of TiO_2. In the catalyst with 10 mole % V_2O_5 only TiO_2 (rutile) and V_2O_5 were found. For the identification of V_6O_{13} the data given by Wilhelmi and coworkers (28) was used. The presence of the other oxides was ascertained by comparison with ASTM data (15).

To obtain relatively more information on the composition of the outer parts of the catalyst particles, the analysis was carried out on undivided particles in a rotating sample holder. The composition of the vanadium oxide phase as a function of the TiO_2 content is given in Figure 9. The composition is expressed as relative intensities of the strongest X-ray reflection of the phases, $I(i)/\Sigma I(i)$ x

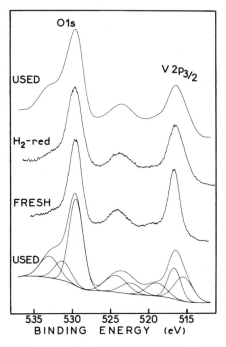

Figure 6. O 1s and V 2p electron spectra for some V_2O_5 catalysts.

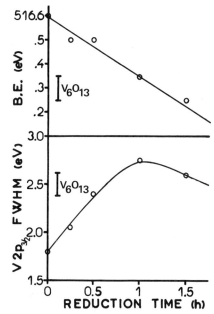

Figure 7. V $2p_{3/2}$ binding energy and half width versus H_2-reduction time at 450°C for a 50 mole % V_2O_5/TiO_2 catalyst.

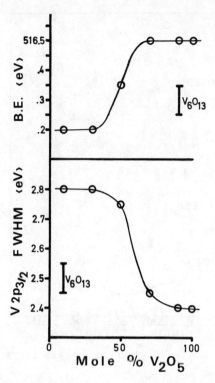

Figure 8. V $2p_{3/2}$ binding energy and half width versus mole % V_2O_5 in original preparation for V_2O_5/TiO_2 catalysts reduced with H_2 at 450°C for 1 hour.

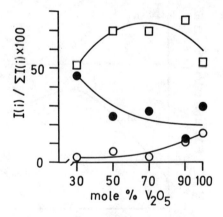

Figure 9. Vanadium oxide composition in V–Ti–O catalysts after reduction in H_2 at 450°C for 1 hour measured by XRD on unground samples. o V_2O_5, □ V_6O_{13} and ● V_2O_4.

100. The lines with the d-values 4.38, 3.32 and 3.20 were used for
V_2O_5, V_6O_{13} and V_2O_4, respectively. It was found that this composition approximately corresponds to the composition in mole %. It
can be seen that V_6O_{13} is the major vanadium oxide phase at the
outer parts of the catalyst particles.

<u>IR</u>. The infrared spectra of the pre-reduced catalysts in the range
950-1050 cm^{-1} are shown in Figure 10. A shoulder can be seen on the
low wavenumber side of the V_2O_5 peak. No significant change in
position of the V_2O_5 peak at 1023 cm^{-1} can be observed, in contrast
to the case in the V_2O_5-MoO_3 (<u>29</u>) and the V_2O_5-SnO_2 (<u>30,31</u>) systems.
In Figure 10(a) the V_2O_5 peak has been normalized. After subtraction
of this peak, the difference spectra presented in Figure 10(b) were
obtained. This figure clearly illustrates the existence of a band
between 960 and 1020 cm^{-1} with a maximum around 995 cm^{-1}. The 10
mole % V_2O_5 catalyst also exhibits the same band, but no V_2O_5 peak.
The only phases in the catalysts which could be identified by XRD
were V_2O_5, V_6O_{13}, V_2O_4 and TiO_2. However, the band around 995 cm^{-1}
has not been found to be due to any of these phases (<u>32</u>). Neither
has it appeared in the spectra for VO_2(A), VO_2(B), V_4O_9 or vanadium
oxide hydrates (<u>33</u>).
    The band at 995 cm^{-1} has also been found in an intimate
V_2O_5/V_6O_{13} mixture (<u>32</u>) as well as in an intimate V_6O_{13}/V_2O_4 mixture
(<u>34</u>), both obtained by decomposition of NH_4VO_3. From these results
it seems most probable that the band observed is connected with the
presence of V_6O_{13}. Pure V_6O_{13} crystals, however, do not absorb in
this region (<u>32</u>). Therefore, it has been suggested that the 995 cm^{-1}
band can be due to a defect structure of V_6O_{13} (<u>34</u>). In contrast to
pure V_6O_{13} this defect structure must have V=O bonds in its lattice.
This conclusion is based on the fact that peaks in the range 900-
1100 cm^{-1} have been ascribed to stretching vibrations of Me=O bonds
(<u>35</u>). The defect structures of non-stoichiometric transition metal
oxides can be divided into four basic types (<u>36</u>). These are crystals
with an excess of metal, resulting from the presence of either
anionic vacancies or interstitial cations, crystals with a de-
ficiency of metal due to the presence of either excess anions in
interstitial positions or cationic vacancies. When considering the
V_6O_{13} structure, the last mentioned type of defect would most easily
result in the formation of V=O units if the cationic vacancies were
localized to the positions of vanadium ions surrounded by 2-coordi-
nated anions. Such cation positions exist between the shear planes
in V_6O_{13} (<u>37</u>). Extended defects of this type would result in an
"amorphous" like V_6O_{13} structure. Amorphous V_6O_{13} has been imaged
by HRTEM and has been found to be very selective in the ammoxidation
of 3-picoline (<u>38</u>).

<u>ESR</u>. These measurements were performed on the reduced catalysts
with 10-100 mole % V_2O_5 in the original sample. All samples showed
the same single broad resonance as for the unreduced catalysts. The
spin concentration was calculated from the spectra at R.T. and is
presented in Figure 11 as the mole % of the total vanadium content
that is present as V^{4+}. It is necessary to add that this is paramagnetic V^{4+} in the vanadium oxide phase as shown above. The XRD
results showed that V_6O_{13} and V_2O_4 were formed upon reduction. It
has been mentioned earlier that pure V_2O_4 does not give any ESR

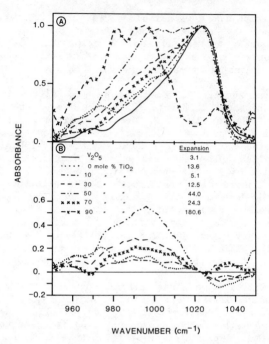

Figure 10. A: Infrared spectra of V_2O_5 and pre-reduced catalysts. 1 mg sample. The strongest peak has been expanded so that
the absorbance is equal to 1.0. B: Resulting absorbance after
subtraction of the V_2O_5 absorbance. (Reproduced with permission
from Ref. 29. Copyright 1982, Academic Press Inc.).

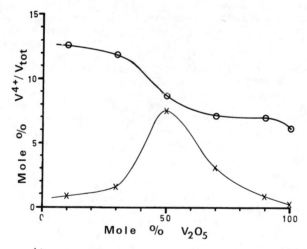

Figure 11. V^{4+} concentration versus mole % V_2O_5 in original
preparation of V_2O_5/TiO_2 catalysts measured by ESR. o after
reduction in H_2 at $450°C$ for 1 hour, x before reduction.

spectra. Furthermore, we have found that the signal for pure V_6O_{13} is much weaker than expected. Evidently, at higher V^{4+} concentrations the signals disappear from the spectra, probably due to spin-spin relaxation effects. Consequently, the observed spectra arise from V^{4+} in the vanadium pentoxide phase and to some extent from the V_6O_{13} phase. Any formation of V^{3+} to account for the loss in intensity for reduced V_2O_5 (<u>39</u>) seems to be an unreasonable and unnecessary assumption. By looking in Figure 11 it can be seen that the % paramagnetic V^{4+} increases with decreasing V_2O_5 content of the initial mixture. This is the same trend as obtained by ESCA, although in that case the total V^{4+} concentration in the surface layers is obtained.

<u>Adsorption.</u> To study the adsorption of 3-picoline with ESCA, the N 1s core line is the most useful core line from the 3-picoline to observe since both the C 1s and O 1s lines are obscured by core lines from the substrate and pump oil contamination. One would expect the 3-picoline molecule to adsorb in at least two different modes: one π-bonded planar adsorption, and one nitrogen lone pair bonded either perpendicular or at an angle to the surface. The former has been shown to be the more strongly retained species in pyridine adsorption on the Ag (111) surface at least (<u>40</u>). A third possibility is that an abstraction of hydrogen from the methyl group results in the formation of a σ-bonded species, which could be preceded by a weaker adsorption as mentioned above. In observing the N 1s B.E. it is possible to determine whether the nitrogen atom is involved in any bonding of the adsorbed species.

In Figure 12 N 1s core line spectra are shown for adsorbed 3-picoline on Pt and on a V_2O_5 catalyst. A N 1s B.E. of 399 eV is obtained for pure 3-picoline condensed in a thick layer on Pt at 140K. At 673K all 3-picoline has desorbed from the Pt surface. To investigate the effect of hydrogen bonding of the N atom on the N 1s B.E., H_2O and 3-picoline were condensed at 190K on the catalyst, forming a thick layer. No substrate core lines could then be observed. From the spectra it is evident that a second species with a N 1s B.E. of 401 eV has appeared which most likely is 3-picoline with water hydrogen bonded to the N atom. Heating this sample to 298K results in a decreased intensity for both peaks and a simultaneous appearance of the substrate core lines in the spectra. At higher temperatures the 401 eV species almost disappear.

For 3-picoline adsorbed on the V_2O_5 catalyst at 250K without H_2O addition a smaller peak at 401 eV and a stronger one at 399 eV are observed. The higher B.E. species may be caused by adsorption on a vanadium cation or -OH groups through the N lone pair. This species disappear upon heating whereas the lower B.E. species is retained to a significant extent even at 673K in vacuum. The strongly retained species with a N 1s B.E. of 399 eV is either the π-bonded or dissociatively bonded 3-picoline. From measurements on the adsorption of pyridine it was observed that only a few per cent was retained compared to the amounts found for 3-picoline at high temperatures. Thus it is concluded that it is the dissociatively bonded species that are observed at 673K. On the Pt surface all 3-picoline was desorbed at 673K, but not on the catalyst, which points to some specific interaction forming the dissociation on the vanadium oxide surface. It is tentatively suggested that through a H abstraction from the methyl group a σ-bond between the carbon atom and a terminal oxygen of the vanadium oxide may be formed. It is possible that it is a homolytic process whereby a hydroxide group is formed simultaneously as well as two V^{4+} ions. This species would be bonded rather strongly to the surface and would be retained at high temperatures. By repeating the same H abstraction mechanism, pyridinecarbaldehyde will be formed and simultaneously desorb from the surface. The initially formed V^{3+} cation would probably have a strong tendency to adsorb an oxygen in a dissociative process reforming the original structure. Measurements of the V $2p_{3/2}$ and O 1s core lines reveal the formation of V^{4+} and the appearance of high O 1s B.E. oxygen species on these catalysts, which supports the above mentioned model.

Pulse chromatographic studies of the adsorption of benzene, pyridine and 3-picoline at 250°C on a V_6O_{13} catalyst showed that the amount adsorbed decreaed in the order 3-picoline>pyridine>>benzene (34). Thus it is obvious that the principal modes of adsorption of 3-picoline are either an adsorption to the nitrogen atom or a chemisorption of the methyl group after abstraction of hydrogen. These results are in line with those obtained from the ESCA studies.

Catalytic Performance

Ammoxidation of 3-picoline. ESCA results indicated that the surfaces of the catalysts were composed mainly of V_2O_5 and V_6O_{13} at low TiO_2 contents. The X-ray diffraction patterns of these catalysts also revealed the presence of V_2O_4 which may be present as separate small crystallites on the V_2O_5/V_6O_{13} surface, since no coherent interface can exist between V_2O_4 and V_2O_5 or V_6O_{13} because of their different structures. Thus a relatively small V_2O_4 area will contribute to the ESCA analysis.

Figure 13 shows the activity, the binding energy of the V $2p_{3/2}$ line, and the FWHM as a function of the composition. The following statements can be made: i) at a high vanadium oxide content the activity is relatively low and the surface composition is close to V_2O_5 ii) around 30-50 mole % V_2O_5 the activity is high and the surface composition is close to V_6O_{13} iii) at low vanadium oxide contents the activity is very low and the V_2O_4 content on the surface increases. These conclusions agree with the published results that V_6O_{13} is more active than V_2O_5 in the ammoxidation of 3-picoline. V_2O_4 was found to have a relatively low activity (7).

Figure 14 illustrates the variation of the selectivity for formation of nicotinonitrile with catalyst composition. The selectivity increases with decreasing vanadium oxide content, but also has a local maximum at 90 mole % V_2O_5. Data obtained from ESR measurements are also given in Figure 14. The V^{4+}/V_{tot} ratio plotted is a bulk property, but it seems that in general an increasing amount of isolated V^{4+} ions can be correlated to increasing selectivity. It is reasonable to suggest that a great amount of isolated V^{4+} species in the bulk also implies a relatively high concentration of oxygen vacancies on the catalyst surface. It has been found (32) that a correlation exists between selectivity and the surface concentration of oxygen vacancies on these catalysts. Oxygen vacancies on the surface are thought to be beneficial for the selectivity

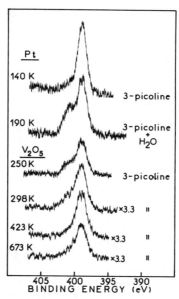

Figure 12. N 1s electron spectra from 3-picoline adsorbed on Pt-foil and V_2O_5 catalyst at various temperatures.

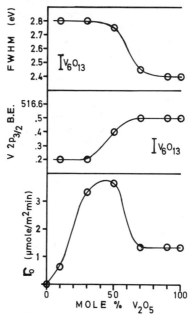

Figure 13. Activity in ammoxidation of 3-picoline, V $2p_{3/2}$ binding energy and half width as a function of catalyst composition. Reaction parameters: mole ratio air/3-picoline/NH_3/H_2O = 245/1/14/60; 320°C; 101 kPa.

because the possibility of complete oxidation to carbon oxides
diminishes.

Oxidation of 3-methyl-5-ethylpyridine. As was shown in Figure 6,
the used catalysts show significant increase in the high B.E. side
of the O 1s line. This is interpreted as being due mainly to an
accumulation of adsorbed intermediates or products on the surface of
the catalyst. Measurements on the increase in the C 1s and N 1s core
lines after use of the catalyst give atom ratios of C:N:O of approximately 7:1:1. The B.E. of the N 1s line for the used catalyst
is 399.3 eV equal to that obtained in the adsorption measurements
after heating as shown above. It was then suggested that 3-picoline
is adsorbed through the substituent.

   In Figure 15 the relative increase of the C 1s and O 1s of
adsorbed species on used catalysts are shown as a function of the
composition of the catalysts in mole % V_2O_5. A maximum is obtained
around 70 mole % V_2O_5 of the initial catalyst formulation. The half
width of the $V\ 2p_{3/2}$ core line also appears to have a maximum at
roughly the same position. Apparently, the more adsorbed species,
the more reduced catalyst and vice versa. The increase in the half
width is caused by the presence of V^{4+} in increased amounts in
addition to V^{5+} and these lines are too close in B.E. for separate
peaks to appear in the spectra. Finally, a comparison with the
selectivity and conversion reveals that maxima in these are also
obtained at the same catalyst composition. Yield data on the oxidation of MEP was also given earlier (41). Evidently, under similar
process conditions, the higher conversion and selectivity obtained
around 70 mole % V_2O_5 leads to a more reduced catalyst with larger
amounts of adsorbed intermediates. The FWHM at the maximum is about
2.5 eV in correspondence with measurements on V_6O_{13}.
   In Figure 16 the selectivity in the oxidation of MEP (calculated from yield data presented earlier (14)) over a 67 mole % V_2O_5
catalyst (which corresponds to the maximum in Figure 15) is shown as
a function of the reaction temperature at otherwise identical
process conditions. The degree of conversion would then naturally
increase with the temperature. Also shown are the $V\ 2p_{3/2}$ core line
B.E. and FWHM measured on the used catalysts. There is a minimum in
the B.E. and a maximum in the half width, both in combination
corresponding to V_6O_{13}, at the same position as the maximum in the
selectivity. Apparently, this phase gives the highest selectivity.
At lower and higher temperatures with a low selectivity the composition corresponds to V_2O_5. There seems to be a dynamic relation
between the selectivity and the oxidation state of the catalyst. As
was discussed elsewhere (14) a catalyst may be cycled between, for
example, intermediate and high temperatures repeatedly, giving the
same steady state selectivity and composition changes.

Conclusions

Catalyst precursors prepared by heating V_2O_5 and TiO_2 powder mixtures at 1150-1250°C, intermediate between the melting point of both
components, were found to contain two phases at higher concentrations of V. One is TiO_2 (rutile), slightly non-stoichiometric and
containing Ti^{3+} and V^{4+} ions. At low V concentrations V^{4+} ions are
found mainly in substitutional positions, but also occur as VO^{2+}

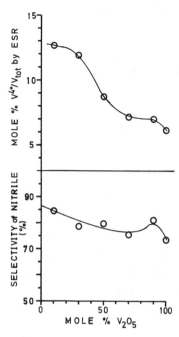

Figure 14. Selectivity in ammoxidation of 3-picoline (320°C, 90 % conv.) and V^{4+}/V_{tot} as a function of catalyst composition.

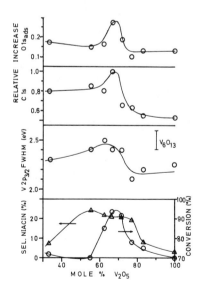

Figure 15. Conversion and selectivity in oxidation of MEP at 400°C and FWHM of V $2p_{3/2}$ and relative increase of the C 1s and O 1s lines from adsorbed species as a function of catalyst composition.

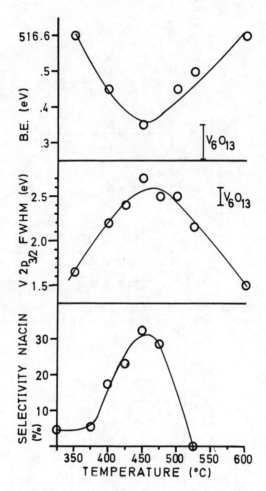

Figure 16. Selectivity in oxidation of MEP over the 67 mole %
V_2O_5 catalyst and B.E. and FWHM of the V $2p_{3/2}$ core line as a
function of reaction temperature.

ions on the surface of the TiO_2 particles. The size of these are approximately 25 μm for V containing catalysts, whereas the pure TiO_2 particles are much smaller indicating the effect of V^{4+} on sintering. At 10 mole % V_2O_5 an agglomeration of the rutile particles is observed, a phenomenon which is absent at higher V/Ti ratios. At increasing V_2O_5 concentrations of the initial mixtures V is dissolved up to 6 atom % in the rutile lattice after fusion. Clusters of V^{4+} or VO_2 islands are then formed in the rutile lattice. Excess vanadium forms V_2O_5 crystals on the surface. At higher V_2O_5 concentrations the rutile particles are completely embedded in the V_2O_5 phase, which is non-stoichiometric with a maximum in V^{4+} concentration when the initial mixture contains 50 mole % V_2O_5. At high V/Ti ratios a small quantity of anatase is also present. Activated catalysts are obtained by H_2 or upon stream reduction of the V_2O_5 phase, whereby V_6O_{13} and V_2O_4 appear in the catalysts. The degree of reduction increases with decreasing V/Ti ratios, giving more V_2O_4 at low V/Ti ratios and predominantly V_6O_{13} at intermediate values. Also the concentration of isolated V^{4+} ions in the vanadium oxides increases with decreasing V/Ti ratios. The V_6O_{13} formed is non-stoichiometric, probably containing cation vacancies between shear planes giving rise to adjacent V=O bonds in the V_6O_{13} lattice. The activity at the initial state of the reduced catalysts in the ammoxidation of 3-picoline has a maximum at intermediate V/Ti ratios, where a maximal amount of defective V_6O_{13} is obtained in the H_2 reduced catalysts. The selectivity seems to be more dependent on isolated V^{4+} centers, reflecting oxygen vacancies. In the oxidation of MEP at steady state, both conversion and selectivity are maximal with intermediate V/Ti ratios giving dominantly defective V_6O_{13}. Adsorption studies reveal strongly bound species which are probably bonded through the methyl group after an initial H-abstraction.

Literature Cited

1. Gasior, I.; Gasior, M.; Grzybowska, B.; Kozłowski, R.; Słoczyński, J. Bull. Acad. Pol. Sci. Ser. Sci. Chim. 1979, 27, 829.
2. Vanhove, D.; Blanchard, M. Bull. Soc. Chim. France 1971, 9, 3291.
3. Cole, D.J.; Cullis, C.F.; Hucknall, D.J. J. Chem. Soc. Faraday Trans. 1 1976, 72, 2185.
4. Bond, G.C.; Sárkány, A.J.; Parfitt, G.D. J. Catal. 1979, 57, 476.
5. Vejux, A.; Courtine, P. J. Solid State Chem. 1978, 23, 93.
6. Simard, G.L.; Steger, J.F.; Arnott, R.J.; Siegel, L.A. Ind. Eng. Chem. 1955, 47, 1424.
7. Andersson A.; Lundin, S.T. J. Catal. 1979, 58, 383.
8. Sembaev, D.Kh.; Suvorov, B.V.; Saurambaeva, L.I.; Suleimanov, Kh.T. Kinet. Katal. 1979, 20, 750.
9. Andersson A.; Lundin, S.T. J. Catal. 1980, 65, 9.
10. Clauws, P.; Vennik, J. Phys. Stat. Sol. (b) 1974, 66, 553.
11. Clauws, P.; Vennik, J. Phys. Stat. Sol. (b) 1975, 69, 491.
12. French Patent 2 119 935, 1972.
13. German Patent 2 423 032, 1975.
14. Andersson, S.L.T. J. Chem. Soc. Faraday Trans. 1 1979, 75, 1356.

15. "ASTM Powder Diffraction File"; Joint Committee on Powder
     Diffraction Standards: Philadelphia, 1974.
16. Klug, H.P.; Alexander, L.E.   "X-Ray Diffraction Procedures";
     Wiley: New York, 1967.
17. Abrahams, S.C.; Bernstein, J.L. J. Chem. Phys. 1971, 55, 3206.
18. Westman, S. Acta Chem. Scand. 1961, 15, 217.
19. Ueda, H. Bull. Chem. Soc. Japan 1979, 52, 1905.
20. Piechotta, M.; Ebert, I.; Scheve, J.   Z. Anorg. Allg. Chem.
     1969, 368, 10.
21. Maksimov, N.G.; Nesterov, G.A.; Zakharov, V.A.; Stchastnev,
     P.V.; Anufrienko, V.F.; Yermakov, Yu.I. J. Molec. Catal. 1978,
     4, 167.
22. Gerritsen, H.J.; Lewis, H.R. Phys. Rev. 1960, 119, 1010.
23. Hörlin, T.; Niklewski, T.; Nygren, M.   J. Phys. Coll. Sect. C4
     1976, 10, 69.
24. Mériaudeau, P.; Vedrine, J.C. Nouv. J. Chim. 1978, 2, 133.
25. Hörlin, T.; Niklewski, T.; Nygren, M.   Acta Chem. Scand. Sect.
     A 1976, 30, 619.
26. Marill, J.-L.; Cornet, L. J. Chim. Phys. 1973, 70, 336.
27. Yoshida, S.; Iguchi, T.; Ishida, S.; Tarama, K.   Bull. Chem.
     Soc. Japan 1972, 45, 376.
28. Wilhelmi, K.-A.; Waltersson, K.; Kihlborg, L.   Acta Chem.
     Scand. 1971, 25, 2675.
29. Bielański, A.; Dyrek, K.; Kozłowska-Róg, A.   Bull. Acad. Pol.
     Sci. Ser. Sci. Chim. 1972, 20, 1055.
30. Yoshida, S.; Murakami, T.; Tarama, K.   Bull. Inst. Chem. Res.
     Kyoto Univ. 1973, 51, 195.
31. Andersson, A. J. Catal. 1981, 69, 465.
32. Andersson, A. J. Catal. 1982, 76, 144.
33. Théobald, F. Rev. Roum. Chim. 1978, 23, 887.
34. Andersson, A.; Wallenberg, R.; Lundin, S.T.; Bovin, J.-O.
     Proc. 8th Int. Congr. Catalysis, 1984; Vol. 5, p. V-381.
35. Barraclough, C.G.; Lewis, J.; Nyholm, R.S. J. Chem. Soc. 1959,
     3552.
36. Haber, J.   In "Catalysis Science and Technology"; Andersson,
     J.R.; Boudart, M., Eds.; Springer: Berlin, 1981; Vol. 2, p. 13.
37. Andersson, A. J. Solid State Chem. 1982, 42, 263.
38. Andersson, A.; Bovin, J.-O.; Lundin, S.T., to be published.
39. Bielański, A.; Dyrek, K.; Serwicka, E.   J. Catal. 1980, 66,
     316.
40. Demuth, J.E.; Christmann, K.; Santa, P.N.   Chem. Phys. Lett.
     1980, 76, 201.
41. Järås, S. Dr. Tech. Thesis, Lund Institute of Technology, Lund,
     1977.

RECEIVED October 4, 1984

9

The Synthesis and Electrocatalytic Properties of Nonstoichiometric Ruthenate Pyrochlores

H. S. HOROWITZ[1], J. M. LONGO[2], H. H. HOROWITZ[3], and J. T. LEWANDOWSKI

Corporate Research Science Laboratories, Exxon Research & Engineering Company, Annandale, NJ 08801

A new series of conductive, mixed metal oxides with the pyrochlore structure has been discovered and tested as electrocatalysts. They can be described by the general formula $A_2[Ru_{2-x}A_x]O_{7-y}$ where A = Pb or Bi, $0 < x < 1$ and $0 \leq y \leq 0.5$. These oxides are prepared by precipitation/crystallization in an aqueous alkaline reaction medium, and a detailed discussion of the pertinent synthesis parameters is given. In aqueous alkaline electrolyte, near ambient temperature, these materials display significantly lower polarizations than any other catalyst for O_2 evolution and are among the best for O_2 reduction. These same materials are found to catalyze the selective electro-oxidative cleavage of olefinic and secondary oxygenated organic compounds. It is suggested that the reaction path for both oxygen electrocatalysis and electro-oxidative cleavage of organics involves a cyclic oxygen-ruthenium intermediate on the surface similar to those formed with osmium tetroxide.

While metal oxides and mixed metal oxides have often been considered for various electrocatalytic applications, they are often limited by low electronic conductivity and/or low surface area. One series of mixed metal oxides with the pyrochlore structure has been discovered (1) that has demonstrated high catalytic activity for electro-reduction and electroevolution of oxygen (2) and the selective electrooxidation of certain organics (3). These materials, which are characterized by high electronic conductivity and can be prepared in high surface area form, are described by the general formula:

[1]Current address: Experimental Station, E. I. du Pont de Nemours and Company, Wilmington, DE 19898
[2]Current address: Exxon Production Research, P.O. Box 2189, Houston, TX 77001
[3]Current address: Exxon Research & Engineering Company, Florham Park, NJ 07932

$$A_2[Ru_{2-x}A_x]O_{7-y} \qquad\qquad (1)$$

where A = Pb or Bi, $0 < x \leq 1$ and $0 \leq y \leq 0.5$. The face-centered cubic oxide pyrochlore structure can crystallographically best be described by the general formula:

$$A_2[B_2]O_6O' \qquad\qquad (2)$$

where A and B represent the two cation sites in the structure. The B-cations are located at the center of oxygen (O) octahedra which share only corners so as to form interconnected cage-like holes. The A-cations are located at the intersection of the cages, and the O' oxygen sites are located at the center of these cages. These "special" oxygens (O') may be partially or totally absent and are the basis for the anion nonstoichiometry observed in pyrochlores. Excellent descriptions of this structure have been given by Sleight (4) and McCauley (5).

Experimental

The synthesis method (6,7) involves reacting the appropriate cations to yield a pyrochlore oxide by precipitation, and subsequent crystallization of the precipitate in a liquid alkaline medium in the presence of oxygen. The alkaline solution serves both as a precipitating agent and as a reaction medium for crystallizing the amorphous precipitate to pyrochlore, thus obviating the need for subsequent heat treatment.

The reactants used were the more soluble cation sources, generally the nitrates. The aqueous solutions of these cation sources were combined in a post transition metal to noble metal ratio appropriate for the ultimately desired pyrochlore stoichiometry. For lead ruthenate syntheses the reactant Pb:Ru ratio was required to be slightly higher than the intended final Pb:Ru ratio because of the high solubility of lead relative to ruthenium. For bismuth ruthenate syntheses the Bi:Ru ratio was required to be slightly lower than the intended final Bi:Ru ratio because of the higher solubility of ruthenium.

The combined cation solutions were stirred for 10 minutes and then added to the alkaline reaction medium (usually KOH or NaOH) where precipitation occurred. The pH of the reaction mixture was adjusted to be between 10.0 and 14.0. Crystallization of the precipitate to the pyrochlore structure was achieved in a period ranging from 8 hours to 5 days by maintaining the stirred, oxygen sparged reaction medium at a temperature of 50-80°C. The reactions were carried out in polymethylpentene beakers to avoid corrosion by the alkaline synthesis medium.

In some cases the precipitate was recovered by conventional filtration. Separation of the solid from the alkaline precipitate medium by this method was followed by repeated washings with hot distilled water. In other instances, the precipitated solid was freeze dried. When this method was employed, the precipitated solid was allowed to settle overnight, and the alkaline precipitation medium was removed by vacuum pipette, leaving behind only enough liquid to keep the solid totally submerged. The removed liquid was then replaced with fresh distilled water, and the precipitate was again allowed to settle. This procedure was repeated three or four times with the precipitate kept completely submerged throughout the entire procedure. The resultant aqueous slurry was then sprayed into a vessel of liquid nitrogen by means of a pneumatic atomization nozzle. The extremely fine droplets constituting the atomized solution were immediately frozen into fine particles as they contacted the liquid nitrogen. The ice was subsequently sublimed from this finely divided frozen powder using a freeze dryer. Characterization procedures for determining physical and chemical properties of these materials have been described in reference (1).

Oxygen electrocatalysis properties were measured on Teflon-bonded porous electrodes (5 cm^2 in area). The electrodes always contained 20 weight percent Teflon and the catalyst loading employed was approximately 60 mg/cm^2. The Teflon-bonded material was hot pressed onto a gold screen which served as the current collector. The Teflon content provided sufficient hydrophobicity so that the electrode, when placed in the electrochemical cell for testing, was able to maintain an interface between the liquid and gas phases. Further details of the experimental methods for evaluating the electrocatalytic oxidation and reduction of oxygen in aqueous alkaline electrolytes are given in reference (2).

Electroorganic oxidation reactions were monitored on electrodes that were fully immersed in the electrolyte. The organic reactants were directly added to the electrolyte. In some of these experiments, the electrodes described above were used. In other experiments, non-Teflon-bonded catalyts were used by employing a specially constructed Teflon cell. The sample was placed in this cell between a piece of gold foil and a piece of gold screen. The gold screen served as the current collector. An outer layer of Celgard microporous film, placed over the screen, was used to prevent loss of sample into the solution. Additional experimental details may be found in reference (3).

Results

General Chemical and Physical Characterization. The x-ray diffraction data, chemical analyses by x-ray fluorescence and the effects of various synthesis parameters explored in this study lead to the conclusion that a new series of pyrochlores represented by formula 1 has been synthesized. The substitution of the larger post transition element cation for the noble metal cation on the octahedrally coordinated B-site leads to a considerable enlargement of the pyrochlore's cubic unit cell dimension. The relationship between lattice parameter (a_0) and extent of substitution of ruthenium by either lead or bismuth is linear as shown in Figure 1.

X-ray diffraction patterns of the lead-substituted lead ruthenates support the conclusion that they are pyrochlores of cubic symmetry. It must be noted, however, that since these materials were prepared at relatively low temperatures, the peaks in their x-ray spectra were broadened. X-ray line broadening was significant [half-height peak width at 50-60° (2θ) was equal to 0.5-0.8°(2θ)]; thus any subtle evidence of distortion to lower symmetries would be difficult to observe. The x-ray diffraction intensity data is also consistent with substitution of the B site ruthenium cations by Pb^{4+}. As the degree of lattice expansion increases the all-odd hkl reflections decrease markedly in intensity since their intensities are proportional to the difference in scattering between the A and B site cations.

Anion stoichiometry measurements made by thermogravimetric reduction in hydrogen show half occupancy of the O' (or seventh anion) site independent of the extent of lead substitution on the B site. Thus, the appropriate general formula for these lead ruthenate phases may be expressed as $Pb_2[Ru_{2-x}Pb_x^{4+}]O_{6.5\pm0.1}$ where $0 \leq x \leq 1$. Recent neutron diffraction results (8) confirm the half-occupancy of the seventh anion site for the phase, $Pb_2Ru_2O_{6.5}$. This oxygen stoichiometry suggests that half of the ruthenium can be formally considered as 5+ and the other half as 4+. This is consistent with the inability to substitute more than half of the ruthenium by Pb^{4+} and obtain a single phase pyrochlore.

In the case of the $Bi_2[Ru_{2-x}Bi_x]O_{7-y}$ series, it is possible to explain the bismuth substituted pyrochlore either as resulting from the substitution of ruthenium by Bi^{3+}, with subsequent vacancy formation on the anion lattice, or by substitution of ruthenium by pairs of Bi^{3+} and Bi^{5+}, with no anion vacancy formation necessary. Any of the above valence distributions for the bismuth ruthenates could account for the expanded lattice parameter since they all involve average B-site ionic radii larger than Ru^{4+}.

The thermal stability of all of these nonstoichiometric pyrochlores is limited and is inversely dependent upon the extent of substitution of noble metal cations on the B-site by post transition element cations (1). For example, $Pb_2[Ru_2]O_{6.5}$ is stable to 850°C in air while $Pb_2[RuPb]O_{6.5}$ is only stable to about 400°C.

The Alkaline Solution Synthesis Route. The precipitation/reaction to form pyrochlores within the alkaline reaction medium appears to involve the gradual evolution of a crystalline pyrochlore from a largely amorphous precipitate. The x-ray diffraction pattern of this "incipient pyrochlore," immediately following precipitation, shows only one discernible feature: a very broad, diffuse peak centered at ~30° (2θ). With increased time in the reaction medium, the major peaks of the pyrochlore begin to become evident in the x-ray pattern of the precipitate, with the original diffuse peak at 30° (2θ) gradually evolving into the (222) reflection of the pyrochlore. An alternative, low temperature solid state approach to synthesizing the substituted, crystalline pyrochlore is to recover the precipitate while still in the amorphous state and subject it to a mild heat treatment. Figure 2 shows the evolution of x-ray diffraction patterns for such a heat treatment carried out at 300°C. Figure 3 is a DTA trace, run in air, on an amorphous or "incipient pyrochlore" precipitate. The very distinctive exotherm accompanying crystallization reinforces the concept that the incipient pyrochlore is a very effective precursor which, with sufficient thermal activation, can undergo a rapid, cooperative transformation to a structurally well-resolved pyrochlore.

Since our interest was in maximizing surface area, attention was directed primarily at the evolution of crystalline pyrochlore directly from the alkaline solution media.

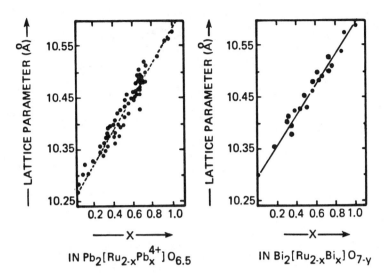

Figure 1.   Lattice parameter vs. composition. Reproduced with permission from "Preparation and Characterization of Materials"; J. M. Honig and C. N. R. Rao, eds.; Academic Press Inc., New York, 1981, p. 38 and 44.

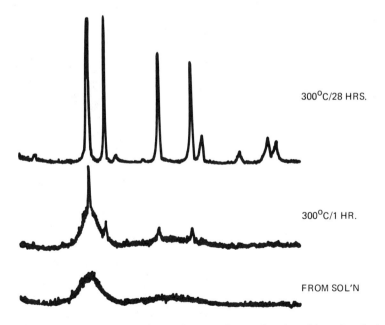

Figure 2.   Transformation of amorphous lead ruthenate into crystalline pyrochlore.

One of the important parameters controlling the alkaline solution synthesis is solubility. The pyrochlores of the present study are found to have a finite solubility in alkali. Experimental observations suggest that the lead-substituted lead ruthenate will always remain in equilibrium with a certain concentration of lead in alkaline solution. If the concentration of lead in solution is increased, thereby exceeding the equilibrium concentration, the system responds by incorporating more lead into the solid, resulting in an increased lattice parameter for the pyrochlore. If the concentration of lead in solution drops below the equilibrium level, the system responds by giving up more lead to the solution resulting in a pyrochlore with a smaller lattice parameter. Thus, there is observed to be a facile transfer of lead between the solid and the solution aiding the synthesis.

The $[OH^-]$ must be maintained at a certain minimum value (pH 10) in order to synthesize crystalline pyrochlore from solution. This implies that a minimum solubility is required for crystallization and that this crystallization of the pyrochlore directly out of alkaline solution may involve a solution-reprecipitation mechanism. Once the restriction of minimum pH has been satisfied, $[OH^-]$ does not seem to have a significant effect on crystallinity as long as oxidizing conditions are maintained. The solubility of lead rapidly increases with hydroxide concentration; therefore, when all else is held constant, the lattice parameter of the product pyrochlore decreases as the pH of the synthesis medium increases.

Increasing the temperature of synthesis results in enhanced crystallinity as would be anticipated because of improved reaction kinetics. However, this observation is also consistent with a crystallization mechanism involving solubility. Furthermore, as the temperature increases so does the equilibrium concentration of lead in solution; thus with all else held constant, increased temperature of reaction results in a smaller lattice parameter for the product lead ruthenate pyrochlore.

One additional parameter that affects the solubility of lead ruthenate pyrochlores in alkali is the extent of lead substitution on the B-site. The greater the substitution (i.e. the larger x is in formula (2), the higher the solubility of the pyrochlores is in alkali (2). If it is assumed that a solution-reprecipitation mechanism of synthesis is operative, the stoichiometry-dependent solubility could explain why it becomes significantly more difficult to crystallize lead ruthenate directly out of alkaline solution when x <0.3.

Perhaps the most important synthesis parameter affecting the crystallization of pyrochlore from alkaline solution is the oxidizing potential within the reaction medium. It is observed that this parameter has dramatic effects on both crystallinity and extent of ruthenium substitution by Pb^{4+}. For example, a synthesis in which O_2 is bubbled into the reaction medium will yield a well crystallized pyrochlore 2-3 times faster than a synthesis where O_2 sparging is not provided. Crystalline pyrochlores cannot be obtained under any synthesis conditions (except those that are electrochemically assisted) when N_2 sparging is used. The necessity for relatively oxidizing conditions in order to yield crystalline expanded pyrochlores is consistent with the hypothesis that these pyrochlores do contain Pb^{4+}, as well as Ru^{4+} and Ru^{5+} cations on the B-site. Thus, oxidation of the divalent lead and trivalent ruthenium starting species is required.

Highly oxidizing potentials can also be achieved by employing an anodic electrode within the reaction vessel. For example, the solution synthesis of crystalline expanded pyrochlore can be done under conditions (16 hours, 50°C, 0.72M KOH, N_2 sparging) that would not yield crystalline pyrochlore if there were no anodic electrode present. Employing a Pb^{4+} source (such as PbO_2 or $Pb(Ac)_4$) rather than a Pb^{2+} source has much the same effect, the kinetics of pyrochlore formation being measurably quicker even though the lead source may have been added as a solid powder. It appears that a minimum electrochemical potential of 1.0V vs. a reversible hydrogen electrode in the same electrolyte (RHE) is required for the synthesis of crystalline pyrochlore. If the potential of the system is appreciably lower than 1.0V RHE an amorphous product results. If the potential is higher than 1.1V RHE, significant amounts of PbO_2 are produced along with pyrochlore. The sparging of O_2 through an alkaline reaction medium (pH = 13) at 75°C develops a natural open circuit potential (OCV) of 1.0V RHE, thus allowing for the formation of crystalline pyrochlore. Nitrogen sparging of the same solution develops an OCV of less than 0.9 V RHE which explains the amorphous product obtained under these conditions. Air sparging at pH 13 gives an OCV well below 1.0V RHE and an amorphous product. However air sparging at greater than pH 14 increases the oxidizing potential of the reaction medium high enough to give crystalline pyrochlore.

The above observations are consistent with the Pourbaix (pH vs. potential) phase diagram for the system lead-water (9). This phase diagram, shown in Figure 4, indicates that in an aqueous alkaline environment of pH 13-14 (the optimum ranges for pyrochlore synthesis), Pb_3O_4 is stabilized in the potential range of 0.9 - 1.0V RHE. Thus, this fairly narrow electrochemical potential range, which coincides with the potential range required to obtain crystalline pyrochlore directly out of solution, is where Pb^{2+} and Pb^{4+} are predicted to coexist in the solid state as is the case in lead-substituted lead ruthenate. Consistent with experimental evidence cited earlier, potentials $> \approx 1.1V$ RHE stabilize only Pb^{4+}, thereby resulting in the formation of PbO_2.

While the discussion of oxidizing potential has centered on its effect in terms of lead, it is important to note that a relatively oxidizing potential is also required to convert the trivalent ruthenium of the starting solution to a higher $(4^+/5^+)$ oxidation state. The Pourbaix diagram (9) for the ruthenium-water system, shown in Figure 5, indicates that at potentials lower than 0.9V RHE, Ru^{3+} is the only stable solid state ruthenium species. At higher potentials Ru^{4+} is stable in the solid state and is in equilibrium with the ruthenate (Ru^{6+}) solution species. Although it is recognized that Pourbaix diagrams are not designed to describe mixed metal systems, in this particular case they help provide a rational explanation for the required solution synthesis conditions of the nonstoichiometric pyroclores.

Synthesis of the bismuth-substituted bismuth ruthenates is, in most respects, similar to that of the lead ruthenate series. Precipitation/crystallization is effected in a relatively oxidizing,

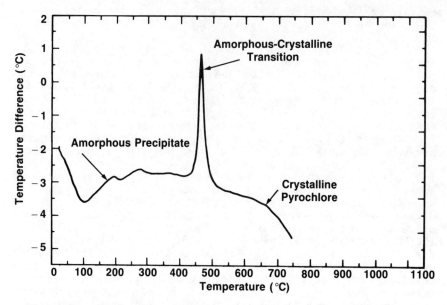

Figure 3.   DTA Trace, run in air at 20°C/min., on amorphous lead
ruthenate precipitate.

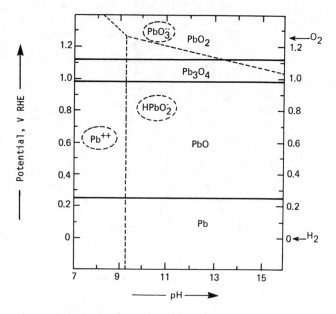

Figure 4.   Potential -pH diagram for the system lead-water at
25°C, adapted from Pourbaix (9).

alkaline medium. As with the lead ruthenates, low levels of electro-chemical potential, pH and temperature all lead to an amorphous product. The synthesis of these materials is further complicated by the existence of a competitive phase, $Bi_2Ru_2O_{7.33}$, which has the $KSbO_3$ structure (10). Figure 6 shows that single phase bismuth-substituted bismuth ruthenate pyrochlores can be synthesized using the alkaline solution route; however, single phase pyrochlores are restricted to a fairly narrow pH/temperature window. The bismuth ruthenate system also differs from the lead ruthenate system in one other respect. In the lead ruthenate series a mild heat treatment (e.g. 400°C) in air will convert any of the amorphous, "incipient pyrochlore" precipitates to pure crystalline pyrochlore. In the bismuth ruthenate system on the other hand, only those amorphous pre-cipitates which have been subjected to a digestion period, within the alkaline medium, of adequate duration will yield pure pyrochlore upon heat treatment. Furthermore, unless the heat treatment is carried out in a low pO_2 atmosphere (e.g. flowing Ar or He) the product will contain substantial amounts of the $KSbO_3$ phase.

The low temperature of synthesis afforded by the alkaline solution route has resulted in the pyrochlore oxides of this study having surface areas ranging from 50-200 m^2/g. Thus, one of the primary requirements for materials being investigated for catalytic applications has been achieved. However, even with these relatively high surface area materials, we have observed that the state of agglomeration can critically affect the electrocatalytic activity. Figure 7 illustrates how a simple variation in the synthesis procedure can drastically affect the state of agglomeration of powders produced via the alkaline solution route. The bulk volume of the powder is seen to increase approximately tenfold when it is recovered from the alkaline synthesis medium by the spray-freeze/freeze dry technique relative to the case where conventional vacuum filtration is employed. Electrodes fabricated from the high bulk volume powder showed a clear improvement in catalytic activity for both reduction and evolution of oxygen.

Oxygen Electrocatalytic Properties: Oxygen Reduction. Figure 8 compares steady-state polarization curves for the electroreduction of O_2 on a typical pyrochlore catalyst, $Pb_2(Ru_{1.42}Pb_{0.58})O_{6.5}$, and 15 w/o platinum on carbon. The latter was considered representative of conventional supported noble metal electrocatalysts. The activities of both catalysts are quite comparable. While the electrodes were not further optimized, their performance was close to the state of the art, considering that currents of 1000 ma/cm^2 could be recorded, at a relatively moderate temperature (75°C) and alkali concentration (3M KOH). Also, the voltages were not corrected for electrolyte resistance. The particle size of the platinum on the carbon support was of the order of 2 nanometers, as measured by transmission electron microscopy.

Figure 9 illustrates the O_2 electroreduction activity of a number of lead ruthenate pyrochlores, $Pb_2(Ru_{2-x}Pb_x)O_{6.5}$, where $0 \leq x < 1.0$. These data demonstrate that catalysts of roughly equivalent activity can be synthesized over the entire compositional range. In two examples where the activity was noticeably lower (x = 0.04 and x = 0.98), the synthesis conditions were such that the surface areas of

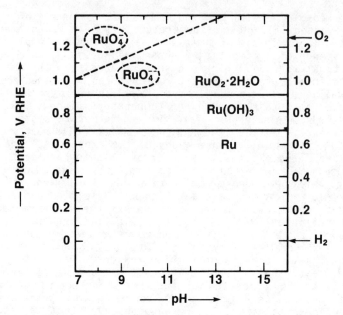

Figure 5. Potential -pH diagram for the system ruthenium-water at 25°C, adapted from Pourbaix (9).

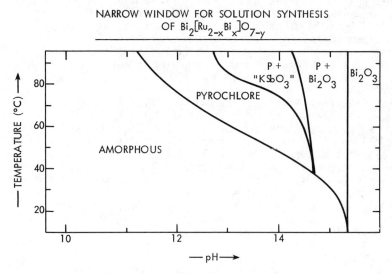

Figure 6. Synthesis products as a function of temperature and pH for the Bi-Ru-O system in KOH reaction media. Time of reaction held constant at 7 days. "KSbO$_3$" signifies the KSbO$_3$-related phase, Bi$_2$Ru$_2$O$_{7.33}$. "P" signifies the pyrochlore phase, Bi$_2$[Ru$_{2-x}$Bi$_x$]O$_{7-y}$.

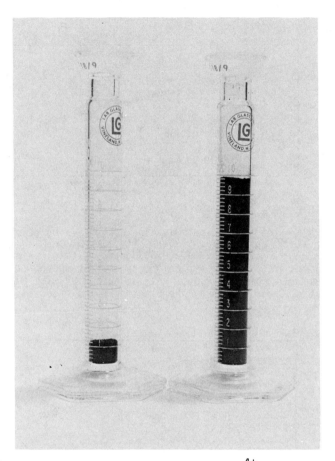

Figure 7. Equal weights of Pb_2 $[Ru_{1.62}Pb_{.38}^{4+}]$ $O_{6.5}$.
Left - Recovered by conventional vacuum filtration and drying.
Right - Produced by the spray-freeze/freeze dry technique. Both
have surface areas of 78 m^2/g.

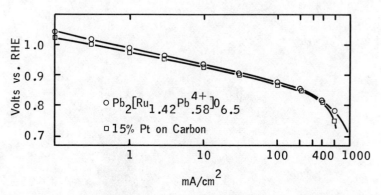

Figure 8. Steady-state polarization curves for O_2 reduction in 3M KOH, 75°C on $Pb_2[Ru_{1.42}Pb_{.58}^{4+}]O_{6.5}$, and 15 w/o Pt on carbon. Catalyst loadings are 60 and 50 mg/cm^2, respectively. The surface areas of the pyrochlore and carbon-supported Pt are 67 m^2/g and 25 m^2/g, respectively. Reproduced with permission from Ref. 1,Copyright 1983, The Electrochemical Soc. Inc.

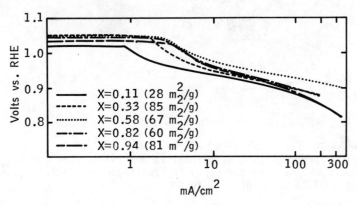

Figure 9. O_2 reduction in 3M KOH, 75°C on $Pb_2[Ru_{2-x}Pb_x^{4+}]O_{6.5}$ as a function of x. Scans at 0.2 mV/sec. Catalyst loading is 60 mg/cm^2. Reproduced with permission from Ref. 1. Copyright 1983, The Electrochemical Soc. Inc.

the catalysts were quite low (<15 m^2/g). Thus, the qualitative conclusion drawn from these activity curves is that the catalyst activity does not appear to be significantly dependent upon composition; surface area does appear to be the dominant factor. The constant polarization at low current densities observed for the pyrochlores run under dynamic scan conditions is a consequence of pseudo-capacitance charging currents (lead ruthenate with surface areas of 60-80 m^2/g has a pseudocapacitance in the range of hundreds of farads per gram; thus, voltammograms under N_2 at 0.2 mV/sec showed capacitance currents of 3-8 mA/cm^2).

Oxygen Electrocatalytic Properties: Oxygen Evolution. Figure 10 shows polarization curves for oxygen evolution on a number of pyrochlores, $Pb_2(Ru_{2-x}Pb_x)O_{6.5}$, where $0 < x < 1$. It can be seen from these curves that catalysts of roughly equivalent activity can be synthesized over very nearly the entire compositional range; surface area again appears to be the primary factor affecting activity. Note again the dominant effect of pseudocapacitance at low current densities.

The data in Figure 10 illustrate the fact that $Pb_2(Ru_{2-x}Pb_x)O_{6.5}$ pyrochlores are even more effective catalysts for oxygen evolution than they are for oxygen reduction. A typical good $Pb_2(Ru_{2-x}Pb_x)O_{6.5}$ catalyst evolves oxygen at 100 mA/cm^2 at a polarization of approximately 120 mV from the theoretical potential (1.18V vs. RHE, 3M KOH, 75°C) as compared to a polarization of approximately 260 mV from theory for oxygen reduction at the same current density. This extraordinarily low anodic polarization is further demonstrated by the oxygen evolution activity comparisons, shown in Fig. 11 between $Pb_2(Ru_{1.67}Pb_{0.33})O_{6.5}$ and some competitor electrocatalysts. Data for Pt black, RuO_2, 1% Mg-doped $NiCo_2O_4$ supported on Ni sheet, and Ni sheet are included along with the pyrochlore. It should be noted that RuO_2, while included in the comparison, is found not to be stable in alkali under anodic polarization.

The performance of Ni could probably have been substantially improved by using a higher surface area form of the metal, but there are no reports in the literature of achieving oxygen evolution activities with nickel that are any better than can be obtained with $NiCo_2O_4$.

It has been established that reasonable activity maintenance is achievable over relatively long testing periods (500-1800 hours). The details of this life testing are given in reference (2). The principal difficulties associated with activity maintenance appear to be related to electrode structure. As noted earlier, evidence of finite catalyst solubility was obtained; however, no indications of gross chemical/electrochemical instability were noted. Interestingly, the observed corrosion appeared to have no significant effect on catalytic activity.

In the case of oxygen electro-reduction, the irreversibility (or polarization from the theoretical voltage) that one almost always encounters can be considered to arise primarily from the extremely high activation energy involved in breaking the very stable O=O bond.   As a result, the electro-reduction of oxygen is typically rate limited by the reduction of an intermediate hydrogen peroxide species.   The potential of this two electron reduction to the

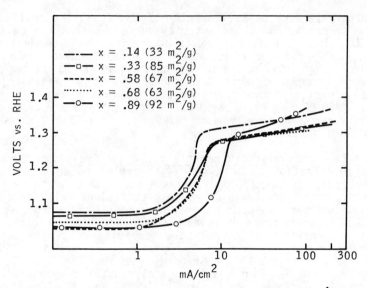

Figure 10. O_2 evolution in 3M KOH, 75°C $Pb_2[Ru_{2-x}Pb_x^{4+}]O_{6.5}$ as a function of x. Catalyst loading is 60 mg/cm^2. Reproduced with permission from Ref. 1. Copyright 1983, The Electrochemical Soc. Inc.

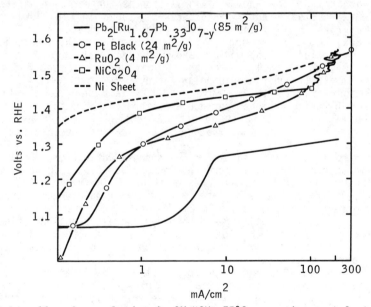

Figure 11. O_2 evolution in 3M KOH, 75°C on various catalysts. Catalyst loadings were: 60mg/cm^2 for the pyrochlore; 33 mg/cm^2 for the Pt black; and 62 mg/cm^2 for the RuO_2. The catalyst loading for $NiCo_2O_4$ spinel was not recorded. Reproduced with permission from Ref. 1. Copyright 1983, The Electrochemical Soc. Inc.

peroxide decreases with the peroxide concentration; unless this concentration is reduced to the extremely low equilibrium level (i.e. 10^{-15} M HO_2^- in 1 M caustic), a potential is obtained which is significantly lower than the theoretical potential for the direct four electron reduction of oxygen to hydroxide.

Rotating ring disk experiments on the lead-substituted lead ruthenates (2), at potentials where significant oxygen reduction current is evident, failed to detect any soluble H_2O_2 as an intermediate. The disk electrode used was a well compacted mass of Teflon and lead ruthenate powders which was run under diffusion limited conditions to minimize any penetration of H_2O_2 into the possibly porous interior. Under similar conditions carbon paste electrodes show significant hydrogen peroxide production in alkaline solutions. The addition of hydrogen peroxide to an open-circuited lead ruthenate electrode, in alkaline electrolyte, caused the formation of O_2 bubbles and a small but definite reduction in potential. This suggests that lead ruthenate catalyzes the rapid disproportionation of peroxide to oxygen and OH^- thereby preventing the formation of soluble peroxide and consequently reducing the mixed potential effect.

While the oxygen electrocatalysis results reported here have been those obtained specifically on the lead ruthenate series, essentially equivalent results were obtained on the bismuth ruthenate series.

Electro-Organic Oxidation Properties. Table I lists some results for the electro-oxidation of primary alcohols and propylene on lead-substituted lead ruthenate. Propylene was cleaved with nearly 100% selectivity to acetic acid and CO_2. In borate buffer at pH 9 the oxidation of propylene also occurred, and the selectivity to acetate and CO_2, based on the amount of carbonate isolated, was also close to 100%. Dissolved ethanol and propanol were both converted with high selectivity to the corresponding carboxylic acid salts in alkaline electrolyte. In contrast, Pt black (also shown in Table I) oxidized ethanol to CO_2 and then rapidly deactivated.

Table I

LEAD RUTHENATES OXIDIZE ALCOHOLS AND CLEAVE PROPYLENE*

Feed	Product	Coulombic Yield % Product	$CO_3^=$	Electrons Per Molec.
$H_3CCH=CH_2$	CH_3COO^-	100	94	10
" at pH 9	CH_3COO^-	N.A.[+]	100	10
CH_3CH_2OH	CH_3COO^-	>94	0	4
$CH_3CH_2CH_2OH$	$CH_3CH_2COO^-$	100	0	4

On PT

CH_3CH_2OH	$CH_3COO^-(?)$	25,27	48	Deactivated At 1 hr.

*Potentiostatic Operation at 1.2 V RHE, 25 or 50°C,0.5-1M KOH
[+]N.A. = Not analyzed for

Table II shows results for the electro-oxidation of secondary alcohols and ketones. In alkaline electrolyte, secondary butanol was not oxidized to methyl ethyl ketone but was cleaved to acetate. Similarly methyl ethyl ketone was cleaved to acetate, although some CO_2 and propionate formed, indicative of cleavage on the other side of the carbonyl group. Butanediol (2,3) went to acetate yielding less CO_2. At pH 9 in borax buffer 2-butanol went exclusively to methyl ethyl ketone at 89% conversion, suggesting that enolization in alkali is a necessary part of the cleavage process. Cyclohexanol and cyclohexanone were both converted to adipic acid. Figure 12 summarizes the various types of electro-organic oxidations, thus far discussed, which are observed to occur on lead ruthenate in alkaline electrolyte.

In order to confirm the reactivity and selectivity of lead ruthenates for the oxidation of isolated double bonds, two soluble, unsaturated carboxylic acids were chosen that contain a double bond far removed from the solubilizing carboxylate group. The two olefinic compounds, 1-undecylenic acid and 2-cyclopentene-1-acetic acid were both cleaved at the double bond as shown in Figure 13.

The initial rates of oxidation for a variety of reactants are shown as a function of temperature in Figure 14. The soluble compounds containing oxidizable oxygen functionalities achieved rates of 100 or more ma/cm^2 at 50°C whereas the oxidation of the olefinic compounds was a factor of 5 or 10 slower. The data on propylene showed a benefit for higher pH values and a diffusion limit above 50°C in strong alkali.

Table II

SECONDARY ALCOHOLS AND KETONES ARE CLEAVED*

Feed	Product	Coulombic Yield % Product	$CO_3^=$	Electrons Per Molec.
$CH_3CH_2\overset{OH}{\underset{H}{C}}-CH_3$	$2CH_3COO^-$	67	8-25	8
$CH_3-CH_2\overset{O}{C}CH_3$	$2CH_3COO^-$	81	14-24	6
$CH_3-\underset{OH}{C}H-\underset{OH}{C}H-CH_3$	$2CH_3COO^-$	84	3-4	6
⬡ = O	$CH_2CH_2COO^-$ $CH_2CH_2COO^-$	87	1	6

+Potentiostatic Operation at 1.2 V RHE, 25 or 50° C, 1M KOH
*Range of Values Depends on e's/MOL of $CO_3^=$ Assumed

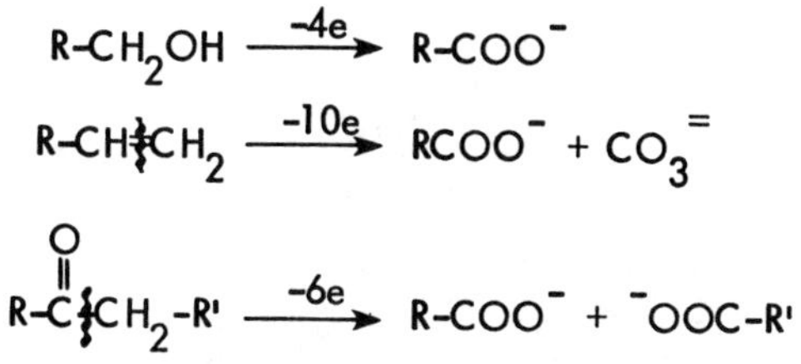

$$R\text{-}CH_2OH \xrightarrow{-4e} R\text{-}COO^-$$

$$R\text{-}CH\text{=}CH_2 \xrightarrow{-10e} RCOO^- + CO_3^=$$

$$R\text{-}\overset{\overset{O}{\|}}{C}\text{=}CH_2\text{-}R' \xrightarrow{-6e} R\text{-}COO^- + {}^-OOC\text{-}R'$$

Figure 12. Typical Electro-Oxidations on Lead Ruthenate in Alkali.

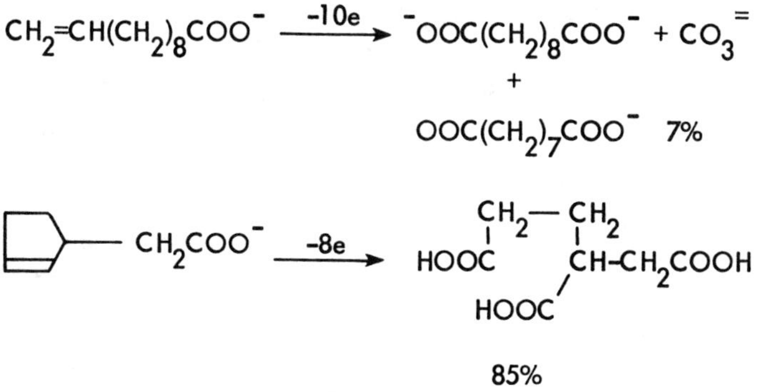

$$CH_2\text{=}CH(CH_2)_8COO^- \xrightarrow{-10e} {}^-OOC(CH_2)_8COO^- + CO_3^=$$

$$+$$

$$OOC(CH_2)_7COO^- \quad 7\%$$

$$\text{[cyclopentene]}\text{—}CH_2COO^- \xrightarrow{-8e} \begin{array}{c} CH_2\text{—}CH_2 \\ \ | \qquad \ | \\ HOOC \qquad CH\text{-}CH_2COOH \\ \ \ \diagup \\ HOOC \end{array}$$

$$85\%$$

Figure 13. Oxidation of Solubilized Unsaturates.

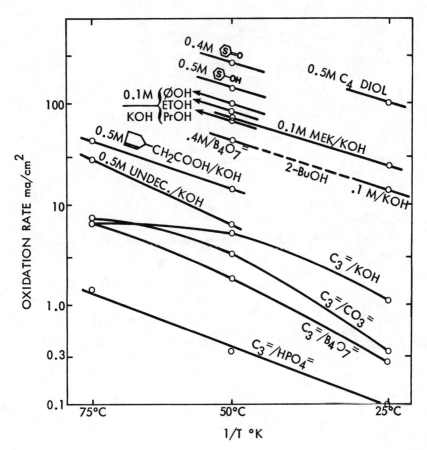

Figure 14.   Rates of Oxidation of Various Organics on Lead
Ruthenate.

Summary and Discussion of Results

A new family of high conductivity, mixed metal oxides having the pyrochlore crystal structure has been discovered. These compounds display a variable cation stoichiometry, as given by Equation 1. The ability to synthesize these materials is highly dependent upon the low temperature, alkaline solution preparative technique that has been described; the relatively low thermal stability of those phases where an appreciable fraction of the B-sites are occupied by post transition element cations precludes their synthesis in pure form by conventional solid state reaction techniques.

The data summarized in this paper have established that the oxide pyrochlores under discussion substantially reduce the activation energy overvoltages associated with oxygen electrocatalysis. Specifically, it is found that these catalysts, in aqueous alkaline media near ambient temperature, are superior to any other oxygen evolution catalyst and are equal in performance to the best known oxygen reduction catalysts. As bidirectional oxygen electrocatalysts, they appear to be unmatched.

It has been demonstrated that these same materials are able to perform the relatively rare reaction of cleavage of olefins and secondary alcohols, ketones and glycols to the corresponding carboxylates with high yields and selectivities.

There are several physical and chemical characteristics of these oxide pyrochlores which may contribute to their high electrocatalytic activity. The previously described alkaline solution synthesis technique (6,7) has provided these materials with surface areas typically ranging from 50 to 200 m^2/g. Thus, one of the basic requirements for an effective electrocatalyst has been satisfied: the electrocatalytic activity is not limited by the unavailability of catalytically active surface sites, as is so often the case with metal and mixed metal oxides.

Since all the catalysts investigated in this study display metallic or near metallic conductivity, the additional basic requirement of minimizing ohmic losses within the catalyst and between the catalyst and current collector has also been satisfied.

While high surface area and metallic conductivity are beneficial to electrocatalysis, they do not alone explain the high catalytic activity. We speculate that the variable oxygen stoichiometry of the pyrochlore lattice, and the multiple valence states of the cations, particularly the ruthenium, are essential to the catalytic activities of these pyrochlores.

Noble metal pyrochlores were originally considered by the authors as prime oxygen electrocatalyst candidates because of their ability to accommodate oxygen vacancies in up to one seventh of the anion sites. Changes in the average valence of the ruthenium can be expected to accommodate any such stoichiometry changes in $Pb_2(Ru_{2-x}Pb_x)O_{6.5}$. The existence of higher and lower valent oxides (either surface or bulk) in the potential range of interest appears to be a characteristic of many oxygen electrocatalysts such as Pt, Ag, Ni (for oxygen evolution), Au, etc. The reasons for this have never been explained exactly, although the ability of the surface to interact with or adsorb the potential intermediate peroxide or hydroperoxide ion on its surface is often invoked.

In the case of lead ruthenate, the oxygen non-stoichiometry concept can be developed further by combining it with the known reactions of the variable valence ruthenium. It has been shown in this work that these same catalysts can cleave carbon-carbon double bonds (3) in a manner analogous to that of osmium and ruthenium tetroxide (11). It is known (12) that OsO_4 (and presumably RuO_4) cleave olefins via complexes with the structure:

$$
\begin{array}{c}
\text{O} \\
\text{||} \\
\text{O} = \text{Os}
\begin{array}{c}
\diagup \text{O} \longrightarrow \text{C} \longrightarrow \text{R'} \\
\diagdown \text{O} \longrightarrow \text{C} \longrightarrow \text{R}
\end{array}
\end{array}
$$

The fact that these catalysts can carry out the very same cleavage suggests that they can form moieties similar to the tetroxide on their surfaces. By analogy, it is possible that the ruthenium atoms at the surface of the crystal lattice can react with oxygen molecules so as to form these same surface complexes:

$$ \text{Ru} + O_2 \longrightarrow [\text{Ru}] \longrightarrow \text{Ru} $$

These can be reduced stepwise to regenerate the starting structure:

$$ \text{Ru} + 2H_2O \xrightarrow{2e^-} \text{Ru} \xrightarrow[+2OH^-]{2e^-} \text{Ru} + 2OH^- $$

This or a similar set of reaction steps would avoid the formation of hydrogen peroxide consistent with the results of the rotating disk electrode experiment.

While the mechanism above is speculative, we prefer it to other possibilities involving -Ru-O-O-Ru- moieties, because (1) the elecro-organic oxidations must involve single Ru atoms, due to the steric requirements of the cleavage reaction; (2) the catalytic activity of non-stoichiometric lead ruthenates is surprisingly insensitive to dilution of the Ru atoms with Pb, whereas high sensitivity would be expected if the mechanism involved pairs of neighboring Ru atoms; and (3) the nearest Ru-Ru distance in the lattice is about 3.5 Å -- probably too long for a bridging peroxide group.

Acknowledgments

The authors would like to thank B. Bowling and H. Brady for assistance in the synthesis and materials characterization of the catalysts used in this study; J. Haberman, K. Strohmaier, and L. Yacullo for the electrochemical measurements; and G. Dupre and K. Rose for organic product analyses.

Literature Cited

1. Horowitz, H.S.; Longo, J.M.; Lewandowski, J.T., Mat. Res. Bull. 1981, 16, 489-496.

2. Horowitz, H.S.; Longo; J.M.; Horowitz, H.H., J. Electrochem. Soc. 1983, 130, 1851-1859.

3. Horowitz, H.H.; Horowitz, H.S.; Longo, J.M., In "Proceedings of the Symposium on Electrocatalysis" O'Grady, W.E.; Ross, P.N. Jr.; Will, F.G., Eds.; The Electrochemical Society, Inc.: Pennington, N.J., 1982; pp. 285-290.

4. Sleight, A.W., Inorg. Chem. 1968, 7, 1704.

5. McCauley, R.A., J. Appl. Phys. 1980, 51, 290.

6. Horowitz, H.S.; Longo, J.M.; Lewandowski, J.T., U. S. Patent 4 129 525, 1978.

7. Horowitz, H.S.; Longo, J.M.; Lewandowski, J.T., In "Inorganic Syntheses"; Holt, S.L. Ed.; Wiley-Interscience: N.Y., 1983; pp. 69-72.

8. Beyerlein, R.A.; Horowitz, H.S.; Longo, J.M.; Leonowicz, M.E.; Jorgensen, J.D.; Rotella, F.J., J. Solid State Chem. 1984, 51, 253.

9. Pourbaix, M.; "Atlas of Electrochemical Equilibria in Aqueous Solutions"; Pergamon Press: New York City, 1966.

10. Abraham, F.; Nowogroki, G.; Thomas, D. C. R. Acad. S. Paris 1974, 278C, 421.

11. Djerassi, C.; Engle, R.R., J. Am. Chem. Soc. 1953, 75, 3838.

12. Criegee, R., Annalen 1936, 522, 75.

RECEIVED December 17, 1984

Solid State Chemistry of Tungsten Oxide Supported on Alumina

S. SOLED, L. L. MURRELL, I. E. WACHS, G. B. MCVICKER, L. G. SHERMAN, S. CHAN, N. C. DISPENZIERE, and R. T. K. BAKER

Exxon Research & Engineering Company, Annandale, NJ 08801

The strong interaction between WO_3 and a γ-Al_2O_3 support is monitored under high temperature reducing and oxidizing conditions by a combination of physical and spectroscopic techniques. Below monolayer coverage a difficult to reduce highly dispersed surface tungsten oxide complex exists, whereas at higher coverages a more easily reduced bulk like WO_3 species is also present. Dynamic structural changes of the supported phase occur during high temperature treatment.

Supports can no longer be considered inert carriers which act solely to disperse a metal or metal oxide and thereby increase effective surface area. In many cases the reactivity and the catalytic properties of supported and bulk phases differ dramatically. A plethora of work has appeared over the last few years describing the strong metal support interaction (SMSI) of Group VIII metals with a titania support (1). In the "SMSI" state, metals display a dramatically reduced H_2 and CO chemisorption ability. Controversy exists about the basis of SMSI and such diverse explanations as electron transfer and TiO_2 migration onto the metal are being argued (1,2).

Examples of supports modifying the properties of transition metal oxides have also appeared in the literature. Recent work points to iron oxide phases as important species in Fischer-Tropsch synthesis (3). Iron oxide supported on SiO_2 (4) and TiO_2 (5) resist reduction under conditions in which bulk iron oxide easily reduces. Thus supported iron oxide catalysts are potentially interesting Fischer-Tropsch catalysts. The extensive studies on ethylene polymerization catalysts suggests that chromium (VI) species exist on a SiO_2 surface at temperatures above which bulk chromic anhydride (CrO_3) decomposes (6).

Recent evidence points to a strong interaction between WO_3 and γ-Al_2O_3 (7-10). The interaction alters the physical and chemical properties of both WO_3 and γ-Al_2O_3. In this review, we

0097-6156/85/0279-0165$06.00/0

describe studies of WO_3 on γ-Al_2O_3 using such diverse techniques
as controlled atmosphere electron microscopy (CAEM) (11), x-ray
photoelectron spectroscopy (XPS or ESCA) (12), thermal gravimetry
(TG) (13,14), and laser Raman spectroscopy (15) to examine the
nature of the tungsten oxide-alumina interaction.

Experimental

In these studies, both powder samples and films were prepared.
Powder samples of nominal 4, 6, 10, 25 and 60 wt.% tungsten oxides
on γ-Al_2O_3 (Engelhard Inc., reforming grade, 180 m^2/gm, 325 mesh)
were prepared by the incipient wetness impregnation method by
adding an aqueous solution of ammonium meta-tungstate to the
alumina powder, drying at 100°C and calcining in air at 500°C for
16 hrs. For the Raman experiments, γ-Al_2O_3 obtained from Harshaw
(Al-4104E, 220 m^2/gm) or Engelhard, Inc., (reforming grade, 180
m^2/g) were used as supports. The impact of calcination and
steaming as a function of temperatures was systematically
studied. Samples of pure WO_3 and $Al_2(WO_4)_3$ were obtained from
Cerac. For the CAEM experiments, film specimens of alumina,
approximately 50 nm in thickness, were prepared according to the
method described previously (16). Electron diffraction
examination of selected areas of the alumina film showed the
predominant phase to be γ-Al_2O_3. Tungsten was introduced onto the
alumina as an atomized spray of a 0.1% aqueous solution of
ammonium meta-tungstate. The tungsten loadings ranged between 4-
20 micromoles/m^2 (which corresponds to 10 to 50 wt.% tungsten on a
γ-Al_2O_3 of 100 m^2/g).

TG measurements were conducted on a Mettler TA-2000C as
described elsewhere (13). For TG reduction studies, samples of
WO_3 on γ-Al_2O_3 were heated to 970°C (at 10°/min) in He and then
held isothermally until constant weight was obtained. This pre-
calcination step minimizes overlapping reduction and
dehydroxylation weight losses. After cooling to room temperature,
H_2 was introduced, and the samples were reheated to a temperature
between 600° and 900°C (at 10°/min) and held isothermally for two
hours. The sensitivity and stability of the thermobalance (0.05
mg) establishes a detection limit of 1 to 2% WO_3 reduction to W.
Slight gray discolorations indicate small amounts of WO_3 reduction
below the TG detection limit.

In situ x-ray photoelectron spectra (XPS or ESCA) were
collected on a modified Leybold Heraeus LHS-10 electron
spectrometer. A moveable stainless steel block allowed sample
transfer in vacuum from a reactor chamber to the ESCA chamber.
The intensities and binding energies of the
W $4f_{5/2,7/2}$ signals (Al Kα radiation) were monitored and
referenced to the Al 2p peak at 74.5 eV. The 10% WO_3 and 60% WO_3
on γ-Al_2O_3 powder samples were calcined in air at 500°C and at
950°C respectively for 16 hrs and then pressed (at 30 Mpa) onto a
gold screen, which in turn was mounted on a moveable stainless
steel block. These samples were calcined in situ at 500°C to
clean the surfaces prior to analysis. For the reduction
treatments, the samples were heated for five minutes at the
desired temperature in flowing H_2 (25 cc/min.), cooled to 250°C in

H_2, evacuated and then transferred into the ESCA chamber. Three samples were investigated; a bulk WO_3 foil, 60 wt.% WO_3 on γ-Al_2O_3, and 10 wt.% WO_3 on γ-Al_2O_3.
The Raman spectrometer consisted of a triple monochromator (Instruments SA, Model DL203) equipped with holographic gratings and F4 optics. The spectrometer was coupled to an optical multichannel analyzer (Princeton Applied Reseach, Model OMA2) equipped with an intensified photodiode array detector cooled to 15°C. Each spectrum reported here was accumulated for about 100 sec or less. The digital display of the spectrum was calibrated to give $1.7 cm^{-1}$/channel with the overall spectral resolution at about 6 cm^{-1}. An argon ion laser (Spectra Physics, Model 165) was tuned to the 514.5 nm line for excitation. A prism monochromator (Anaspec Model 300S) with a typical band width of 0.3 nm removed the laser plasma lines (16). A 0.2 gm sample was pelletized under 60 MPa pressure into a 13 mm diameter wafer for mounting on a spinning sample holder. The laser power at the sample location was set in the range of 1-40 mW. The scattered light was collected by a lens (F/1.2, f/55 mm) held at about 45° with respect to the excitation.

Results

CAEM. Controlled atmosphere electron microscopy (11) was used to observe the behavior of tungsten oxide particles supported on γ-Al_2O_3 films when heated at temperatures up to 1150°C in 0.7 kPa oxygen. The two specimens described in the Experimental section were heated at increasing temperatures and the specimen changes were recorded in real time on video tape (15). The detailed observations of the dynamic behavior of the different tungsten oxide phases on the γ-Al_2O_3 film as a function of temperature and tungsten oxide content will be described in the Discussion section.

Thermalgravimetry. γ-Al_2O_3 on programmed heating (10°/min) to 1100°C in the presence of oxygen, continuously lost weight as a result of dehydroxylation: the weight lost between 200 and 1100°C equaled about 3.5%. In addition, a weak exotherm with an onset near 1050°C occurred during the transition of γ-Al_2O_3 to α-Al_2O_3. A 10% WO_3 on γ-Al_2O_3 sample showed different behavior. When this sample was heated in an oxygen atmosphere, a larger exotherm occurred at 1050°C as a fraction of the Al_2O_3 support reacted with WO_3 to form $Al_2(WO_4)_3$. The formation of $Al_2(WO_4)_3$ was confirmed by X-ray diffraction measurements. Alumina not utilized in tungstate formation transformed predominantly to θ-Al_2O_3: only a trace of α-Al_2O_3 was produced. Thus, the presence of the tungsten oxide surface phase inhibits the transition of θ-Al_2O_3 to α-Al_2O_3.

TG experiments indicate that a surface tungsten oxide phase
on alumina is difficult to reduce. Table I shows the degree of
reduction (expressed as percent WO_3 reduced to W^o) as a function
of WO_3 loading after two hour reductions at 600 or 900°C. A 10%
WO_3 on alumina sample was reduced at several intermediate
temperatures as well. Ambiguities resulting from simultaneous
weight loss due to water were minimized by initially calcining
these samples in He to 970°C. This pre-treatment yields tungsten
oxide on a transitional alumina (mostly θ) possessing a surface
area of ~80 m^2/gm. The retardation of WO_3 reduction depends on
loading levels. At low loading levels (<6%), little or no
reduction occurs. The 10% WO_3 on Al_2O_3 showed the first sign of
reduction at 800°C. Although no weight loss was detected by TG, a
slight greyish discoloration indicated some reduction had
occured. In contrast, bulk WO_3 is completely reduced after 2
hours at 600°C. Following 850 and 900°C reductions, 10% WO_3 on
Al_2O_3 was extensively reduced and the presence of tungsten metal
was confirmed by x-ray diffraction measurements.

ESCA. ESCA measurements also reveal the reduction resistance of
the tungsten oxide surface phase on Al_2O_3. An oxidized tungsten
foil serves as a standard for the ESCA reduction experiments. The
ESCA W $4f_{5/2,7/2}$ spectra for the oxidized, partially reduced and
fully reduced tungsten foil are presented in Figure 1. The ESCA W
$4f_{7/2}$ binding energy for the oxidized foil occurs at ~36 eV and
corresponds to tungsten in the +6 oxidation state (17-19). The
completely reduced foil exhibits an ESCA W $4f_{7/2}$ peak at ~32 eV
corresponding to metallic tungsten (17-19). The partially reduced
tungsten foil displays a very broad W $4f_{5/2,7/2}$ ESCA spectrum.
Deconvolution of the ESCA W 4f signal from the partially reduced
sample reveals the presence of five oxidation states of tungsten
(W^{+6}, W^{+5}, W^{+4}, W^{+2} and W^o) (17,20). Thus, the reduction of bulk
tungsten oxide to metallic tungsten proceeds through ESCA
observable intermediate oxidation states of W^{+5}, W^{+4}, and W^{+2}.
     The reduction behavior of a 60% WO_3 on Al_2O_3 sample as shown
in Figure 2 was very similar to that observed for the oxidized
tungsten foil. The ESCA W $4f_{7/2}$ binding energy for the oxidized
sample occurs at ~36eV and reveals that tungsten is present as
W^{+6}. During partial reduction of the 60% WO_3 on Al_2O_3 sample the
ESCA W 4f signal broadens indicating that in addition to W^{+6} and
W^o other oxidation states are present. However, the tungsten
oxide in the 60% WO_3 on Al_2O_3 sample requires higher temperatures
to completely reduce the tungsten oxide than tungsten oxide on the
foil.
     Figure 3 presents the ESCA W 4f spectra for the 10% WO_3 on
Al_2O_3 sample. Note the higher temperatures required to initiate
reduction for the supported tungsten oxide compared to the
oxidized tungsten foil or the 60% WO_3 on Al_2O_3 sample. The W
$4f_{7/2}$ binding energy for the oxidized sample occurs at ~36 eV and
reveals that tungsten is present as W^{+6}. The tungsten oxide
completely reduces to metallic tungsten at 900°C. Following this
reduction the total intensity of the ESCA W 4f signal decreases by
about 70%. X-ray diffraction shows the growth of large tungsten
metal particles consistent with the decrease in the ESCA signal

Table I. Reduction Behavior of Tungsten Oxide on Alumina

After 900°C H_2 Treatment (2 hr)					
% WO_3	4	6	10	25	100
% Reduction ($WO_3 \rightarrow W$)	-0-	2.5%	49%[a]	85%[b]	Black
Color		White	Lt. Gray	Black	Black	Black

After 600°C H_2 Treatment (2 hr)					
% WO_3	4	6	10	25	100
% Reduction ($WO_3 \rightarrow W$)	-0-	-0-	-0-	~22%[a]	-100-
Color	White	White	White	Black	Black

10% WO_3/γ-Al_2O_3 (2 hr at Reduction Temperature)		
Temperature °C	% Reduction $WO_3 \rightarrow W$	Color
600	-0-	White
700	-0-	White
800	-0-[c]	Tint of Gray-Slight Discoloration
850	17%[a]	Gray
900	49%[a]	Black

a. Reduction still continuing after 2 hr.
b. Some α-Al_2O_3 present.
c. Detection limit of 1 to 2%.

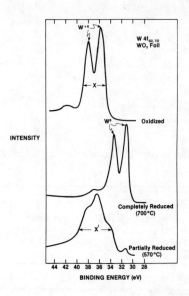

Figure 1. ESCA W $4f_{5/2, 7/2}$ spectra for a tungsten foil, oxidized and reduced. X indicates half width of oxidized sample; X' indicates half width of partially reduced sample.

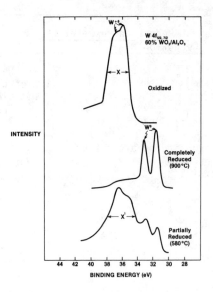

Figure 2. ESCA spectra for 60 Wt.% WO_3 on γ-Al_2O_3, oxidized and reduced. X indicates half width of oxidized sample; X' indicates half width of partially reduced sample.

intensity. For the partially reduced 10% WO_3 on Al_2O_3 sample the ESCA W 4f$_{7/2}$ peak does not broaden suggesting the absence of W^{+5}, W^{+4}, and W^{+2} states. Deconvolution of the ESCA W 4f spectra for the partially reduced 10% WO_3 on Al_2O_3 sample only reveals the presence of W^{+6} and W^0. Thus, the high temperature reduction of 10% WO_3 on Al_2O_3 does <u>not</u> proceed through ESCA <u>observable</u> intermediate tungsten <u>oxidation</u> state of W^{+5}, W^{+4} and W^{+2}.

RAMAN. The Raman spectra of WO_3, $Al_2(WO_4)_3$, and 10% WO_3 on Al_2O_3 are presented in Figure 4. Crystalline WO_3 contains a distorted octahedral WO_6 network with the major vibrational modes at 808, 714 and 276 cm^{-1}. These modes have been assigned to W=0 stretching, W=0 bending and W-0-W deformation, respectively (21). Minor bands appear at 608, 327, 243, 218, 185 and 136 cm^{-1}. $Al_2(WO_4)_3$ (defect $CaWO_4$ structure) contains distorted, isolated tetrahedral tungsten groups. The major Raman peaks of $Al_2(WO_4)_3$ were assigned by comparison with tetrahedrally coordinated tungsten in an aqueous solution of WO_4^{2-} as well as with solid Na_2WO_4 (22). In Na_2WO_4, the WO_4^{2-} groups are required to sit at crystallographically constrained tetrahedral sites (symmetry 42m). WO_4^{2-} (aq.) and Na_2WO_4 exhibit major vibrational modes at 933 and 928 cm^{-1} (symmetric W=0 stretch), 830 and 813 cm^{-1} (antisymmetric W=0 stretch), 324 and 312 cm^{-1} (W=0 bending), respectively. Thus, the $Al_2(WO_4)_3$ peak at 1052 cm^{-1} is attributed to the W=0 stretching mode and the doublet at 378-394 cm^{-1} is assigned to the W=0 bending mode. Distortion in the tetrahedra dramatically affects the position of the bands.

The major Raman peak for 10% WO_3 on Al_2O_3 occurs around 970 cm^{-1}, and has been assigned to the W=0 symmetrical stretch of the surface tungsten oxide species (19,23). The intensities of the major Raman band for WO_3 (808 cm^{-1}), $Al_2(WO_4)_3$ (1052 cm^{-1}), and 10% WO_3 on Al_2O_3 (970 cm^{-1}) were compared after normalization with respect to the laser power applied. The relative Raman intensity ratios for these peaks are 1600:40:1 for normalized laser power. These Raman intensity ratios were further scaled for the different tungsten oxide contents and yielded relative ratios of 160:5:1 (15).

The states of tungsten oxide on alumina depend on the tungsten oxide content and the temperature of calcination. The effect of tungsten oxide content is shown in Figures 5 and 6 for 15 and 25% WO_3 on Al_2O_3, whereas the effect of changing temperatures with constant WO_3 content is shown in Figure 7. Figures 5a and b show Raman spectra of 15% and 25% WO_3 on Al_2O_3 calcined (with 8% steam present) at 760°C. Both materials exhibit surface areas of 120 m^2/gm. The 15% WO_3 on Al_2O_3 sample, Figure 5a, exhibits Raman bands of both a surface tungsten oxide species on the alumina support and a trace amount of crystalline WO_3. The 25% WO_3 on Al_2O_3 sample, Figure 5b, however, shows very intense crystalline WO_3 Raman bands which dominate the spectrum due to the large Raman cross-section of this phase. The intensities of the Raman bands resulting from surface tungsten oxide species are similar for both of these samples as shown in Figure 6 after scaling for the different applied laser powers over the region 850-1150 cm^{-1} (15).

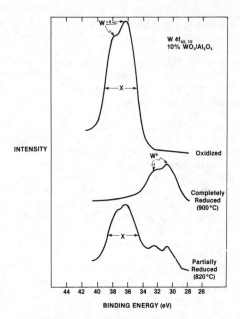

Figure 3. ESCA spectra for 10 Wt.% WO₃ on γ-Al₂O₃, oxidized and
reduced. X indicates half width of oxidized sample; X' indicates
half width of partially reduced sample.

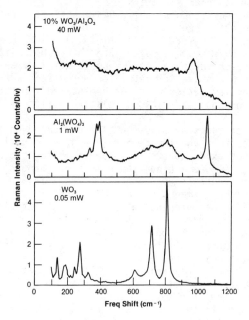

Figure 4. Laser Raman spectra of 10% WO₃ on Al₂O₃, aluminum
tungstate and tungsten oxide.

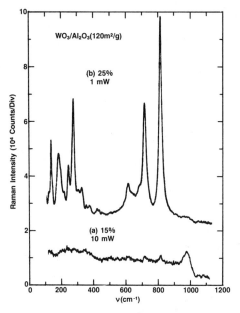

Figure 5. Laser Raman spectra of 25 Wt.% WO_3 on γ-Al_2O_3 and 15 Wt.% WO_3 on γ-Al_2O_3 de-surfaced to 120 m^2/g surface area.

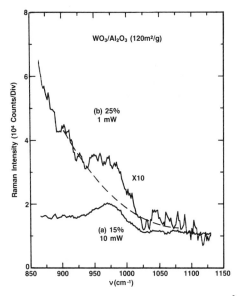

Figure 6. Laser Raman spectra from 850 to 1150 cm^{-1} of 25 Wt.% WO_3 on γ-Al_2O_3 and 15 Wt.% WO_3 on γ-Al_2O_3 de-surfaced to 120 m^2/g surface area.

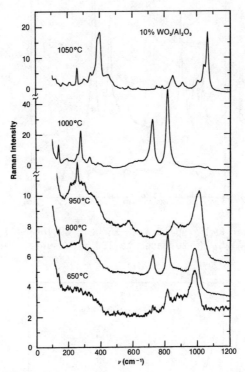

Figure 7. Laser Raman spectra of 10% WO_3 on γ-Al_2O_3 calcined at 450, 800, 950, 1000 and 1050°C.

The states of tungsten oxide on alumina were investigated over a wide temperature range (650-1050°C) for 10% WO_3 on Al_2O_3. The 10% WO_3 on Al_2O_3 sample calcined at 650°C exhibits Raman peaks at 972, 809 and 718 cm^{-1}. The peak near 970 cm^{-1} is associated with a tungsten oxide surface complex (8,19,23). The position of this Raman peak shifts monotonically from 970 to about 1000 cm^{-1} as the calcination temperature is increased to 950°C. Similar shifts are observed when tungsten oxide loading is increased for samples calcined at 500°C (7,19).

Samples calcined at 1000 and 1050°C display bulk tungsten oxide WO_3 and $Al_2(WO_4)_3$ phases. Raman peaks at 811, 717, 273 and 137 cm^{-1} are characteristic of crystalline WO_3. These peaks decrease in intensity as the calcination temperature increases from 650-950°C, so that at 950°C the crystalline WO_3 Raman peaks at 811, 717 and 273 cm^{-1} are absent. At 1000°C these peaks appear again, and then disappear following a calcination at 1050°C. The major Raman peak of the $Al_2(WO_4)_3$ at 1055 cm^{-1} is first observed following a calcination at 1000°C and dominates the Raman spectra after a 1050°C calcination. The θ-Al_2O_3 Raman peak at 253 cm^{-1} is present in the spectra of samples calcined at 950-1050°C and agrees with X-ray diffraction. This series of Raman spectra reveals the dynamic nature of the WO_3 on Al_2O_3 system and its dependence upon calcination temperature.

Discussion

The strong interaction of WO_3 with a γ-Al_2O_3 surface modifies the behavior of both tungsten oxide and alumina. γ-Al_2O_3 will transform from a series of closely related transitional alumina phases possessing a defect spinel structure, containing both tetrahedral and octahedral aluminum ions, to α-Al_2O_3, a corundum structure containing only octahedral aluminum ions (24). The γ to α-Al_2O_3 transformation occurs by the condensation of surface hydroxyl groups and the elimination of H_2O. TG studies on powder samples, as well as, CAEM studies on model film systems indicate that WO_3 inhibits the γ to α-Al_2O_3 phase transformation. The presumed bonding of WO_3 with the hydroxyl surface of γ-Al_2O_3 blocks the transformation to α-Al_2O_3 (8). At sufficiently high temperatures ($\sim$1150°C) and high WO_3 concentrations, approximately 3X monolayer coverage (19), CAEM detects the reaction of tungsten oxide with alumina forming $Al_2(WO_4)_3$. Monolayer coverage is defined as 4.3×10^{18} W atoms/m^2 or 7 micromoles/m^2. A monolayer of WO_3 on γ-Al_2O_3 simply refers to the surface phase oxide structure at maximum packing of the alumina surface before crystallites of WO_3 are formed (15,19). The observation of the sequential formation of WO_3 followed by $Al_3(WO_4)_3$ is supported by the Raman studies (10,15,19).

Bulk WO_3 crystallizes in a distorted version of the ReO_3 structure with WO_6 octahedra linked via corner-sharing with neighboring WO_6 octahedra. Some thirty years ago, Magneli discovered that during reduction, WO_3 behaves in a nonclassical manner. As oxygens are removed, the WO_3 structure rearranges maintaining the metal coordination at six. During this process, the octahedra restructure along crystallographic shear planes to

share edges in place of corners. This tightened packing allows
the structure to retain its metal coordination number at six even
though the cation-to-anion ratio increases. In this way,
partially-reduced WO_3 exists over a wide compositional range (25)
with formal tungsten oxidation states of 5 and 6 and tungsten
coordination of 6. Further reduction produces a discrete (WO_2)
phase which displays a distorted rutile structure. WO_2, which
contains a distorted octahedral coordination about tungsten also
reduces via a series of shear planes with tungsten assuming formal
valence states of 3 and 4. Thus during the reduction of bulk WO_3,
intermediate oxidation states occur. Consistent with this
mechanism (20), ESCA observes several intermediate oxidation
states (+5, +4 and +2) during the reduction of bulk WO_3, see Fig.
1.

The reduction behavior of tungsten oxide supported on γ-Al_2O_3
differs significantly from that of bulk WO_3. TG studies of WO_3 on
γ-Al_2O_3 show that below a coverage of 6% WO_3, the surface tungsten
oxide phase is essentially irreducible (10). At intermediate
loadings (~10%), WO_3 partially reduces; while, at higher loadings
(25%), the additional WO_3 behaves like bulk WO_3, see Table I. We
would suggest that the isolated surface tungstate groups
(8,16,19,26) on the low loaded samples do not reduce through
intermediate structures but, as our ESCA results indicate, the
reduction proceeds directly to tungsten metal. Since reduction
also occurs only at high temperature, the tungsten metal formed
rapidly sinters into large particles.

Our high temperature reduction experiments using ESCA agree
with both our TG studies and the recent low temperature reduction
experiments reported by Salvati, et al. (19). Salvati and co-
workers found a loading level dependence on reduction and the
presence of bulk-like WO_3 species above a critical coverage. Our
study (see Fig. 3) indicates that at the temperature necessary to
reduce the surface phase of tungsten oxide on Al_2O_3 the reduced
tungsten rapidly sinters to metallic particles (10). Apparently,
the highly-dispersed state of the tungsten oxide complex on the
alumina surface (8,19,26) precludes the formation of
nonstoichiometric tungsten oxide phases which form during the
reduction of unsupported WO_3.

For WO_3 on Al_2O_3 samples containing more than a monolayer the
additional tungsten oxide is present as WO_3 crystallites. These
WO_3 crystallites are not in direct contact with the alumina
support and are indistinguishable from bulk WO_3 in their reduction
behavior (19,20). The WO_3 crystallites reduce at mild
temperatures and exhibit ESCA observable intermediate
tungsten oxidation states. Detailed analysis (20) of the 10%
WO_3 on Al_2O_3 ESCA W $4f_{5/2,7/2}$ spectra in Figure 3 only reveals the
presence of W^{+6} and W^0 on the alumina support after partial
reduction. Within the experimental limits of ESCA, the high
temperature reduction of the tungsten oxide surface complex on the
alumina support does not proceed through observable intermediate
tungsten oxidation states of W^{+5}, W^{+4} and W^{+2}, but rather directly
from W^{+6} to W^0. The different reduction behavior of the WO_3
crystallites and the tungsten oxide monolayer can be used to
distinguish between these two forms of tungsten oxide on

alumina. Salvati et al. also used this approach to distinguish between the tungsten oxide surface complex and the WO_3 crystallites (19). Below monolayer coverage, tungsten oxide on alumina is not reduced after 12 hours at 550°C, but above monolayer coverage reduction of W^{+6} to W^0 at this temperature is observed with ESCA.

The TG and ESCA technique have shown evidence for a strong metal oxide-support interaction between WO_3 and γ-Al_2O_3 under high temperature reducing conditions. As we will now discuss CAEM and Raman spectroscopy suggest a strong interaction between WO_3 and γ-Al_2O_3 under high temperature oxidizing conditions, as well. Dynamic studies by CAEM of WO_3 on γ-Al_2O_3 at high temperature have been carried out for one-half and 3X monolayer loadings (19) of WO_3 on a model γ-Al_2O_3 film support. Following an in situ decomposition in oxygen of the ammonium meta-tungstate at 500°C on the 4 micromole/m^2 loaded film (about one-half monolayer) electron diffraction confirmed the presence of only transitional alumina phases (no α-phase was present). Heating this film to 1050°C shows no crystallization of the support; whereas with pure alumina films CAEM detects restructuring of the film to form α-Al_2O_3 at this temperature. The particle size of the tungsten oxide phase for the one-half monolayer covered film at 500°C lies below the resolution limit (2.5 nm) of the microscope. Even upon continued heating to 1050°C WO_3 particles are still not visible, and the support does not transform to α-Al_2O_3. This result suggests the presence of a highly-dispersed WO_3 phase stabilizes the alumina support from restructuring to α-Al_2O_3.

When the more highly loaded WO_3 on alumina specimen, i.e. 3X monolayer coverage, was heated in oxygen, particles were detected by CAEM at 500°C. Because the size of these particles (3 nm) are near the resolution limit of the CAEM, we could not determine a detailed particle size analysis. As the temperature was raised to 700°C the WO_3 particles grow in diameter to between 5 and 10 nm. Detailed examination shows a nearly uniform distribution of the particles across the support. The particles have irregular angular shapes, but in many cases remained thin enough to avoid masking the structural features of the underlying support. This morphology is similar to that proposed for FeO "raft" structures on SiO_2 (27). The particles do not change size, shape or position upon further heating at 1050°C for 1 hr. Maintenance of particle identity indicates a strong interaction between particles and support (27,28).

As the temperature was raised from 1050°C to 1150°C the specimen with the higher WO_3 loading changes dramatically in appearance. Initially, the electron scattering density of the particles increases. While maintaining their 5 to 10 nm lateral dimensions, the particles apparently become thicker. As the change proceeds, the area of the alumina support surrounding the tungsten oxide particles becomes progressively more transparent to the electron beam, suggesting that Al_2O_3 preferentially leaves these areas and forms $Al_2(WO_4)_3$. Subsequent examination of specimens in the high resolution transmission electron microscope (where defocusing and tilting experiments were performed) confirm this phenomena and eliminate any question of over-focus or phase contrast artifacts (29,30).

In summary, at sufficiently high temperatures and high WO_3 concentrations, the CAEM follows the agglomeration of WO_3 into thin oxide cluster structures (26,27), which subsequently react with Al_2O_3 forming $Al_2(WO_4)_3$. Figure 8 outlines the proposed stepwise interaction of tungsten oxide with transitional γ-Al_2O_3 films.

Laser Raman spectroscopic studies of alumina-supported WO_3 catalysts have shown that three different tungsten oxide phases are present: WO_3, $Al_2(WO_4)_3$, and a surface tungsten oxide species (8,19), Figure 4. The concentrations of these phases in WO_3 on Al_2O_3 catalysts depend on tungsten oxide loading and temperature of calcination (8). Previous studies have shown that Raman spectroscopy is more sensitive to WO_3 and $Al_2(WO_4)_3$ than to the surface tungsten oxide complex (8,19). No attempt, however, had been made to estimate the relative Raman cross sections of these tungsten oxide phases. This information would be useful to develop a model for the WO_3 on Al_2O_3 system (15).

The dynamic changes of tungsten oxide that occur on the surface of γ-Al_2O_3 as a function of calcination temperature and tungsten oxide content were followed by Raman spectroscopy for 10%, 15% and 25% WO_3 on γ-Al_2O_3 as described in the Results section. The Raman spectra of 15% WO_3 and 25% WO_3 on γ-Al_2O_3 calcined at 760°C, Figure 5 and 6 exhibits similar intensities for the bands of the surface tungsten oxide complex although the band intensities for crystalline WO_3 differ dramatically. Since the 15% WO_3 on γ-Al_2O_3 sample calcined at 760°C contains near monolayer coverage (19,26) the 25% WO_3 on γ-Al_2O_3 sample at this same temperature must contain crystallites of WO_3. Raman spectroscopy confirms that the surface phase tungsten oxide complex will form WO_3 crystallites as the tungsten oxide content increases (at a constant surface area) beyond monolayer coverage (19,26).

The study of the 10% WO_3 on γ-Al_2O_3 as a function of temperature also reveals changing states of tungsten oxide, Figure 7. Initially this sample contains tungsten oxide below monolayer coverage, but as the temperature is raised and the surface area collapses, the tungsten oxide concentration exceeds monolayer coverage (19,26). The crossover point of approximate monolayer coverage occurs between 60 to 100 m^2/gm and is generated by calcination temperatures between 850 and 950°C.

Below monolayer coverage (less than ~25-30% WO_3 on Al_2O_3 of 200 m^2/g) tungsten oxide is primarily in a highly dispersed and amorphous state on the alumina surface and remains so at low calcination temperatures (500-800°C) (8,19,23,26). For 10% WO_3 on Al_2O_3, Figure 7, the surface tungsten oxide complex is present up to 950°C. In addition, Raman peaks for crystalline WO_3 are also observed in this temperature range. The WO_3 crystallite concentration is low since they are not detected by X-ray diffraction. The amount of tungsten oxide present as crystalline WO_3 for 10% WO_3 on γ-Al_2O_3, Figure 7, is estimated to be less than 1% of the total tungsten oxide content present in the 10% sample calcined at 650 and 800°C (16). As the calcination temperature increases, the relative amount of crystalline WO_3 initially decreases, as measured by the intensity ratio I (811 cm^{-1})/ I (965-1000 cm^{-1}), (15). Thus, at the higher calcination temperatures a substantial decrease in the surface area of the alumina support occurs and simultaneously the WO_3 particles disperse to form the tungsten oxide surface complex. As the surface area decreases the distance between the tungsten oxide surface species decreases and the tungsten oxide surface density on the alumina support increases (19,26). The increase in the intensity of the ESCA W $4f_{5/2,7/2}$ signal (19,26) and the shift from ~965 to ~1000 cm^{-1} in the Raman band associated with the tungsten oxide surface complex (16) reflect this change. These structural changes in the WO_3 on Al_2O_3 system are depicted in Figures 9a and 9b. Thus, Raman spectroscopy confirms that the surface phase tungsten oxide complex will form WO_3 crystallites as the alumina desurfaces (at a constant tungsten oxide content), and monolayer coverage is exceeded.

A close-packed monolayer of tungsten oxide on alumina apparently forms when the minimum distance between non-polymeric tungsten oxide centers is achieved (8,19,26). The formation of the close-packed tungsten oxide monolayer, however, does not preclude the alumina from additional loss in surface area at still higher temperatures. The close-packed tungsten oxide monolayer accommodates further de-surfacing by forming bulk tungsten oxide phases WO_3 and $Al_2(WO_4)_3$ (see Figures 9c and 9d). The formation of WO_3 and $Al_2(WO_4)_3$ crystallites at higher temperatures is detected by laser Raman spectroscopy, Figure 7.

The CAEM studies indicate that at high temperatures the surface tungsten oxide phase transforms to thin WO_3 particles. These in turn react at high temperatures with the underlying alumina support to form $Al_2(WO_4)_3$. Thus both CAEM and Raman spectroscopy point to the same model for the transformation of the tungsten oxide surface phase; first to form tungsten oxide particles and then to form subsequently $Al_2(WO_4)_3$ at high temperature calcination conditions.

Conclusions

The strong interaction between WO_3 and γ-Al_2O_3 manifests itself under both high temperature reducing and oxidizing conditions. Under reducing conditions, TG and ESCA demonstrate that the critical coverage for virtual non-reducibility of WO_3 on an alumina surface (which has been exposed to high temperature and de-surfaced to ~80 m^2/gm) is 6-8 wt.%. Above this monolayer coverage more easily reduced bulk-like WO_3 species are present. For loadings of 10 wt.%, where partial reduction occurs at high temperature, the tungsten oxide appears to reduce directly from W^{+6} to W^0 without accessing the intermediate oxidation states that bulk WO_3 passes through.

Under high temperature oxidizing conditions, laser Raman spectroscopy and CAEM demonstrate the dynamic behavior of the amorphous and crystalline structural transformations occuring in the WO_3 on Al_2O_3 system. Below monolayer coverage of tungsten oxide on alumina, the tungsten oxide phase is present as a highly dispersed and amorphous surface complex on the support. Above

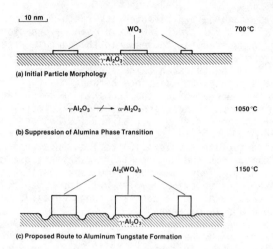

Figure 8. Model of the transformations observed for tungsten oxide on an alumina film by controlled atmosphere electron microscopy.

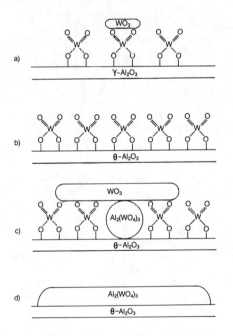

Figure 9. Model of the structure/transformations of tungsten oxide on high surface area γ-Al_2O_3 as a function of calcination temperature.

monolayer coverage, both a surface complex and discrete WO_3 crystallites are present. During high temperature de-surfacing a portion of the surface complex converts first to WO_3 and then reacts with the alumina to form $Al_2(WO_4)_3$.

Literature Cited

1. Tauster, S. J.; Fung, S. C.; Baker, R. T. K.; Horsley, J. A. Science 1981, 211, 1121.
2. Chen, B. H.; White, J. M.; Brostrom, L. R.; Deviney, M. L. J. Phys. Chem. 1983, 87, 2423.
3. Reymond, J. P.; Meriaudeau, P.; Teichner, S. J. J. Catal. 1982, 75, 39.
4. Topsoe, H.; Dumesic, J. A.; Morup, S. In "Applications of Mossbauer Spectroscopy"; Cohen, R. L., Ed.; Academic Press: New York, 1980; Vol. II.
5. Murrell, L. L.; Garten, R. L. (in press).
6. Hogan, J. P. J. Poly. Sci. 8(A-1) 1970, 2637.
7. Thomas, R.; Kerkhof, F. P. J. M.; Moulijn, A. J.; Medema, J.; deBeer, V. H. J. J. Catal. 1980, 61, 559.
8. Tittarelli, P.; Iannibello, A.; Villa, P. L. J. Sol. St. Chem. 1981, 37, 95.
9. Thomas, R.; deBeer, V. H. J.; Moulijn, J. A. Bull. Soc. Chim. Belg. 1981, 90(12), 1349.
10. Soled, S.; Murrell, L.; Wachs, I.; McVicker, G. Am. Chem. Soc. Div. Pet. Chem. Prepr. 1983, 28, 1310.
11. Baker, R. T. K.; Harris, P. S. J. Sci. Instrum. 1972, 5, 793.
12. Wagner, C. D.; Riggs, W. M.; Davis, L. E.; Moulder, J. F.; Muilenberg, G. E. "Handbook of X-Ray and Photoelectron Spectroscopy"; Physical Electronics Industries 1979.
13. Mettler Technical Bulletin, TA-2000C, Hightstown, N.J., 1980.
14. Soled, S.; McVicker, G. B.; DeRites, B. Proceedings 11th NATAS Conf., 1981, 417.
15. Chan, S. S.; Wachs, I. E.; Murrell, L. L. (in press).
16. McVicker, G. B.; Garten, R. L.; Baker, R. T. K. J. Catal. 1978, 54, 129.
17. Haber, J.; Stock, J.; Ungier, L. J. Sol. St. Chem. 1976, 19, 113.
18. Salje, E.; Carley, A. F.; Roberts, M. W. J. Sol. St. Chem. 1979, 29, 237.
19. Salvati, L.; Makovsky, J. M.; Stencel, J. M.; Brown, F. R.; Hercules, D. M. J. Phys. Chem. 1981, 85, 3700.
20. Wachs, I. E.; Chersich, C. C.; Hardenbergh, J. H. (in press).
21. Anderson, A. Spectr. Lett. 1976, 9, 809.
22. Busey, R. H.; Keller Jr., D. L. J. Chem. Phys. 1964, 41, 215.
23. Thomas, R.; Moulijn, J. A.; Kerkhof, F. P. J. M. Recl. Trav. Chim. Pays-Bas 1977, 96, M134.
24. Knozinger, H.; Ratnasamu, P. Catal. Rev-Sci. Engr. 1978, 17(1), 31.
25. Wells, A. F. Structural Inorganic Chemistry, Oxford Press, London 1962.
26. Murrell, L. L.; Grenoble, D. C.; Baker, R. T. K.; Prestridge, E. B.; Fung, S. C.; Chianelli, R. R.; Cramer, S. P. J. Catal. 1983, 79, 203.

27.   Yuen, S.; Chen, Y.; Kubsh, J. E.; Dumesic, J. A.; Topsoe, N.;
      Topsoe, H. J. Phys. Chem. 1982, 86, 3022.
28.   Baker, R. T. K.; Prestridge, E. G.; Garten, R. L. J. Catal.
      1979, 56, 390.
29.   Flynn, P. C.; Wanke, S. E.; Turner, P. S. J. Catal. 1974, 33,
      233.
30.   Treacy, M. M. J.; Howie, A. J. J. Catal. 1980, 63, 265.

RECEIVED October 4, 1984

High Resolution Electron Microscopy
The Structural Chemistry of Bismuth-Tungsten-Molybdenum Oxides and Bismuth-Tungsten-Niobium Oxides

D. A. JEFFERSON

Department of Physical Chemistry, University of Cambridge, Cambridge, CB2 1EP, United Kingdom

The application of high resolution electron microscopy to the determination of structures in systems with actual or potential use in selective oxidation catalysis is discussed. Problems of image interpretation, arising from instrumental aberrations and multiple scattering, are outlined, and examples of its use are given in the systems Bi-Mo-W-O, Bi-Mo-Nb-O and Bi-W-Nb-O. In all three, intermediate phases are revealed which show either potential or actual structural adaptability to varying stoichiometry, particularly with regard to oxygen content.

Heterogeneous catalysis has, until recently, been exclusively the preserve of the surface chemist. Detailed study of the bulk structural features has become important with the advent of shape selective catalysts, notably zeolites, where the distinction between external and internal surface is difficult to make, but surface studies have been considered most appropriate for other systems. However, in many real catalysts, where the catalytic action undoubtedly occurs on the external surface, it does so by means of intermediate structural states, and the catalytic efficiency is then dependent upon the relative stability and interactions of such intermediate states with the bulk material. Consequently an understanding of the structural chemistry and structural modification possible in the parent catalyst phase is still essential to understanding the catalytic action. This is especially true in the case of oxidation catalysts, where it can be shown (1) that lattice oxygen plays a part in the catalytic process.

Unfortunately, many of the oxide systems used as catalysts are extremely complex structures and phase relationships in them are not well understood. This arises partly from the difficulties of applying traditional methods of structure determination to materials which are complex but poorly crystalline, and are frequently disordered or polyphasic. In such cases the average structure resulting from a diffraction experiment is of little help and an attempt must be made to determine the specific structure of each individual crystallite. Only in this way can the structure be characterised completely.

0097–6156/85/0279–0183$06.25/0
© 1985 American Chemical Society

At present the only technique available for specific (sometimes called real-space) structure determination is high resolution electron microscopy (HREM). At first sight this appears to be an ideal method, as the direct imaging of the structure avoids the phase problem normally associated with diffraction methods, and can be applied to all materials, whatever their state of long range order. Compared with diffraction methods, however, the accuracy is relatively poor, as the available resolution is limited to not much better than 2Å, well above the theoretical limit. Furthermore, severe problems of image interpretation occur, but within certain limitations, these can be overcome and the technique applied successfully. The object of this paper is to illustrate the use of these direct imaging methods in systems with possible catalytic application.

Experimental

In principle, HREM is a very simple technique for structure determination and with electron wavelengths currently used, atomic detail should, in theory, be easily resolvable, without any "phase problem" However, the practical difficulties are by now well known and characterised (2) and a brief summary of the relevant points will be given here.
    Although there is no inherent phase problem in the HREM technique it is replaced by an "instrumental" phase problem arising from the relatively imperfect nature of electromagnetic lenses. Spherical aberration is the chief limitation, and its effect can be coupled with that of deviation from the Gaussian focus into a phase factor which can be considered to act upon the diffracted amplitudes before re-combination by the lens to produce the initial image amplitude, namely:

$$\chi(\alpha) \;=\; \frac{2\pi}{\lambda}\left\{\; \tfrac{1}{2}\Delta F \,\alpha^2 \;-\; \tfrac{1}{4}C_s \,\alpha^4 \right\}$$

where λ = electron wavelength, ΔF = deviation in focus from the Gaussian position, C_s = spherical aberration coefficient, and $\alpha(=1/d_{hkl})$ is the spatial frequency of each diffracted beam. For a given instrument, C_s is usually fixed, but ΔF can be varied continually, the shape of this composite function being shown in Fig. 1. Consequently the diffracted beams will, in general, be recombined into an image with incorrect phases, and the resulting detail will be confused. It is customary, therefore, to adjust ΔF until the two terms in the above expression partly cancel over a range of spatial frequencies, as shown in Fig. 1c, when the diffracted beams within the "plateau region" of the function are all recombined with the correct phase shifts. The image detail corresponding to these beams will then be essentially correct and the d-spacing corresponding to the outer edge of the plateau, d_{lim}, is usually referred to as the point resolution of the instrument.
    It should be noted that spherical aberration does not, in itself, limit the extent of the image detail available, but merely confuses interpretation. Other aberrations, such as chromatic factors, arising from the energy spread of the electron beam and an uncertainty in the focal length of the objective lens due to current instabilities, do

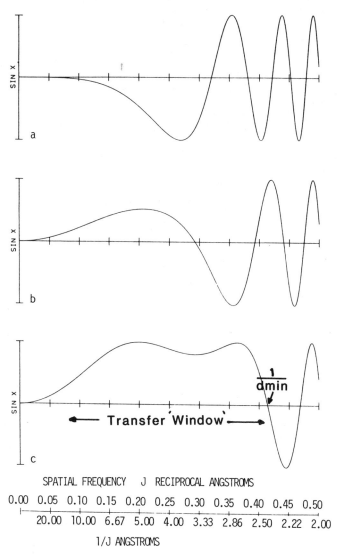

SPATIAL FREQUENCY   J RECIPROCAL ANGSTROMS

0.00 0.05 0.10 0.15 0.20 0.25 0.30 0.35 0.40 0.45 0.50

       20.00 10.00 6.67 5.00 4.00 3.33 2.86 2.50 2.22 2.00

            1/J ANGSTROMS

Fig. 1. The imaginary part of the phase contrast transfer function, sinχ, plotted as a function in reciprocal space for a microscope operating at 200 kV with a spherical aberration coefficient of 1.2 mm   a) Gaussian focus;   b) 325Å underfocus;   c) 650Å underfocus.

however set an ultimate limit on the available detail imaged. The finite divergence present in the electron beam has a similar effect. These factors create a blurring of detail in the image intensity, which can be represented mathematically by convolution with a blurring function, and, for very weak scatterers, this can be visualised as a damping envelope superimposed on the aberration function. This is shown in Fig. 2.

In addition to these instrumental factors, specimen dependent effects must also be considered, and of these, by far the most important is multiple elastic scattering. Given the above instrumental limitations, it is still possible to observe a direct correspondence between image contrast and potential density in the specimen out to a defined resolution limit, or at least to reconstruct the latter from the observed image. However, this is only true if the electron wavefront emerging from the specimen bears a simple relationship to the crystal potential, and this implies that only single-scattering events can occur. If multiple scattering is present, the observed image may never resemble the true specimen potential density, no matter how the instrumental conditions are varied. The effect of multiple scattering is shown in Fig. 3 in the case of a tungsten bronze structure which contains hexagonal tunnels of WO_6 octahedra filled with bismuth. As tungsten and bismuth scatter equally strongly, the tunnels are virtually invisible in the potential density, and this is also evident at the edge of the crystal where multiple scattering is negligible. Further away from the crystal edge, multiple scattering becomes evident, and then the observed image reveals the tunnels as large white dots. This is particularly confusing as it would suggest, if interpreted directly in terms of potential, that the tunnels were in fact empty. Indeed, if the thin edges were not present, a logical interpretation of the image would be in terms of a model with empty tunnels.

Although the factors described above impose severe limitations on the interpretation of image contrast it is possible to calculate their effects. The phase-contrast transfer theory of imaging is long established, and multiple scattering of electrons can be simulated numerically using the multislice formulation (3,4). Consequently, although there is, in general, no analytical method of recovering the specimen structure from the observed image of a thicker crystal, model structures can be constructed and their images simulated over a wide range of experimental conditions. These can then be compared with the observed image and the structure determined on a trial and error basis. Consequently a trial structure can be refined to a limited extent depending upon the instrumental resolution and on the computing resources available. Obtaining a trial structure can however only be carried out when very thin regions of the specimen are present, as then some of the atomic positions can be observed directly. Once this has been done, however, multiple scattering can actually be used as an aid to refinement, because the image detail in thicker regions is extremely sensitive to small shifts in atomic positions. Both instrumental and computational aspects of any HREM structure determination are therefore equally important, and just as the experimental side is limited by the available point resolution (ca. 2.4Å at 200 kV, 1.8Å at 400-500kV), so too are limitations imposed by computational resources. particularly as regards storage requirements when large unit cells are involved.

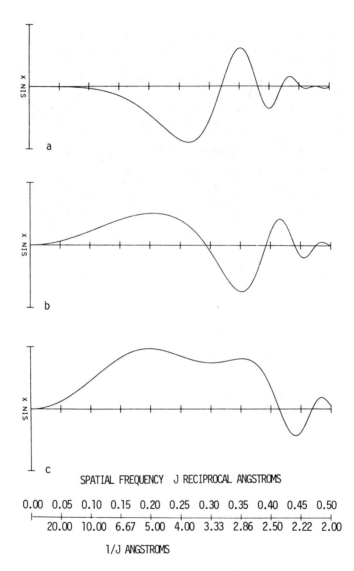

SPATIAL FREQUENCY J RECIPROCAL ANGSTROMS

0.00 0.05 0.10 0.15 0.20 0.25 0.30 0.35 0.40 0.45 0.50

20.00 10.00 6.67 5.00 4.00 3.33 2.86 2.50 2.22 2.00

1/J ANGSTROMS

Fig. 2. The functions of Fig. 1, after modification by a damping
envelope to take account of chromatic factors and beam
divergence. For a very thin crystal, the reduction in amplitude of sinχ represents a reduction in the intensity of the
fourier components in the image.

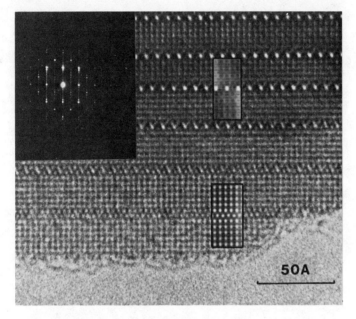

Fig. 3. Illustration of the effects of multiple scattering in an image of an intergrowth bismuth-tungsten oxide bronze, nominally $Bi_{0.1}WO_3$. The simulated images shown correspond to thicknesses of 20Å (near the crystal edge) and 60Å respectively, with an underfocus of 1220Å.

Examples :   1) Bismuth molybdo-tungstates

Although the bismuth molybdates, $Bi_2Mo_nO_{3n+3}$, which are widely used in selective oxidation catalysis, and their tungsten analogues share a common structure for the n=1 member of the series (5), this is not the case for $n \geq 1$. In particular, $Bi_2Mo_3O_{12}$ has a defective scheelite-like structure (6) with tetrahedrally co-ordinated Mo, whereas the tungstate phase $Bi_2W_3O_{12}$ adopts a layer-like structure similar to, but more complex, than Bi_2WO_6, and usually occurs in a severely disordered form (7). The relationship between these two structural types is not clear and in particular, it is uncertain whether intermediate phases can exist. Consequently materials of the general composition $Bi_2(Mo,W)_3O_{12}$ were prepared by quenching from the melt and annealing for various periods at $600°C$. In the more tungsten rich members a new phase of unknown structure was detected (8), an outline of the structure determination being given below.

   Although X-ray powder diffraction showed samples of nominal composition $Bi_2W_2MoO_{12}$ to be a single phase, indexable on a modified $Bi_2Mo_3O_{12}$ unit cell, examination in the electron microscope immediately revealed a biphasic assemblage. Typical crystals are shown in Fig. 4, with portions of the corresponding X-ray emission spectra. The majority phase did have the $Bi_2Mo_3O_{12}$ structure, with slightly altered unit cell dimensions, but a very high W content as revealed by X-ray emission spectrometry (Fig. 4a). This represents considerable solid solution of W in $Bi_2Mo_3O_{12}$. The minority phase, however, showed a very complex electron diffraction pattern, typical of a layered structure with stacking disorder and correspondingly complex image contrast, (Fig. 4b). This phase contained very little Mo. The principle features of the image contrast, shown at higher magnification in Fig. 5, suggest a compound layer as the basic structural unit containing six structural sub-layers separated by some $3.8Å$, with an interlayer region showing fringe contrast reminiscent of the simple layered tungstates. The number of sub-layers was variable, and some stacking defects were noted, these being responsible for the streaking observed in the electron diffraction patterns. Attempts to interpret these images in terms of 1:1 correspondence between image contrast and crystal potential failed, the main difficulty being the periodicity of some $5.9Å$ within the layers, which was not reconcilible with any known arrangement of corner-sharing WO_6 octahedra or isolated WO tetrahedra. A model based on W_2O_5 rather than WO_3 layers was therefore constructed.

   The structural model proposed, and the computer-simulated images generated from it, are shown in Fig. 5. Because the cation-cation distances for edge-sharing WO_6 octahedra are less than the point resolution of the instrument at the optimum focus of Figs. 1 and 2, it was necessary to go beyond this point of focus and image the cations with reversed contrast, as shown in Fig. 5a. The simulated images match very well, although it was necessary to refine the idealised model of Fig. 5c by allowing the cations in adjacent edge-sharing octahedra to increase their separation by some 10-15%. Although the final structure remains a trial model rather than a fully refined one, it is nevertheless extremely significant as it proves conclusively that complex reduced phases can exist in this

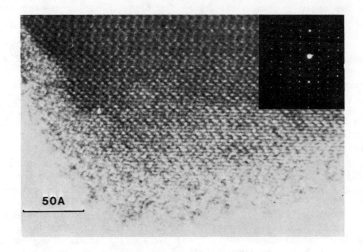

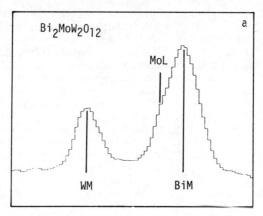

Figure 4a. Electron micrographs of the phases in a sample of
nominal composition $Bi_2MoW_2O_{12}$. $Bi_2Mo_3O_{12}$-like phase with
tungsten-substitution. Portions of the x-ray emission spectra of
each are also shown.

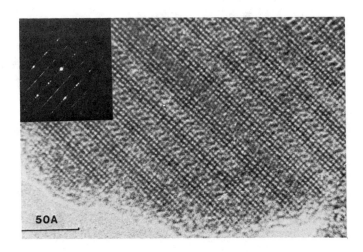

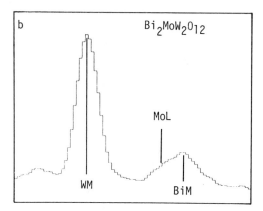

Figure 4b. Electron micrographs of the phases in a sample of nominal composition $Bi_2MoW_2O_{12}$. The new phase discussed in the text. Portions of the x-ray emission spectra of each are also shown.

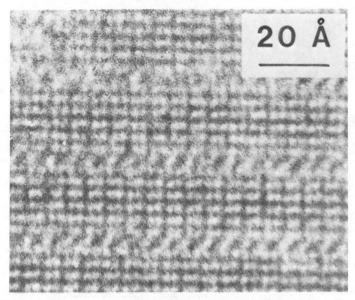

Figure 5a. Enlargement of the image contrast of the phase shown in Figure 4b.

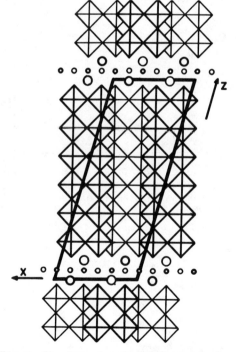

Figure 5b. Structural model proposed.

c

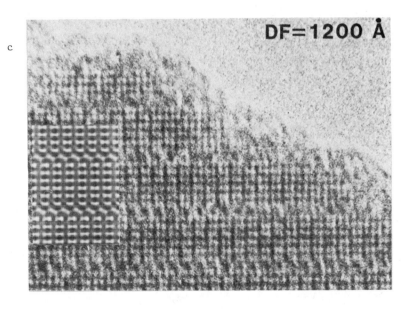

d

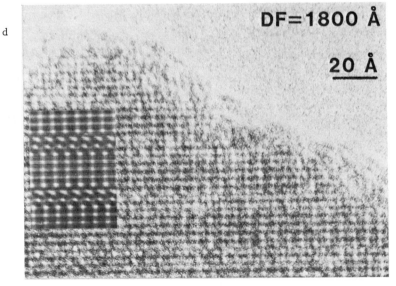

Figure 5c and 5d. Comparisons of the simulated images with experimental ones. The crystal thickness is 20Å and underfocus is 1200Å and 1800Å respectively.

system. Furthermore, by varying both the number of sub-layers used
and their composition (altering the spacing of the rows of edge-
sharing octahedra) such phases can be made both compositionally adap-
table and compatible with varying conditions of reduction. Further
studies on these, and related, structures are in progress.

2) Bismuth-Niobium-Molybdate, $Bi_2(Mo,Nb)O_{6-x}$

Although no unusual intermediate phases of the above type were found
in the system $Bi_2(Mo,W)O_6$ the manner in which these simple layered
structures adapt to reduction is nevertheless, of great practical
interest. One relatively simple way of testing whether reduced structures are feasible, in structural, if not thermodynamic terms is by
substituting the octahedrally co-ordinated hexavalent cation (either
Mo or W) by a pentavalent cation such as niobium. Specimens of this
type have been prepared, using the same technique as that described
above, but with the annealing being carried out in sealed quartz
tubes, and the preliminary results have been reported (9). A more
detailed discussion of the structural models and their derivation will
be given here. For the niobium-substituted Bi_2MoO_6, little difference
was observed in the X-ray powder diffraction pattern until approximately 50% replacement of Mo was achieved. At this point, some extra
lines became visible but most of the original ones remained unaffected.
Further Nb-substitution only served to increase the intensity of the
additional maxima and alter their positions slightly. Although these
results suggested the presence of two phases, no significant compositional variations could be observed in x-ray emission analysis of
this sample in the electron microscope, using previously characterised
specimens of Bi_2MoO_6 and $Bi_5Nb_3O_{15}$ as standards. Examination by
electron diffraction indicated that most crystals appeared to correspond to Bi_2MoO_6, but for those examined in [100] projection a pronounced superlattice was observed. A typical diffraction pattern
from a sample of nominal composition $Bi_2Mo_{0.5}Nb_{0.5}O_{5.75}$ is shown in
Fig. 6, together with the lattice image, where the superlattice fringes are clearly visible. Apart from these characteristic fringes, the
image contrast is identical to that obtained from pure Bi_2MoO_6, and
the spacing of the superlattice was found to decrease with increasing
Nb content. As can be seen in the electron diffraction pattern, the
superlattice was incommensurate and the overall periodicity was
approximately 8.3 times that of the basic lattice.
        Features such as long-period superlattices are immediately remi-
niscent of effects such as crystallographic shear (CS) in structures
based upon corner-sharing WO_6 networks (10), and this phenomenon has
been postulated as a mechanism for catalyst reduction in molybdates
(11). Although experimental verification of this hypothesis has not
been forthcoming (12), CS structures would seem to be the most likely
explanation of the images described above. However, in Bi_2MoO_6, CS
planes would be expected to be either (101) (10$\bar{1}$) within the plane
of the layers of MoO_6 octahedra, whereas the effect is observed on
(001). Other possibilities could be (011) or (110), but these would
involve "steps" in the layer planes which are not observed in the
experimental images.
        If the features observed are based upon edge-sharing of octahedra
within the MoO_6 component, they can only be reconciled with the known

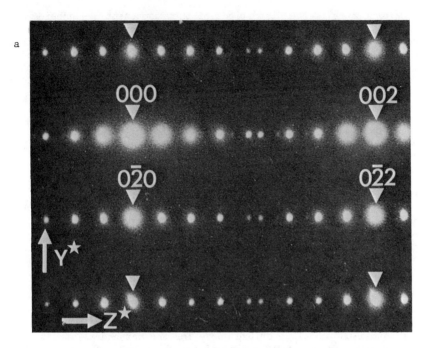

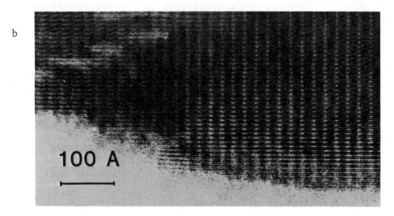

Fig. 6. a) Enlarged portion of the electron diffraction pattern of a crystal of nominal composition $Bi_2Mo_{0.5}Nb_{0.5}O_{5.75}$. Maxima indexable on the normal Bi_2MoO_6 unit cell are indicated. b) The corresponding lattice image, showing the superlattice fringes. The electron beam is parallel to $[100]$.

Bi_2MoO_6 structure if the distortions of that component are considered.
In common with most ReO_3-based oxides, octahedra are rotated and
tilted within the layer such that the actual arrangement is more like
that of Fig. 7a, rather than the idealised form normally represented.
It is then possible to introduce <u>pairs</u> of edge-sharing octahedra into
the array, as shown in Fig. 7b, and these pairs can lie on the (001)
planes, as observed experimentally. When viewed in [100] projection,
there will be a local excess of cation density at the positions of
these edge-sharing pairs generating the observed superlattice. Separation of these pairs by seven normal octahedra will produce the
correct stoichiometry, and furthermore, their presence will slightly
alter the dimensions of the octahedral component such that it no
longer exactly matches that of the Bi-O layer. Consequently, the
possibility of uncorrelated intergrowth and an incommensurate superlattice will arise.

Image simulations using the model of Fig. 7b proved most encour-
aging. It was possible to simulate the gradual, near-sinusoidal
appearance of the superlattice fringes almost exactly, as is shown
in Fig. 8. Furthermore, some limited refinement was possible, by
allowing the Mo atoms around the edge-sharing pairs to relax, and
thereby improving the image match. However, no data can be obtained
on the oxygen atom positions. This latter point is important, as
recent work by Gai (12) has suggested that the high temperature form
of Bi_2MoO_6, with tetrahedrally-coordinated Mo, can exist at a lower
temperature than originally thought. In an exactly analogous manner
to the "quasi-CS" model described above for the octahedral structure,
a similar arrangement can be envisaged for the layer with tetrahedral
Mo, involving edge-sharing of pairs of tetrahedra. Schematically,
this is shown in Fig. 7c, and comparison with the octahedral variant
indicates that, in [100] projection, the two models are identical in
terms of cation positions. Image simulations at 2.5Å resolution
from the tetrahedral model are correspondingly almost identical to
those from the octahedral case, and therefore, at this resolution, it
is impossible to determine which model applies. The tetrahedral case
is possibly the most appropriate, as the revised oxygen lattice implies reduced correlation between the Bi- and Mo-containing components,
making an incommensurate superstructure more likely. However, until
greater structural resolution can be achieved, both models remain
equally possible, and this illustrates one of the principal difficulties of the lattice imaging technique.

3) Bismuth-Niobium-Tungstate $Bi_2(W,Nb)O_{6-x}$

Similar studies on the niobium substituted tungstate have produced
equally complicated, but different structural configurations. Typical
electron diffraction patterns and medium resolution lattice images of
material of nominal composition $Bi_2W_{0.5}Nb_{0.5}O_{5.75}$ are shown in Fig. 9.
As with the corresponding molybdate, x-ray powder diffraction shows
only faint extra lines on increasing niobium-substitution, whereas
the superlattice structure is immediately apparent in electron diffraction. In this case the superlattice lies on the (103) planes and
is always commensurate, and furthermore it is evident from the (001)
fringes in experimental images that the layers are "stepped", or
displaced along [001] where they intersect the superlattice planes,

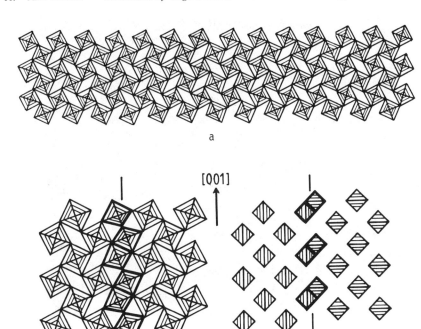

Fig. 7. a) A portion of the octahedral sub-layer in Bi_2MoO_6, with exaggerated octahedral distortion b) The same layer, but with pairs of edge-sharing octahedra inserted. c) The alternative model with tetrahedral co-ordination, having pairs of edge-sharing tetrahedra.

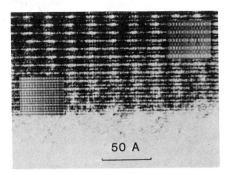

Fig. 8. Comparison of observed and simulated images for the composition $Bi_2Mo_{0.5}Nb_{0.5}O_{5.75}$. The crystal is a sharply tapering one and the simulations are for a thickness of 35Å (near the edge) and 105Å and underfocus settings of 20Å.

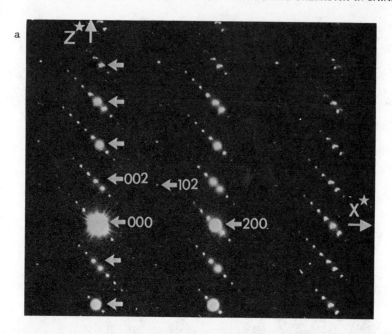

Fig. 9. a) Enlarged portion of the electron diffraction pattern of a
crystal of nominal composition $Bi_2W_{0.5}Nb_{0.5}O_{5.75}$. Maxima
indexable on the Bi_2WO_6 unit cell are indicated. b) The
corresponding lattice image, showing the superlattice fringes.
The electron beam is parallel to [010].

such that the displacement is one-half of an individual layer thickness. This "stepping" of layers did not occur during electron microscopic examination, nor was it merely a contrast effect caused by increasing crystal thickness. The resemblance to CS planes in ReO_3-based oxides (13) is very striking, and it is possible to envisage several CS structures which will reproduce the observed unit cell.

However, CS structures involving edge-sharing of octahedra within the W-containing sub-layer suffer from two major disadvantages, in that they fail to give the stoichiometry determined experimentally by x-ray emission analysis, and furthermore, to reproduce the observed unit cell it is necessary always to insert the CS planes in pairs, otherwise the half-layer displacement is not achieved. There appears to be no reason why this should be so, particularly as many samples are disordered and satisfactory co-ordination for interlayer bismuth can be found if only single CS planes are involved. In addition, multislice simulations of CS-based models give grossly exaggerated contrast along the superlattice planes. Consequently, an alternative picture of layer stepping was necessary.

One alternative which fits the requirements is an actual overlapping of layers at the superlattice planes, and is illustrated schematically in Fig. 10a. With this model, layer offsets of one half are obligatory, and where the layers overlap, additional perovskite-like sites are produced which can accommodate extra bismuth atoms and so achieve the correct composition. Image-matching studies on this model have been successful, a comparison of experimental and simulated images being made in Fig. 10b, and the essential features are undoubtedly correct. Once again, however, the oxygen positions cannot be defined, and the distortions of the perovskite-like region remain undetermined.

Further substitution of niobium results in exceedingly complex structures, and a micrograph of a typical crystal of the niobium endmember of the series, $Bi_4Nb_2O_{11}$, is illustrated in Fig. 11. The unit cell of this material is extremely large, approximately 115 x 80 x 5.5 Å, and the x-ray powder diffraction diagram is impossible to interpret. From the micrograph shown, it would appear that the structure is based upon different principles from the one described above, but it can nevertheless be derived from it by repeated overlap of layers on (112) and (113) planes in a very complex sequence. Why such a complex sequence should be employed, and whether either of these phases are true "phases", or merely certain compositions in a quasi-continuous solid solution series, is not yet certain. What is however, demonstrated, is the remarkable ability which these simple layered structures show to variations in stoichiometry.

Conclusions

The degree of success of HREM in elucidating the structural chemistry of the systems described above can only be assessed in relative terms. In none of the structures has a true refinement been attempted, particularly in the case of the oxygen atom positions. Furthermore, the possibilities of subsequent diffraction refinement of the trial structures obtained is not high, owing to their disordered nature, which, combined with their complexity, would make even neutron powder refinement impracticable. Consequently the nature and dimensions of

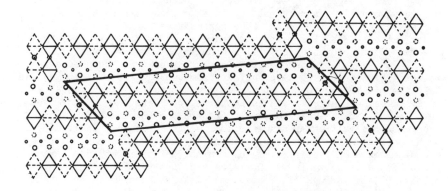

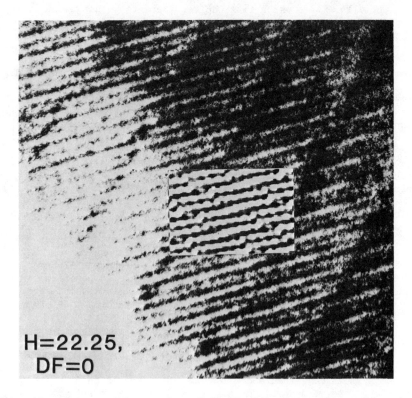

Fig. 10. Top: possible overlap of Bi_2WO_6-like layers to produce the "steps" observed in the experimental images. Bottom: comparison of experimental image and one simulated using this model, showing good agreement. The simulation is for a crystal of thickness 23Å, imaged at the Gaussian focus.

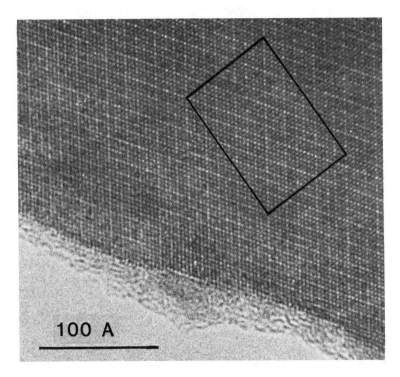

Fig. 11 High resolution electron micrograph of a crystal of
$Bi_4Nb_2O_{11}$. The true unit cell is indicated. The direction
of the electron beam corresponds to the $[1\bar{1}0]$ axis of Bi_2WO_6.

metal-oxygen linkages, which are of crucial importance to the exact
mechanism of catalysis, remain undetermined. However, the results
obtained show beyond all doubt that even simple ternary oxides such
as Bi_2MoO_6 have an inherent structural adaptability to oxygen content
similar to, but based on different principles from other systems such
as the calcium-manganese oxides (14). In addition, the way in which
variable stoichiometry can be accommodated in either ordered or disordered arrangements suggests that the introduction of small regions
of differing stoichiometry will be a relatively facile process, and
that such regions are a very real possibility in actual catalytic
activity.

The difficulties of establishing the existence of compositional
fluctuations of this type in a working catalyst should not, however,
be underestimated. In the examples discussed, any disorder present
has been either one- or two-dimensional : in a real catalyst, it is
liable to be three-dimensional. Consequently observation of such
fluctuations might be expected to be extremely difficult, particularly
as the HREM can only give structural data in projection. However, it
is possible to examine extremely thin specimens, and when the dimension parallel to the beam falls below 20Å, the bulk structural features examined must inevitably show some surface characteristics.
Greater structural resolution is certainly available, with point
resolution at 500kV or greater being of the order of 1.8Å, but even
at this level anion positions remain undetermined. The real value
of the HREM technique, however, lies in its ability to characterise
grossly defective structural systems. If the data obtained in this
way can be used as an aid to the interpretation of other, indirect
evidence such as x-ray photoelectron-spectroscopic data, then the
results may be extrapolated to deal with materials where defects are
only present at or near the surface, such as real, rather than model
catalyst systems.

Acknowledgments

The author wishes to acknowledge the help and encouragement of
Professor J.M. Thomas and Dr. R.K. Grasselli, without whom this work
would not have been possible. Drs A. Reller, J.L. Lievin and M.K.
Uppal carried out much of the experimental work. The financial
support of the S.E.R.C. is also acknowledged.

Literature Cited

1.   Grasselli, R.K. and Burrington, J.D. In "Selective Oxidation and
      ammoxidation of Propylene by Heterogeneous Catalysis" Pines, H.
      ed; ADVANCES IN CATALYSIS No. 30. Academic Press, New York, 1981.
2.   Spence, J.C.M. "Experimental High Resolution Electron Microscopy"
      Clarendon Press, Oxford, 1981.
3.   Cowley, J.M. and Moodie, A.F. Acta Crystallogr. 1957, 10, 609.
4.   Goodman, P and Moodie, A.F. Acta Crystallogr. Sect. A, 1974, 30,
      280.
5.   Van den Elzen, A.F. and Rieck, G.D. Acta Crystallogr. Sect. B,
      1973, 29, 2436.
6.   Van den Elzen, A.F. and Rieck, G.D. Acta Crystallogr. Sec. B,
      1973, 29, 2433.

7. Jefferson, D.A., Gopalakrishnan, J. and Ramanan, A. Mater Res. Bull. 1982, 17, 269.
8. Jefferson, D.A., Thomas, J.M., Uppal, M.K. and Grasselli, R.K., J. Chem. Soc. Chem. Comm., 1983, 594.
9. Lievin, J.L., Reller, A. and Jefferson, D.A. Mater. Res. Bull. 1984, 19, 571.
10. Tilley, R.J.D. Chemica Scripta 1979, 14, 147.
11. Grasselli, R.K., Burrington, J.D. and Brazdil, J.F. Faraday Discussion Chem. Soc. 1981, 72, 203.
12. Gai, P.L. Journ. Solid State Chem. 1983, 49, 25.
13. Iijima, S. Journ. Solid State Chem. 1975, 14, 52.
14. Reller, A., Jefferson, D.A., Thomas, J.M., Beyerlein, R.A. and Poeppelmeier,K.R. J.Chem.Soc. Chem. Comm. 1982, 1378.

RECEIVED January 10, 1985

Temperature-Programmed Decomposition of 2-Propanol on the Zinc-Polar, Nonpolar, and Oxygen-Polar Surfaces of Zinc Oxide

K. LUI, S. AKHTER, and H. H. KUNG

Chemical Engineering Department and Ipatieff Laboratory, Northwestern University, Evanston, IL 60201

The temperature programmed desorption and decomposition of 2-propanol, acetone, and propene were studied on the Zn-polar (0001), the stepped-nonpolar (5051), and the O-polar (0001) surfaces of ZnO. The desorption temperatures of all the species, except water, were the lowest on the O-polar surface, and the highest on the Zn-polar surface. Propene did not adsorb on the Zn-polar surface. Adsorbed acetone desorbed in two peaks from the nonpolar surface. On the other surfaces, adsorbed propene or adsorbed acetone desorbed in a single peak. 2-Propanol was decomposed into acetone, propene, H_2, and H_2O. The ratios of acetone to propene were the same independent of the surface and the 2-propanol coverage. The evolution of the products except H_2O, was reaction limited. The decomposition activity was the highest on the Zn-polar face, and lowest on the O-polar face.

Studies in recent years on the surface properties of transition metal oxides have demonstrated that the surface structural stability, the surface electronic structure, and the surface chemical reactivity depend on the crystallographic orientation of the exposed surface and the presence of surface imperfection, such as steps and point defects (1). ZnO is one recent example. The natural surfaces of ZnO, which can be prepared in a relatively well-ordered state, include the Zn-polar (0001), the O-polar (0001), and the nonpolar (1010) surfaces. (See Figure 1 for a schematic representation of these surfaces). These surfaces have been shown to possess different chemisorptive properties and reactivities. It was shown that CO_2 was desorbed from a nonpolar surface at about 120°C, but from a Zn-polar surface at 410°C (2). The decomposition of methanol, formaldehyde, and formic acid also differed (3,4). In general, the temperature at which the decomposition products appeared was the highest on the Zn-polar

0097-6156/85/0279-0205$06.00/0
© 1985 American Chemical Society

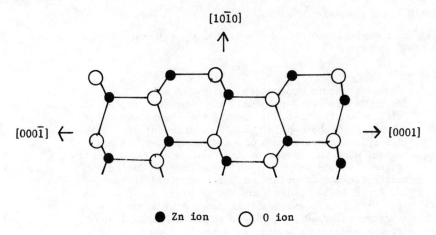

Figure 1.   Schematic drawing of the various ZnO crystal surfaces.

face, intermediate on the nonpolar face, and the lowest on the O-polar face (5). The decomposition activity, measured as the fraction of adsorbed molecules decomposed, decreased in the same order: Zn-polar > nonpolar > O-polar. The reaction products in methanol decomposition also differed. From these observations, it was concluded that on the Zn-polar surface, the reaction of methanol proceeded via both oxidation by the lattice, and dehydrogenation. The ratio of oxidation to dehydrogenation varied according to the condition of the surface. On the nonpolar and the O-polar surfaces, only the oxidation pathway participates. These were interpreted as an indication of the presence of both metallic and oxidic character on the Zn-polar surface, but only the oxidic character on the other surfaces, which is consistent with the surface atomic structures. Such a crystal plane dependence of the adsorption and reaction properties of ZnO has also been shown using powder samples (6).

In view of the interesting results of methanol decomposition, we proceeded to study the decomposition of 2-propanol. It is known that 2-propanol decomposes via two competing pathways.

$$2\text{-}C_3H_7OH \longrightarrow (CH_3)_2CO + H_2 \qquad \text{dehydrogenation}$$

$$2\text{-}C_3H_7OH \longrightarrow C_3H_6 + H_2O \qquad \text{dehydration}$$

The dehydrogenation reaction produces acetone and hydrogen, and is dominant over basic oxides (7). The dehydration reaction produces propene and water, and is dominant over acidic oxides. It would be interesting to see if the competition between these two pathways depend on the exposed crystal planes of ZnO. We report here the results of such an investigation. 2-Propanol was decomposed on ZnO single crystal surfaces by the temperature programmed decomposition technique. To assist the interpretation of data, the temperature programmed desorption of propene and acetone were also studied.

Experimental

Experiments were carried out in a standard ultra high vacuum system described before (2-4). The system was equipped with a low energy electron diffraction unit, an Auger electron spectrometer, a sputter-ion gun, and a UTI quadrupole mass spectrometer. The mass spectrometer was computer-controlled. It had a shield in front of the ionizer, that had a hole of 1 cm in diameter to discriminate the mass spectrometer signal due to direct flux of molecules from the background pressure increase during temperature programmed desorption (TPD). TPD was performed by heating the ZnO crystal at 10 K s^{-1} radiatively with a tungsten filament behind the crystal.

The ZnO crystals were purchased from Atomergic Chemicals. The direction of the c-axis was determined by etching (2). The desired surfaces were about 6 mm x 6 mm. The edge and the back side of the crystal were covered by evaporated gold. Adsorption of molecules was achieved with a doser. Thus the exposures reported here, which were calculated with the chamber pressure indicated by an ion gauge, were ten to one hundred times less than the actual exposure.

Preparation of the (10$\bar{1}$0), (50$\bar{5}$1), and (0001) surfaces has

been described before (2). Briefly, it involved alignment of the
crystal by Laue x-ray back scattering, cutting out the desired
face, mechanical polishing, and repeated Ar-ion sputtering and
annealing. The O-polar (000$\bar{1}$) face was similarly prepared. The
annealing in vacuum was performed at 500°C. A 6-fold "1 x 1" LEED
pattern was obtained. Small but detectable (by Auger electron
spectroscopy) amounts of S, Cl, and K were present. The TPD
results did not appear to depend on the amounts of these impurities
when the amount was small.

As before, three types of control experiments were routinely
performed to help discriminate the desired TPD signals from other
desorption signals. The first type was a blank run in which the
entire procedure of TPD was carried out except that no gases were
introduced onto the surface. The second type was to perform the
TPD procedure but with the sample facing away from the doser during
adsorption. This was used to determine the amount of adsorption on
places other than the desired sample surface. The third type was
also the same as an ordinary TPD experiment except that desorption
was performed with the surface facing away from the mass
spectrometer. This also helped identify which desorption was from
the desired surface. These control experiments normally provided
enough information for the identification of the desired desorption
signals versus desorption from background gas adsorption or from
other places than the desired surface.

Desorption was monitored with mass spectroscopy. The cracking
patterns of 2-propanol, acetone, and propene were individually
determined (8). For quantitative analysis, masses 45, 43, 41, 18,
and 2 were used for 2-propanol, acetone, propene, water, and
hydrogen, respectively, after correction for cracking in a similar
procedure as described (3,4). The mass spectrometer sensitivities
were determined to be 5.26, 7.88, 3.07, 4.74, and 5.20 amp/torr,
and the pumping speeds were 9.5, 13.1, 31.0, 1.7, 36.9 L sec^{-1},
respectively for the five species. These two latter quantities
were used to convert the mass spectrometer readings into molecular
fluxes.

Results

As mentioned in the previous section, control experiments were
performed to discriminate the desired signals from the other
signals. Thus, unless specified, the results reported are the
desired signals. Most of the experiments were performed on the
(0001), the (50$\bar{5}$1), and the (000$\bar{1}$) surfaces, which will also be
referred to as the Zn-polar, nonpolar, and the O-polar surfaces,
respectively. The (50$\bar{5}$1) surface was actually a stepped nonpolar
face. A few experiments were performed on the flat nonpolar (10$\bar{1}$0)
surface at high 2-proponol exposures; the results were similar to
the (50$\bar{5}$1) surface.

2-Propanol Decomposition. Typical TPD spectra for 2-propanol
decomposition on the (0001), (50$\bar{5}$1), and (000$\bar{1}$) surfaces are shown
in Figures 2a, b, and c, respectively. Common to all three
surfaces was the evolution of five products: undecomposed 2-
propanol, hydrogen, acetone, water, and propene. Other species

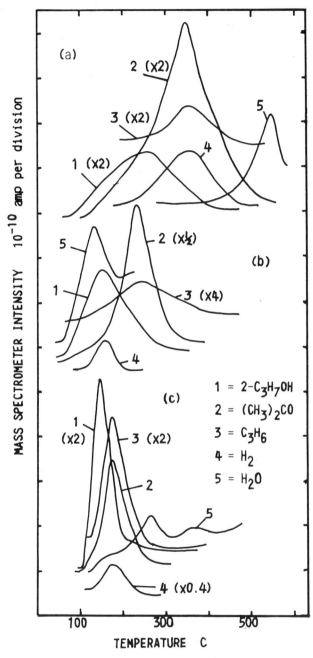

Figure 2. Typical TPD spectra for 2-propanol decomposition on different ZnO surfaces. a) Zn-polar $(000\bar{1})$ surface, 0.225 L exposure; b) nonpolar $(50\bar{5}1)$ surface, 0.03 L exposure; c) O-polar $(000\bar{1})$ surface, 0.0015 L exposure.

that were looked for but were not found included CO, CO_2, methanol,
formaldehyde, acetaldehyde, ethanol, propane, and 1,3-
cyclohexadiene. In general, the temperatures at which these
products were detected were the highest on the Zn-polar plane, and
lowest on the O-polar plane. Table I shows the peak temperatures
of the various products. These temperatures did not shift with the
coverage of 2-propanol.

Table I. Temperature Programmed Desorption and Decomposition
          Peak Temperatures of Various Products

Experiment	Species	Peak Temperature C		
		(0001)	(50$\bar{5}$1)	(000$\bar{1}$)
2-Propanol decomposition	H_2	125	160	350
	H_2O	225,325	130	540
	propene	125	230	350
	acetone	125	225	350
	2-propanol	100	150	250
H_2O adsorption	H_2O	100,225,325	130	230,540
Acetone desorption	acetone	100	100,200	250 to 300
Propene adsorption	propene	90	100	not ads.

Both dehydrogenation and dehydration were observed on these
surfaces. On the three surfaces studied, the acetone peak, which
was a product of dehydrogenation, always appeared at the same
temperature as the propene peak, which was a product of
dehydration. The H_2 peak also appeared at the same temperature as
acetone on the (000$\bar{1}$) and the (000$\bar{1}$) surface, but was at a lower
temperature on the nonpolar surface. The water peaks were small,
and appeared at temperatures different from the peaks of the carbon
compounds.
        Table II lists the peak areas of the carbon products for the
three surfaces at different coverages. The areas of the H_2 peaks
were not shown because their accuracies were low due to the much
higher H_2 background pressures than the carbon compounds. The
water peak areas were also not quantified because adsorption from
background water resulted in water desorption at the same
temperature as water from the reaction. Since the amount of water
adsorbed from the background was   not easily controllable, the
water areas were not used. The location of the water peaks from
the reaction was definitively identified by using deuterated 2-
propanol $((CD_3)_2CDOD)$. The reaction product, which was D_2O, could
be separately measured from the background water, which was H_2O.
H-D exchange in the mass spectrometer (3), however, still made
quantitative determination of the peak area inaccurate.
Nonetheless, within the large uncertainties, the H_2/acetone and the
H_2O/propene ratios were independent of the 2-propanol coverage. No

Table II. Product Quantities in 2-Propanol Decomposition on ZnO

Exp. L[a]	Acetone	Propene	2-PrOH	Ac/2-PrOH[b,d]	Propene/Ac[c,d]
(0001) surface					
0.003	5.63	0.49	1.47	3.86(3.02)	0.087(0.52)
0.015	14.07	0.94	3.90	3.61(2.82)	0.067(0.40)
0.03	16.01	1.03	4.51	3.50(2.72)	0.065(0.39)
0.12	13.81	1.07	3.67	3.76(2.93)	0.077(0.47)
0.225	14.93	1.19	4.02	3.72(2.88)	0.078(0.47)
(50$\bar{5}$1) surface					
0.005	9.23	0.5	2.05	2.88(2.27)	0.053(0.32)
0.015	16.11	0.89	6.45	2.50(1.97)	0.055(0.33)
0.03	21.23	1.31	10.43	2.04(1.60)	0.062(0.36)
0.06	30.24	1.70	17.82	1.70(1.34)	0.057(0.33)
0.12	31.27	1.62	24.15	1.31(1.02)	0.052(0.28)
0.24	38.84	2.31	32.34	1.20(0.95)	0.060(0.36)
0.54	43.41	2.35	44.88	0.97(0.76)	0.055(0.32)
(000$\bar{1}$) surface					
0.0015	7.52	e	4.45	1.69(1.45)	e
0.006	6.14	e	5.93	1.04(0.89)	e
0.3	15.78	e	13.87	1.14(0.98)	e
1.2	19.96	e	17.91	1.11(0.96)	e
3.6	19.51	e	19.27	1.01(0.87)	e

[a] Apparent exposure. The actual exposure was ten to one hundred times higher due to the dosing action.

[b] Ratio of acetone to 2-propanol.

[c] Ratio of propene to acetone.

[d] Values in the brackets are ratios corrected for the mass spectrometer sensitivities and pumping speeds. They represent the ratios of the molecular fluxes for desorbing from the surface.

[e] These values were small. They were not measured because of the high background signals in these experiments.

attempt was made to obtain these ratios using deuterated 2-propanol because the major cracking fragments for deuterated acetone and deuterated propene were both at m/e=46.

Acetone and Propene Desorption. Desorption of adsorbed acetone yielded only acetone on all three surfaces. Other possible products that were searched for but not found included H_2, H_2O, methane, CO and CO_2. The desorption profiles from these surfaces are shown in Figure 3, and the peak temperatures are listed in Table I. A single sharp desorption peak was observed on the O-polar surface. A broad peak was observed on the Zn-polar surface which might be two overlapping peaks. Two distinct peaks were observed for the nonpolar (5051) and (1010) surfaces. The areas of the two peaks increased but at different rates with increasing exposure. The area of the higher temperature peak increased more rapidly initially, but appeared to be saturated at about 0.27 L. This saturation might be real, but it might also be due to increasing competition from background water adsorption with increasing exposure (see next section). The area of the lower temperature peak increased more slowly, but continued to increase until its area was more than five times the area of the higher temperature peak at the highest exposure studied (20 L). The peak temperatures were constant. The dependence on the exposure was not studied on the other two surfaces.

The TPD profiles of adsorbed propene are shown in Figure 4 for the O-polar and the nonpolar surface. A single peak was observed on both surfaces. The peak temperatures are listed in Table I. Only the desorption of propene, and no hydrogen, CO, CO_2, or methane were detected. Only one exposure was studied. On the Zn-polar surface, exposure up to 6 L did not result in detectable propene desorption. H_2, water, methane, CO, CO_2, benzene, and cyclohexene were searched for but were not found. Thus propene appeared not to adsorb on this surface. In fact, ethylene was also found not to adsorb on the Zn-polar surface under these conditions.

Water Desorption. After each TPD or high temperature annealing, background water was adsorbed onto the surface during cooling of the sample. The amount of water adsorbed increased with longer cooling time. Heating of the samples after cooling in vacuo always showed desorbed water at temperatures listed in Table I. The effect of adsorbed water was studied on the (5051) and the (0001) surface. No effect was found in the 2-propanol decomposition. However, in the acetone adsorption on the (5051) surface, it was found that the higher temperature acetone desorption peak decreased with increasing cooling time, for a given acetone exposure. The lower temperature peak, however, was not affected.

Discussion

The TPD results of acetone and propene are discussed first, since they are used for the discussion of 2-propanol decomposition.

Propene adsorption on ZnO powder had been studied by TPD and by infrared spectroscopy (9-12). A reversibly and an irreversibly adsorbed propene were found. Since the reversibly adsorbed form

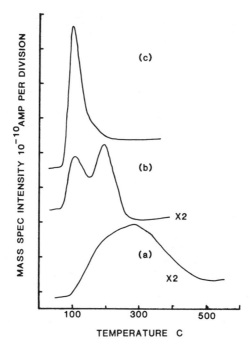

Figure 3.   TPD spectra of acetone in acetone desorption.
a) Zn-polar (000$\underline{1}$) surface, 0.6 L exposure;
b) nonpolar (50$\underline{5}$1) surface, 0.28 L exposure;
c) O-polar (0001) surface, 0.03 L exposure.

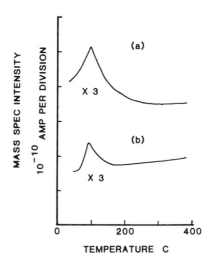

Figure 4.   TPD spectra of propene in propene adsorption.
a) nonpolar (50$\underline{5}$1) surface, 14.4 L exposure;
b) O-polar (0001) surface, 0.72 L exposure.

could be removed by evacuation at room temperature (10), it would
not be detected in our experiments. The irreversibly adsorbed form
was found to exist as a π-allyl anion species based on IR
spectroscopy (11,12). It competes with CO for adsorption sites
(10), indicating that its adsorption involves a Zn ion. TPD of
this species from powder ZnO showed a peak at about 100°C (12).
This temperature corresponds well with the desorption temperature
observed in this study. Therefore the propene desorption peak
observed here must be from the recombination of an adsorbed π-allyl
and hydrogen. Dissociative adsorption of propene requires a Zn-O
pair site which is abundant on the nonpolar surface. That the O-
polar surface, but not the Zn-polar surface adsorbs propene in this
manner suggests that surface defects which expose the nonpolar
surfaces (or other Zn-O pairs) exist more abundantly on the O-polar
than on the Zn-polar surface.

     Adsorption of acetone on ZnO was much less studied. Results
from an infrared spectroscopic study showed that acetone is
adsorbed in the form of an enolate (13). At high coverage, a
polymeric species is formed. The single sharp TPD peak observed
for the O-polar surface indicated only one type of adsorbed
acetone. The two distinct peaks for the nonpolar surface, and the
broad peak for the Zn-polar surface indicated the presence of
different types of adsorption on these surfaces. Since the
coverages were low in this study, the formation of polymeric
acetone was not likely. The different forms may then be due to two
different forms of adsorbed acetone (such as acetone or enol), or
two different kinds of surface sites. Water apparently reduced the
amount of higher temperature form of acetone. It might be that
adsorbed water reduces the enol formation (which is assumed to
result in the higher temperature acetone peak) in the former
explanation, or it competes for adsorption on the higher
temperature sites in the latter explanation. More definite
statements would require additional information.

     The decomposition of 2-propanol showed both similarities and
differences among the surfaces. The most notable similarity is the
fact that propene and acetone were produced at about the same ratio
on all surfaces. Dehydrogenation to form acetone was the dominant
reaction, as has been observed on ZnO powders (7). The desorption
temperatures of the reaction products, acetone, propene, and
hydrogen were always higher than the temperature of desorption of
the adsorbed acetone, propene, and hydrogen (hydrogen does not
adsorb on ZnO under our conditions). Thus the evolution of acetone
and propene are reaction limited in 2-propanol decomposition.
These, together with the observation that acetone and propene were
always evolved at the same temperature suggest that acetone and
propene are formed from a common intermediate on the different
surfaces, the formation or decomposition of which is the rate
limiting step. The evolution of water, however, was at the same
temperature as the desorption of adsorbed water. Thus the process
is desorption limited.

     Three interesting differences can be easily identified. The
first is the difference in the temperature of desorption from the
different surfaces. In general, evolution of the decomposition
products are at the lowest temperature on the O-polar surface,

intermediate on the nonpolar surface, and the highest on the Zn-polar surface. The order is identical to that observed for the decomposition of methanol on the same three surfaces (3-5). Thus, the common intermediate postulated above interacts most weakly with the O-polar surface, and most strongly with the Zn-polar surface. In addition to different bonding characteristics of the reaction intermediate on the different surfaces, at least two other interactions that differ on the different surfaces can be identified. The atomic arrangement of an ideal Zn-polar surface is such that a layer of Zn ions is more outwardly situated than the next layer of O ions. Because the exposed ions are Zn ions which are nonpolarizable, this surface is a hard acid (14). Conversely, an ideal O-polar surface has a layer of O ions more outwardly situated than the next layer of Zn ions. These exposed O ions make the surface a soft base. The intermediate in 2-propanol decomposition is an enolate ion (13). Being a base, it should interact more strongly with a hard acid than a soft base.

The second type of interaction is dipolar interaction. The atomic arrangement of the surfaces is such that a strong dipole pointing outward is present on the Zn-polar surface, and a strong dipole pointing inward is present on the O-polar surface. The nonpolar surface, relatively speaking, does not possess a dipole. The opposite orientation of the dipole on the two polar surfaces would interact with the dipole of the enolate intermediate in an opposite manner. This should contribute to the different temperature of evolution of the products.

The second difference among the surfaces is the fact that, except H_2O, the other three decomposition products, H_2, acetone, and propene, were evolved at the same temperature on the two polar surfaces, but H_2 was evolved at a lower temperature on the nonpolar surface. It is interesting to compare these results with the observations by Koga et al. who studied the decomposition of 2-propanol at 100°C on ZnO powder (13). They found that if the gas phase 2-propanol was suddenly removed from the gas phase, the evolution of hydrogen continued, but the evolution of acetone stopped. The evolution of acetone resumed after readmission of 2-propanol. This behavior can be explained by the fact that the major exposed face of their ZnO powder sample was the nonpolar plane. It is only on this surface that H_2 can be evolved without concurrent evolution of acetone in the absence of gaseous propanol.

From the temperature at which the decomposition products evolved, it would seem that the O-polar surface should be the most active in 2-propanol decomposition. However, a close examination of the temperatures in Table I shows that on the O-polar surface, the desorption temperature of the minor product water was actually rather high - higher than any of the products from the nonpolar surface. Thus in a steady state reaction at temperatures below about 100°C, the O-polar surface could be easily poisoned by adsorbed water, leaving only the nonpolar surface active.

That H_2 was evolved at a lower temperature than acetone and propene on the nonpolar surface, but not on the other surfaces, was interesting. A possible explanation is that the formation of enolate (13) from 2-propanol takes place most readily on the nonpolar face because of the availability of Zn-O pairs such that

the conversion of it into propene and acetone was the slow step.
On the polar surfaces, its formation was the slow step.
     It has been reported that the decomposition of 2-propanol was
a structure-insensitive reaction on ZnO (15). Our results suggest
otherwise. Further work being planned to study steady state
reactions on these single crystal surfaces will provide the answer
to this discrepancy.
     The third difference is the different reactivity of the
surfaces. The fraction of adsorbed 2-propanol being decomposed is
illustrated in Table I as the ratios of acetone/undecomposed 2-
propanol. These ratios are the lowest on the O-polar surface and
the highest on the Zn-polar surface. On the nonpolar surface, the
ratios changed from a high value close to those on the Zn surface,
to a low value close to those on the O surface as the coverage of
2-propanol increased. These data can be explained by the presence
of two types of site on these surfaces: a reactive site on which 2-
propanol decomposes, and an unreactive site on which 2-propanol
simply adsorbs and desorbs. The sticking coefficients of 2-
propanol on these two sites vary in the same manner with coverage
on the two polar surfaces, but differently on the nonpolar surface,
such that the reactive site is populated more easily. An alternate
explanation is that there is only one type of site for each polar
surface. But the sites on the two polar surfaces are different,
yielding different acetone to 2-propanol ratios. Both types of
sites are present on the nonpolar surface. If the initial sticking
coefficient of 2-propanol on the sites similar to those found on
the Zn surface is higher, the variation of the acetone to 2-
propanol ratio is then explained. However, this latter model does
not automatically explain why the products acetone and propene
should desorb at one identical temperature from the nonpolar plane,
that was different than those on the polar surfaces unless this
difference can be accounted for by the different dipolar
interaction between the surface and the adsorbed intermediate.
     In conclusion, the chemical properties of ZnO depend on the
particular surface plane that is exposed. This surface specificity
has now been demonstrated for the decomposition of 2-propanol,
methanol, formaldehyde and formic acid, and adsorption and
desorption of acetone, propene, water, CO, and CO_2. These data
have made possible better understanding of the results using ZnO
powder. It will be intersting to see how different are the
catalytic properties of these surfaces.

Acknowledgment

     Support of this work by the Petroleum Research Fund
administered by the American Chemical Society is gratefully
acknowledged.

Literature Cited

1. V.E. Henrich, Prog. Surface Sci. 1983, 14, 113.
2. W.H. Cheng, and H.H. Kung, Surface Sci. 1982, 122, 21.
3. S. Akhter, W.H. Cheng, K. Lui, and H.H. Kung, J. Catal. 1984,85,
   437.

4. W.H. Cheng, S. Akhter, and H.H. Kung, J. Catal. 1983, 82, 341.
5. S. Akhter, and H.H. Kung, to be published.
6. M. Bowker, H. Houghton, K.C. Waugh, T. Giddings, and M. Green, J. Catal. 1983, 84, 252.
7. O.V. Krylov, "Catalysis by Nonmetals," Academic Press, 1970.
8. K. Lui, MS thesis, Northwestern University, 1984.
9. A.L. Dent, and R.J. Kokes, J. Amer. Chem. Soc. 1970, 92, 6709, 6718.
10. A.A. Efremov, and A.A. Davydov, Kinet. Catal. 1980, 21, 383.
11. R.J. Kokes, Intra-Science Chem. Rept. 1972, 6, 77.
12. A.A. Davydov, A.A. Yefremov, V.G. Mikhalchenko, and V.D. Sokovskii, J. Catal. 1979, 58, 1; R. Spinicci, and A. Tofanari, J. Thermal Analysis, 1982, 23, 45.
13. O. Koga, T. Onishi, and K. Tamaru, J. Chem. Soc. Faraday Trans. I, 1980, 76, 19.
14. R.G. Pearson, editor: "Hard and Soft Acids and Bases," Dowden, Stroudsburg, 1973.
15. G. Djega-Mariadassou, and L. Davignon, J. Chem. Soc. Faraday Trans. I, 1982, 78, 2447.

RECEIVED October 4, 1984

SULFIDES

The Role of Solid State Chemistry in Catalysis by Transition Metal Sulfides

R. R. CHIANELLI

Corporate Research Science Laboratories, Exxon Research & Engineering Company, Annandale, NJ 08801

The Transition Metal Sulfides are a group of solids which form the basis for an extremely useful class of industrial hydrotreating and hydroprocessing catalysts. Solid state chemistry plays an important role in understanding and controlling the catalytic properties of these sulfide catalysts. This report discusses the preparation of sulfide catalysts, the role of disorder and anisotropy in governing catalytic properties, and the role of structure in the promotion of molybdenum disulfide by cobalt.

The Transition Metal Sulfides have been widely used in petroleum upgrading processes for many years, and due to their catalytic and structural stability in feedstocks containing large amounts of sulfur, the demand for better sulfide catalysts will continue as we are forced to upgrade an increasingly heavier feedstock supply (1). Although industrially important for over sixty years, it has been only recently that progress has been made in forming a basis for a fundamental understanding of how these solids catalyze important reactions. An understanding of the solid state chemistry of these solids has played a key role in this recent progress. The catalytically important solid state chemistry of the sulfides includes not only "classical" solid state areas such as the structure of supported and unsupported catalysts, but also areas which are at the forefront of solid state chemistry itself. These areas include novel low temperature methods for producing the solid catalysts at low temperature, the study of disorder and its effect on the catalytic properties of the solids and the importance of crystalline anisotropy in determining and controlling the reactivity of the solid, both to the catalytic environment and to other metals which may cause poisoning or promotion. This report discusses these issues as the basis for understanding the relationship between the properties of the solids and their ability to catalyze a reaction.

0097–6156/85/0279–0221$06.00/0

Binary Transition Metal Sulfide Hydrodesulfurization (HDS) Catalysts

In the period between WWI and WWII, work primarily in Germany focused on MoS_2 and WS_2 as being the best sulfides for hydrogenation and heteroatom removal reactions in the presence of hydrogen and sulfur (2). Originally, these catalysts were used in an unsupported form and without additional transition metals which serve to promote catalytic activity. Recently, the binary transition metal sulfides (binary refers to the simple transition metal sulfides containing one transition metal and sulfur) have been investigated for their activity in a model HDS reaction (3). The model HDS reaction was the desulfurization of dibenzothiophene (DBT) and all the group IV, V, VI, VII and VIII transition metal sulfides, with the exception of Tc, were studied. The active sulfide phases as determined by X-ray diffraction after reactivity measurements, along with the activity of these phases in the HDS reaction are presented in Figure 1.

Most of the phases which were identified after approximately eight hours under catalytic conditions (400°C and 1300 kpa) were poorly crystalline, as determined by broadened Bragg diffraction peaks. However, in some cases (Os and Ir) the diffraction pattern obtained contained only diffuse scattering and no hint of any remaining Bragg peaks. The disorder (discussed further below) which appears in these catalysts contributes (in some cases) to an uncertainty regarding the precise phase which is the stable state of the active sulfide under catalytic conditions. For example, in the cases of Os and Ir amorphous phases of the metal sulfides were obtained. These phases have approximately a 1:1 metal-to-sulfur stoichiometry, but currently their structures are unknown. In the cases of V and Fe the poorly crystalline state of the catalysts did not permit an exact choice of phase based solely on X-ray diffraction data from among several closely related phases (4). Nevertheless, in most cases the crystal structure, upon which the catalyst is based, can be discerned. On the left of the periodic table in group IV, V and VI we find structures which are layered types either cadmium iodide or molybdenite like (TiS_2, ZrS_2, NbS_2, TaS_2, MoS_2, and WS_2) or nickel arsenide related ($V_{1+x}S_2$, Cr_2S_3). All contain only six coordinate metal atoms, but as we move further to the right in the periodic table we find that the structures vary to a greater extent. In the first row we have different structures starting with MnS through to Ni_3S_2, which are dominated by tetrahedral coordination as well as variable stoichiometry. In the second row we find RuS_2 with a pyrite structure (whereas in the first row FeS_2 is not stable under catalytic conditions), Rh_2S_3 with a nickel arsenide related structure and PdS with a unique structure. In the third row we find ReS_2 with a distorted layered structure, Os and Ir with undetermined amorphous structures, PtS with a unique structure and Au not forming a stable sulfide under catalytic conditions. In other words, we find that as we move across the periodic table the state of the sulfide catalyst is constantly changing in regard to crystal structure, stoichiometry and degree of order. Yet the catalytic activity for the model reaction is varying in a continuous fashion as we move from element to element. Furthermore, the activity is varying in a way familiar

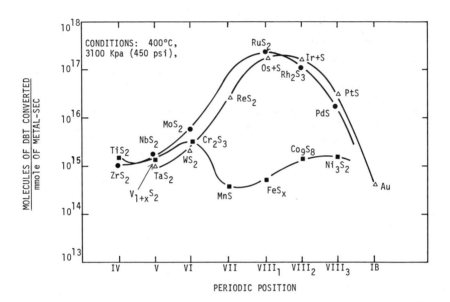

Figure 1. Periodic trend for the HDS of Dibenzothiophene by Transition Metal Sulfides.

in other areas of catalysis: the "volcano" plot (5). From Figure 1
it can be seen that the first row transition metal sulfides are
inactive relative to the second and third row transition metal
sulfides which exhibit maxima in the group VIII metals, at RuS_2 in
the second row and between Os and Ir in the third row. This
behavior points to the secondary role of crystal structure and to
the dominance of the 4 and 5 d electrons in determining catalytic
activity in the transition metal sulfides. A considerable amount of
work has gone into understanding the origin of this effect termed
the "electronic effect" in sulfide catalysis, but a detailed discussion of this work is beyond the scope of this paper (6).

Preparation of Transition Metal Sulfide Catalysts

The catalysts described above were prepared via low temperature
precipitation from non-aqueous solution (7). This technique
involves the precipitation of the transition metal sulfide from a
non-aqueous solvent such as ethyl acetate by dissolving the
appropriate transition metal halide in the solvent and reacting it
metathetically with a sulfiding agent such as lithium sulfide to
precipitate the insoluble sulfide for example:

$$MoCl_4 + 2 Li_2S \xrightarrow{\text{ethyl acetate}} MoS_2 \downarrow + 4 LiCl \qquad (1)$$

The black precipitate is separated from the product LiCl by extensive washing with ethyl acetate. Because the product is formed
rapidly at room temperature, it is completely amorphous to X-rays.
The amorphous MoS_2 which has initially low surface area ($\sim 5m^2/gm$)
may then be converted into higher surface area poorly crystalline
MoS_2 by heat treating in a flowing gas of $H_2/15\%$ H_2S at elevated
temperature (8). For example, if amorphous MoS_2, prepared as
described above, is treated at 400°C or 600°C in a $H_2/15\%$ H_2S mixture, the resulting surface areas will be 63 and 44 M^2/gm, respectively. Thus, by controlling the temperature of the heat treatment
a continuous series of MoS_2 catalysts can be prepared which have
variable surface areas. In a similar manner, all the transition
metal sulfides can be prepared in an oxide free form with sufficient
surface area for catalytic measurements. The lower temperature
precipitation method offers an additional advantage. In preparing
all the transition metal sulfides the chemistry is varied as little
as possible from one sulfide to the next. No other preparative
method currently available offers all these advantages. Furthermore, by changing the solvent to propylene carbonate a homogeneous
colloidal dispersion of MoS_2 can be obtained which is stable over
long periods of time. By slurrying a support material such as
Al_2O_3, SiO_2 or MgO with the colloidal MoS_2, the catalyst can be
selectively adsorbed on the support and thus the effect of supporting the transition metal sulfides can be studied, again keeping the
method of preparation as closely related as possible.
    Another method used in the preparation of the transition metal
sulfides is the thermal decomposition of a suitable precursor in a

sulfiding environment. For example, MoS_2 may be prepared via thermal decomposition of ammonium thiomolybdate which proceeds through the amorphous intermediate MoS_3 (9):

$$(NH_4)_2MoS_4 \xrightarrow{250\,^0C} MoS_3 + H_2S\uparrow + 2NH_3\uparrow \qquad (2)$$

$$MoS_3 \xrightarrow{400\,^0C} MoS_2 + S^o \qquad (3)$$

This method has the advantage of being simpler than the above method and of providing catalysts with surface areas which are generally higher than those obtained by other methods (>100 M^2/gm). A third method of preparation which is convenient is the direct sulfidation of the appropriate ammonium hexachloride in H_2/H_2S (10):

$$(NH_4)_2OsCl_6 + 2H_2S \xrightarrow[H_2/H_2S]{350\,^0C} OsS_2 + 2NH_4Cl\uparrow + 4HCl\uparrow \qquad (4)$$

This method, which has been applied to the noble metal sulfides, has the advantage that all by-products are easily removed in the flowing gas phase. Both the thermal decomposition method and the direct sulfidation method can be applied to specific sulfides only in cases where the precursor material can be easily synthesized. The low temperature precipitation technique is the only method which can be applied to all the transition metal sulfides. All the above methods easily provide reasonable quantities of high surface area catalysts for further study.

The Effect of Crystal Structure in Transition Metal Sulfide Catalysts

In the preceding part of this paper the predominance of the "periodic" effect on HDS by sulfide catalysts was described. Because periodicity dominates, crystal structure is of secondary importance. However, in this section we briefly examine the effect of crystal structure on the catalytic properties of the transition metal sulfides. In the case of catalysts such as MoS_2 and WS_2, the most industrially important catalysts, the effect of crystal structure is quite pronounced. An understanding of the effect of crystal structure in these catalysts is essential to optimizing their catalytic properties for a given application.

The effect of crystal structure may be investigated by preparing catalysts, as described above, at various temperatures which assures a set of catalysts having variable surface areas, pore size distributions, and crystallinity. Measuring the catalytic activity as a function of these physical properties will help to define the role of crystal structure for the particular transition metal sulfide. In general, the HDS is poorly correlated to N_2 BET surface area. This non-correlation can be most easily seen by preparing a series of MoS_2 catalysts by a variety of methods and measuring the HDS activity as a function of BET surface area. There is virtually no correlation between hydrodesulfurization and BET surface area. There is, however, a rather good correlation to a specific chemisorption technique, in this case O_2 chemisorption (11). It is believed that this result arises from the anisotropy of the layered structure (Figure 2). In this structure single layers of transition metals are sandwiched between two layers of close-packed chalcogen atoms. Within these layers the transition metal atoms are bound to six sulfur atoms which are arranged trigonal prismatically about the metal. Each sulfur atom bridges three transition metal atoms within the same layer, forming the only strong intralayer forces, and the layers can be viewed as two-dimensional macromolecules which stack, bound only by van der Waals forces, to form three dimensional crystals (12). As a result of this, the basal planes of MoS_2 catalysts are extremely inert contributing to the N_2 surface area measurements but not to the catalytic activity. O_2, on the other hand, chemisorbs on the MoS_2 edge planes where the catalytically active sites are presumed to be located.

The inertness of the basal planes has been demonstrated in high vacuum studies on MoS_2 single crystals (13). Adsorption and binding studies of thiophene, H_2S and related molecules were carried out. These studies indicated that only physical adsorption of these molecules occur on the basal plane of MoS_2. The probe molecules desorbed without detectable decomposition indicating very low chemical activity of the basal planes of MoS_2. It was further shown that the basal plane of MoS_2 was inert to O_2 exposure at 520 K (14). Only sputtering with He ions, which caused the destruction of the hexagonal feed pattern, would induce reactivity toward O_2. The basal plane could be annealed at 100°K and its inertness recovered. From this study it was concluded that defects may be introduced into the surface by sputtering, and this produced a drastic increase in the rate of oxidation of the surface and in removal of sulfur. This indicates that oxygen chemisorption (and thus HDS) is associated with defect sites in MoS_2 and not ordered basal planes. In a well crystallized catalyst these defects lie on the edge planes of the crystallites.

Because of the anisotropic nature of MoS_2, it tends to grow in very thin crystals which do not permit easy study of edge planes. However, Tanuka and Okuhara showed that the edge planes of MoS_2 were reactive for certain types of reactions by cutting single crystals into pieces and comparing the rates of cut and uncut crystals (15). Further evidence for O_2 interaction at the edge planes of MoS_2 comes from studies on single crystals. When single crystals of MoS_2 are placed in an oxidizing environment at elevated temperatures (>400°C), oxidation of the MoS_2 can be seen to proceed through the edge planes (16). Furthermore, oxygen enrichment at edge planes of MoS_2 single crystals has been demonstrated by scanning Auger studies on single crystals (17).

RuS_2, on the other hand, has a completely isotropic cubic structure identical to pyrite (FeS_2). The isotropic nature of RuS_2 in the HDS reaction can be seen by the linear correlations between HDS activity and O_2 chemisorption and HDS activity and BET surface

Figure 2. Structure of a single layer of MoS$_2$. Reproduced with permission from Ref. 8. Copyright 1982, Taylor & Francis Ltd.

228   SOLID STATE CHEMISTRY IN CATALYSIS

area (18). RuS_2 and MoS_2 exhibit the two extremes which can occur
in HDS by sulfides. Presumably, the effect of crystal structure in
all other transition metal sulfides lies somewhere in between.

The Role of Edge Planes in Promoted Transition Metal Sulfide Catalysts

The importance of edge planes also arises in the industrially important promoted transition metal sulfide catalyst systems. It has
been known for many years that the presence of a second metal such
as Co or Ni to a MoS_2 or WS_2 catalyst leads to promotion (an
increase in activity for HDS or hydrogenation in excess of the
activity of the individual components) (2). Promotion effects can
easily be observed in supported or unsupported catalysts. The
supported catalysts are currently the most important industrial
catalysts, but the unsupported catalysts are easier to characterize
and study. Unsupported, promoted catalysts have been prepared by
many different methods, but one convenient way of preparing these
catalysts is by applying the nonaqueous precipitation method
described above. For example, for Co/Mo, appropriate mixtures of
$CoCl_2$ and $MoCl_4$ are reacted with Li_2S in ethyl acetate:

$$xCoCl_2 + yMoCl_4 + (x + 2y)Li_2S \xrightarrow{\text{ethyl acetate 25}^0\text{C}}$$

$$x\text{"CoS"(amorphous)} + yMoS_2 + 2(x + 2y)LiCl \qquad (5)$$

The amorphous products are then heat treated in 15% H_2/H_2S at the
desired temperature, as described above for the binary systems. By
using this technique, the ratio of Co/Mo in the catalyst can easily
be controlled. It has been well established that somewhere in the
region of a Co/Mo ratio of 0.2 to 0.3 a maximum in catalytic
activity will occur (2). The enhancement in activity can be as much
as an order of magnitude above the unpromoted systems. The precise
shape of the activity vs. promotion curves depends greatly on the
method of preparation of the catalyst. The unsupported catalyst,
after heat treatment, contains both Co_9S_8 and MoS_2, as determined by
x-ray diffraction. A Ni/Mo catalyst prepared in the same fashion
will contain Ni_3S_2 and MoS_2. Thus, from the point of view of the
solid state chemist, this catalyst is a two phase system and the
question arises, what is the origin of the promotion effect? Delmon
et al., initially working with unsupported catalysts, first pointed
out that the Co/Mo catalyst system was phase separated into Co_9S_8
and MoS_2. As a result, the concept of "contact synergy" was
developed, which attributes the promotion effect to the two phases
(Co_9S_8 and MoS_2) being in contact (19). The microscopic origins of
the synergic by contact are not addressed in detail by this
theory. However, Phillips and Fote explained the shape of promotion
curves by invoking surface enrichment of Co or Ni on MoS_2 or WS_2
(20). Thus, the presence of the second phase provides a source of
Co or Ni to surface enrich the MoS_2 or WS_2. Again, no microscopic
origin of promotion is discussed.

Voorhoeve et al., also working primarily with unsupported catalysts and in the Ni_3S_2/WS_2 system, introduced the concept of "pseudointercalation" as the origin of the promotional effect (21). The promotional effect occurs because "pseudointercalation" of Ni in WS_2 creates more W^{+3} sites which are believed by the authors to be the active sites. The term "pseudointercalation" refers to the fact that the WS_2 and MoS_2 do not form bulk intercalates as other layered sulfides do, presumably because of filled conduction bands. Therefore, the intercalation occurs only at edge planes of WS_2 or MoS_2 where the trigonal symmetry is relaxed, allowing "pseudointercalation". The pseudointercalation theory is in agreement with the surface enrichment theory, but the surface enrichment takes place specifically at the edge planes. However, the pseudointercalation theory also provides a microscopic explanation of promotion. "Pseudointercalation" theory and "synergy by contact" are also consistent, if we assume that the second phase is necessary (in unsupported catalysts) to produce the surface enrichment at edge planes. In fact, the Co_9S_8 crystallites may be epitaxially related to the MoS_2 edge planes and thus in direct contact.

Further, insight into the origin of promotion comes through the recognition that the pairs of sulfides (synergic pairs), which in the unsupported form exhibit the promotion effect (Ni_3S_2/MoS_2, Co_9S_8/MoS_2, Ni_3S_2/WS_2, Co_9S_8/WS_2), can be related to the simple binary sulfides through their heats of formation (22). This is done by noting that the average heats of formation of the "synergic pairs" fall in the same range as the heats of formation of the most active sulfides. Whereas, pairs of sulfides which do not exhibit promotion fall outside this range. This suggests that the synergic pairs behave catalytically as second and third row "pseudobinary" sulfides. Since they are phase-separated bulk systems, the averaging of properties must occur at the surface. Thus, the sulfided Co/Mo or Ni/Mo catalysts behave at the surface as sulfides of hypothetical elements of periodic position between those of the members of the pair, hence, the term pseudobinary sulfide. How does this averaging occur? The suggested pseudobinary relationship calls for a Co or Ni atom present at the surface of the MoS_2. This Co or Ni atom is bound to sulfur atoms which are shared with an Mo atom. Somewhere at the edge, sulfur atoms, which upon leaving create vacancies, are shared by Co and Mo leading to average electronic properties of the sulfur atom or vacancy. Thus, Co at the edge of MoS_2 creates a vacancy which has properties similar to a vacancy which occurs on the surface of one of the most active binary sulfides (RuS_2).

In phase-separated promoted sulfides, direct observation of the promoting Co is difficult due to the small amount of it present. Topsøe et al. have shown the presence in both supported and unsupported CoMo catalysts of a unique form of sulfided Co (the CoMoS" phase) which correlates with activity (23,24). In the CoMo system the CoMoS phase is probably the active species, and Topsøe et al. consider that it resides at the edge of the MoS_2 crystallites in both supported and unsupported catalysts. Additional evidence for Co enrichment at MoS_2 edges comes from scanning Auger studies of

single crystals of Co doped MoS_2 crystals, which clearly demonstrates this effect (17). It is again evident that the edge planes
of layered catalysts, such as MoS_2, play an important, perhaps dominant, role in their catalytic behavior.

The Role of Disorder in Layered Transition Metal Sulfide Catalysts

In the above sections the transition metal sulfide catalysts are
described in rather idealized terms. Edge planes and basal planes
exist in single crystals and microcrystals, but in most real
catalysts a great deal of disorder exists. In fact, in the layered
transition metal sulfides disorder in the catalysts is quite important. This disorder must be understood and taken into account for
proper interpretation of the physical characteristics of these
catalysts. When MoS_2 is prepared by the low temperature precipitation method and heat treated as described above, the compound
crystallizes partially in what is termed the "rag" structure (25).
This structure consists of several stacked, but highly folded and
disordered, MoS_2 layers (Figure 3). Because of rapid growth during
precipitation and the anisotropy of the structure, the layers grow
very rapidly in two dimensions but only slowly in the C or stack
direction. The resulting "rags" can be several thousand angstroms
across but only 20 to 30 Å thick. Because of this structure, x-ray
analysis of these materials can be very misleading. The powder diffraction pattern of MoS_2 contains a strong 002 maximum in the low
angle region. From a standard line broadening analysis of this
peak, an estimate of the number of stacked layers can readily be
obtained. However, at higher diffraction angles, a broad envelope
beginning approximately at $2\theta = 30°$ and continuing out to above $2\theta = 60°$ is observed using $Cu_{K\alpha}$ radiation. This envelope contains the
100, 101, 102, 103, 006, 105, 106, 110 and 008 reflections with
well-defined maxima appearing for the 100, 103, and 110
reflections. The asymmetric shape of the envelope in the region of
the 100 is characteristic of random layer lattice structures in
which the layers are stacked randomly with respect to one another.
This random stacking, combined with the folding, makes it impossible
to extract the crystallite or particle size dimension by line
broadening analysis. Large rags and small rags may have the same
order length in the plane, as determined by line broadening
analysis, but vastly different particle sizes, edge areas and,
therefore, catalytic activity. Thus, as opposed to isotropic systems, x-ray diffraction data is only marginally useful in interpret-
ing catalytic properties of these anisotropic catalysts. An extreme
example of this type of disorder is seen in ReS_2, which actually
forms large spherical particles on the order of 0.1-1.0μ but whose
order parameters obtained from x-ray diffraction are 35 x 74 Å(8).
        A more detailed modeling of the diffraction of poorly crystal-
line MoS_2 has recently been presented (26). The entire diffraction
pattern of the MoS_2 was computed using the Debye scattering
equation. The results of the calculations where compared with
experiments and the model refined for a better fit. The size,
stacking, bending and rotation of MoS_2 sheets were varied. The
experimental x-ray patterns could be fit quite well by rotating the

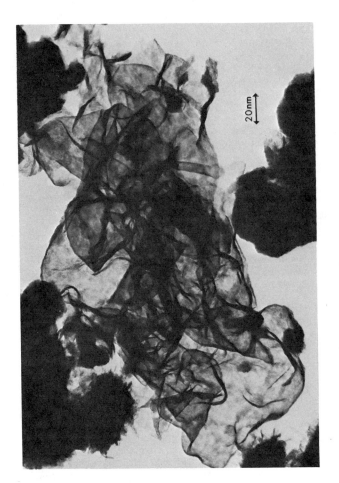

Figure 3. Transmission electron micrograph of highly folded layers of MoS$_2$. Reproduced with permission from Ref. 8.

Copyright 1982, Taylor & Francis Ltd.

layers with respect to each other. A rotation of 0.1 radian enabled
a fitting of the diffuse envelope described above. Thus, the poorly
crystalline MoS_2 catalysts commonly encountered have several
features contributing to its structure: stacked layers, rotation of
the layers with respect to each other, and folding of the layers.
All of these factors must be considered in understanding the
properties of these catalysts.

Finally, we again note that the concept of active sites at the
edge planes is a highly idealized one. In reality this means that
the active sites in MoS_2 or related compounds are edge-like defects
which have the structure and stoichiometry of Mo atoms which occur
at the edge planes of MoS_2 crystals, but which may occur throughout
the surface of the above mentioned disordered catalysts. Although
currently the exact structure of these sites is unknown, their
number may be directly measured by O_2 chemisorption or perhaps by
ESR measurements (27). Recently, in situ UPS measurements have
given a direct look at the surface defects in poorly crystalline
MoS_2 (28). Several poorly crystalline MoS_2 catalysts, prepared as
described above, of differing edge area were studied. The preparation and measurements were carried out in a UHV-compatible cell.
The results showed new electronic states not present in crystalline
MoS_2. These states are observed as a band tail just above the $d_z 2$
band in MoS_2. It was shown that the states are chemically active
surface states by reversible oxygen adsorption – desorption experiments. These experiments are the first to show the direct
connection between surface electronic states and bulk electronic
states in MoS_2 catalysts and should form the basis for developing a
deeper understanding of the nature of these states in MoS_2 and
related catalysts.

Summary

Solid state chemistry plays an important role in the catalysis by
Transition Metal Sulfides; however, it is a role that is somewhat
different than the role usually assigned to solid state chemistry in
catalysis. In catalysis, by sulfides, the chemistry of ternary
phases is not now important and thus, the usual role of solid state
chemistry in preparing ternary phases and systematically studying
the effect on catalytic properties through variation of the composition of these ternary phases is absent. Nevertheless, preparation
of the Transition Metal Sulfides is crucial in controlling the
properties of the catalysts. Low temperature solid state preparations are the key to obtaining good catalysts in reasonable surface
area for catalytic measurements.

Crystal structure plays a secondary role in catalysis by the
Transition Metal Sulfides. As the periodic trends for HDS of the
binary sulfides shows the dominant effect is which transition metal
is present in the reaction, this transition metal takes on the
structure and stoichiometry of the phase which is most stable in the
sulfur containing catalytic environment. The unsupported promoted
catalyst systems can be grouped into "synergic" pairs of sulfides.
Because these pairs are related to the basic periodic trends of the
binary Transition Metal Sulfides through average heats of formation,

they can be thought of as "pseudobinary" sulfides; which, though phase separated bulk systems, operate synergically by forming surface phases which have the average properties of the bulk phases.

Although crystal structure plays a secondary role in the general sense in catalysis by Transition Metal Sulfides, in the particular case of the industrially important layered Transition Metal Sulfides, the catalytic properties are dominated by the anisotropic crystal structure. The edge planes of MoS_2 and WS_2 are the seat of the active sites in the unpromoted systems and the seat of promotion in the promoted systems. An understanding of these edge planes or edge-like defects is central to understanding catalysis by these compounds. Finally, again, because of their anisotropic structure, disorder plays a large role in the catalytic property of these solids.

Acknowledgments

The author would like to acknowledge the contributions of: T. A. Pecoraro, S. Harris, M. B. Dines, S. J. Tauster, M. Salmeron, G. A. Somorjai, A. Wold, M. H. Farias, A. J. Gelman, K. S. Liang, S. K. Behal, B. H. Kear, E. I. Stiefel, T. R. Halbert, W. H. Pan, E. B. Prestridge, J. P. de Neufville, F. Z. Chien, S. C. Moss, B. G. Silbernagel, G. J. Hughes and J. D. Passaretti. Also, a special thanks to D. L. Jocelyn for administrative assistance.

Literature Cited

1. Chianelli, R. R. "Surface Properties and Catalysis by Non-Metals," Bonnelle, J. P.; Delmon, B.; Derouane, E., Eds.; Riedel, Dordrecht, 1982; p. 361.
2. Weisser, O.; Landa, S. "Sulfide Catalysts: Their Properties and Applications;" Pergamon Press, Oxford; 1973.
3. Pecoraro, T. A.; Chianelli, R. R. J. Catal. 1981, 67, 430.
4. Chianelli, R. R. J. Catal. 1982, 71, 228.
5. Sinfelt, J. Progress in Solid State Chem. 1975, 10, 55.
6. Harris, S.; Chianelli, R. R. J. Catal. 1984, 86, 400.
7. Chianelli, R. R.; Dines, M. B. Inorganic Chem. 1978, 17, 2758.
8. Chianelli, R. R. International Reviews in Physical Chemistry 1982, 2, 127. ..
9. Diemann, E.; Muller, A. Coord. Chem. Rev. 1973, 10, 79.
10. Passaretti, J. D.; Kaner, R. B.; Kershaw, R.; Wold, A. Inorg. Chem. 1981, 20, 501.
11. Tauster, S. J.; Pecoraro, T. A.; Chianelli, R. R.; J. Catal. 1980, 63, 515.
12. Schollhorn, R.; Sick, E.; Serf, A. Mater. Res. Bull. 1975, 10, 1005.
13. Salmeron, M.; Somorjai, G. A.; Wold, A.; Chianelli, R. R.; Liang, K. S. Chem. Phys. Letts. 1983, 90, 105.
14. Farias, M. H.; Gelman, A. J.; Somorjai, G. A.; Chianelli, R. R.; Liang, K. S. Surface Sci. 1984, 140, 181.
15. Tanuka, K.; Okuhara, T. Proceedings of the Third International Conference on "The Chemistry and Uses of Molybdenum"; Climax Molybdenum 1979, 170.

16. Bahl, O. P.; Evans, E. L.; Thomas, J. M. Proc. Roy. Soc. A.
     1983, 306, 53.
17. Behal, S. K.; Kear, B. H.; Chianelli, R. R. Materials Letters,
     in press, 1984.
18. Chianelli, R. R. "Fundamental Studies of Transition Metal
     Sulfide Hydrodesulfurization Catalysts," Catal. Rev.-Sci. Eng.,
     1984; 26 (3&4), p. 361.
19. Delmon, B. "Proceedings, 3rd International Conference on the
     Chemistry and Uses of Molybdenum;" Ann Arbor, Michigan, August
     19-23, 1979; p. 73.
20. Phillips, R. W.; Fote, A. A. J. Catal. 1976, 41, 168.
21. Voorhoeve, R. J. H.; Stuiver, S. C. M. J. Catal. 1971, 23, 243.
22. Chianelli, R. R.; Pecoraro, T. A.; Halbert, T. R.; Pan, W. H.;
     Stiefel, E. I. J. Catal. 1983, 86, 226.
23. Topsøe, H.; Clausen, B. S.; Candia, R.; Wivel, C.; Morey, S.
     J. Catal. 1981, 68, 433.
24. Topsøe, H. "Surface Properties and Catalysis by Non-Metals,"
     Bonnelle, J. P.; Delmon, B.; Derouane, E., Eds.; Riedel,
     Dordrecht, 1982; p. 329.
25. Chianelli, R. R.; Prestridge, E. B.; Pecoraro, T. A.;
     DeNeufville, J. P. Science 1979, 203, 1105.
26. Chien, F. Z.; Moss, S. C.; Liang, K. S.; Chianelli, R. R. J.
     de Physique 1981, C4, 273.
27. Silbernagel, B. G.; Pecoraro, T. A.; Chianelli, R. R. J.
     Catal. 1982, 78, 380.
28. Liang, K. S.; Hughes, G. J.; Chianelli, R. R. J. Vac. Soc.
     Tech. 1984, A2(2), 991.

RECEIVED December 17, 1984

14

The Role of Promoter Atoms in Cobalt–Molybdenum and Nickel–Molybdenum Catalysts

H. TOPSØE[1], N.-Y. TOPSØE[1], O. SØRENSEN[1], R. CANDIA[1], B. S. CLAUSEN[1], S. KALLESØE[2], E. PEDERSEN[2], and R. NEVALD[3]

[1] Haldor Topsøe Research Laboratories, DK-2800 Lyngby, Denmark
[2] Department of Inorganic Chemistry of Copenhagen, DK-2100 Copenhagen, Denmark
[3] Department of Electrophysics, Technical University of Denmark, DK-2800 Lyngby, Denmark

By use of analytical electron microscopy (AEM) and infrared spectroscopy (IR) using NO as a probe molecule, it has been shown that the Co-Mo-S structure responsible for the promotion of the hydrodesulfurization activity can be considered as MoS_2 with the Co atoms located in edge positions. Evidence for similar catalytically important Ni-Mo-S structures has also been obtained. Magnetic susceptibility results show that the Co edge atoms interact with the surrounding Mo atoms. The presence of weak antiferromagnetism indicates some interactions between Co atoms. The results show that Co or Ni edge sites play a more important role than Mo edge sites in both hydrodesulfurization and hydrogenation.

The increasing need for efficient treatment of various fossil fuel feedstocks has resulted in many studies (for reviews, see e.g., Refs. (1-8)) devoted to the understanding of the catalytic properties of hydrodesulfurization (HDS) catalysts (e.g., Mo or W based catalysts promoted by Co or Ni). The efforts have been directed towards an understanding of the structural form in which the different atoms are present, and to establish connections between structural information and the various catalytic functions. It has, however, been difficult to make progress since for a long time, information regarding the structural state of the active elements has been limited. In the absence of direct structural information many models have been proposed. For the important Co-Mo catalysts, some of the models most often referred to are: the "monolayer model", where Co is proposed to be associated with the alumina and Mo is present as an oxysulfide bound to the alumina surface; the "contact synergy model" where Co_9S_8 is supposed to be present in contact with MoS_2; and the "intercalation models" where Co is supposed to form intercalation structures with MoS_2. These and other models have been extensively reviewed in the past (see e.g., (1-8)).

0097-6156/85/0279-0235$06.00/0

Recently, we have found that two techniques, Mössbauer emission spectroscopy (MES) (see e.g., Refs (6, 8-13)) and extended X-ray absorption fine structure (EXAFS) (14-16), can provide some of the needed structural information. This has not only resulted in a better description of the structural state of the catalysts but it has also allowed a better understanding of the catalytic properties. In this connection, it should be stressed that both of the above techniques conveniently allow studies to be carried out under in situ conditions.

The MES investigations have shown that the Co promoter atoms may be present in not one but many different configurations. Of particular interest, it was found that one of these is a structure which also contains molybdenum and sulfur atoms (8, 9, 11). This structure was termed Co-Mo-S and since the promotion of the HDS activity was found to be associated with the presence of this structure (6, 12, 13), it is important to understand in detail the properties of Co-Mo-S.

In the present paper different approaches used for investigating the location and the properties of the promoter atoms in Co-Mo-S and Ni-Mo-S type structures will be discussed. In one approach, NO is used as a probe molecule. By studying the adsorption using infrared spectroscopy, it is possible to follow simultaneously the adsorption taking place on Co and Mo surface atoms. In another approach, analytical electron microscopy (AEM) is used to determine the location of Co in Co-Mo-S. By choosing unsupported Co-Mo-S samples with dimensions much larger than the diameter of the electron beam, edge and basal plane regions can be examined independently. Finally, information concerning the chemical nature of the Co atoms has been obtained by means of magnetic susceptibility. This method has not previously been used on Co-Mo catalysts with known Co phase composition.

Experimental

The preparation of the unsupported Co-Mo catalysts with Co/Mo=0.005 was carried out using the homogeneous sulfide precipitation (HSP) method described earlier (11, 17). In short, a hot (335-345 K) solution of a mixture of cobalt nitrate and ammonium heptamolybdate is poured into a hot (334-345 K) solution of 20% ammonium sulfide under vigorous stirring. The hot slurry formed is continuously stirred until all water has evaporated and a dry product remains. Two HSP catalysts were studied. The catalyst (Co/Mo=0.0625) used in the magnetic studies was sulfided at 675 K for 4 hr in a flow of 2% H_2S in H_2, whereas the catalyst (Co/Mo=0.005) used in the AEM study was sulfided in the same gas mixture at 1175 K and kept at this temperature for 24 hr.

The alumina-supported Co-Mo and Ni-Mo catalysts were prepared according to the method described in (12) by incipient wetness impregnation of Mo/Al$_2$O$_3$ catalysts (8.6% Mo impregnated on η-Al$_2$O$_3$ with surface area of 250 m^{2}g^{-1}) with solutions containing cobalt nitrate or nickel nitrate followed by drying and calcining in air at 775 K for 2 hr.

The microscopic analysis was carried out using a JEOL 100 STEM-

SCAN system modified to make it especially suitable for measuring small element concentrations in microcrystals. The unsupported catalysts were sulfided under successively more severe conditions until the resulting MoS_2 crystals were much larger (ca. 4000 Å) than the electron beam diameter (ca. 100 Å). MES confirmed that Co-Mo-S was the only Co-Mo-S phase present. When analyzing the Co concentrations in edge and basal plane regions, 45° tilting of the MoS_2 plates was employed such that the total number of surface atoms is the same in the two positions. The absolute Co concentrations were calculated using an Al film of known thickness as reference. The AEM measurements are presented in detail elsewhere together with HREM results (18).

The Mössbauer emission measurements were performed using the constant-acceleration spectrometer and in situ cell system described in (11).

Infrared spectra were recorded on a Perkin-Elmer 180 grating spectrometer. Self-supporting wafers of the catalyst powder were placed in an in situ quartz IR cell (19) which allowed pretreatments at various conditions. Before the NO adsorption experiments, N_2 (purified by passage through Cu turnings at 523 K and a molecular sieve trap (Linde 5A) kept at 195 K) was passed over the catalyst at 673 K for 16 hr. This was followed by cooling to ambient temperature. Nitric oxide (99% purity) was further purified by freeze-thaw cycles. Further details have been given previously (20).

Most of the magnetic susceptibilities were measured by the Faraday technique at a field strength of 12,000 Oe in the temperature range 5-277 K. No corrections for diamagnetic contributions were applied. Preliminary descriptions of the instrumentation and standard deviations of temperatures and susceptibilities are given in (21). The catalysts were sulfided outside the susceptibility apparatus in a reactor equipped with a side arm tube which could be sealed off after sulfiding. Without contact to air, the catalyst was transferred to the sample cells using a stainless steel high-vacuum glove-box equipped with a turbomolecular pump. Some measurements of the magnetic susceptibilities were also carried out between liquid helium and room temperature in the low frequency vibrating magnetometer described previously (22). These measurements were carried out with fields up to 60,000 Oe. The susceptibilities reported here were calculated from the magnetization curves below saturation.

Results and Discussion

All of the results obtained so far, including the Mo EXAFS studies (14, 16), indicate that the Co-Mo-S structure has a MoS_2-like structure (see e.g., 6, 8).

Figure 1 shows the MoS_2 structure with some of the possible locations. Position (a) represents the position proposed in the original intercalation model, (b) represents an edge or "pseudo-intercalated" Co atom, (c) Co substituting for Mo in the interior, (d) Co located at the edges of the slabs (e.g., at a surface substitutional site), and (e) Co at basal planes.

The combined results from MES (9, 11, 12) and IR (23, 24) of the same Co-Mo/Al_2O_3 catalysts revealed that Co-Mo-S may be present in catalysts where single slab MoS_2 structures dominate. In these structures, intercalation (position (a)) or pseudo-intercalation (position (b)) sites do not exist since at least two S-Mo-S slabs must be stacked on top of each other in order to have intercalation or pseudo-intercalation positions. Thus, the intercalation models do not describe the structure of Co-Mo-S, although the basic idea of an association between Co and MoS_2 appears to be right.

The EXAFS studies of the Co K-edge (15, 16) are not consistent with the substitution of Co in the bulk of MoS_2 (i.e. position (c)) since the number of sulfurs surrounding the Co seems to be less than six and the observed Co-S distance is significantly smaller than the Mo-S distance in MoS_2.

The only Co locations consistent with all of the above measurements on Co-Mo-S are the basal plane (position (e)) or edge positions (position (d)). Such positions are also consistent with the observation by MES (see e.g., (6,9)) and EXAFS (15, 16) that the Co atoms are surface atoms.

It has been difficult to obtain definite proof whether the Co atoms are located at edge or basal plane positions although the former has been favored (see e.g., 5, 6, 8). Considering the MoS_2 structure the edge position would seem more favorable than the basal plane position. The bonding between S-Mo-S slabs comprises of weak van der Waals forces. Consequently, MoS_2 will readily cleave along a van der Waals gap and since none of the covalent bonds are broken by such a cleavage, the exposed basal plane surfaces will be expected to be relatively inert. In contrast, the edge surfaces have coordinatively unsaturated Mo atoms. It could also be mentioned that the two types of surfaces have different adsorption and catalytic properties (25-29).

Direct information on the location of the Co atoms in Co-Mo-S was obtained from analytical electron microscopy (AEM) studies of unsupported Co-Mo catalysts. These exhibited large MoS_2 crystals and MES showed that the Co atoms were exclusively present as Co-Mo-S. Different crystallographically regular MoS_2 crystals were selected. For each plate-like crystal, the relative Co concentration was measured in both interior and edge regions. The results of many such analyses are given in Table I (see also (18)). It was found that with the beam in the interior positions of the MoS_2 crystals, the Co

Table I. AEM measurements of the Co concentration in interior and edge plane regions of large MoS_2 crystals exhibiting Co-Mo-S.

Electron beam positon	Co concentration (expressed as Co/Mo)
interior	below detection limit ($\lesssim 0.001$)
edge planes	ca. 0.8 [a]

[a] referred to the density of Mo atoms in an edge plane.

concentration was below the limit of detection. This allows one to make several conclusions concerning the location of the Co atoms in

Co-Mo-S. First of all, in agreement with earlier results we can exclude the presence of Co in bulk intercalation (position (a) in Fig. 1) and bulk substitutional positions (position (c)). With the present limit of detection, the maximum amount of Co that could be present in these positions corresponds to 0.1 at% Co. It can also be concluded from these measurements that no significant amount of Co is present at the basal plane surfaces (position (e)) (the detection limit corresponds to about 5% coverage of the basal planes).

In view of the above results, the large Co signal measured with the electron beam probing the edge regions must be exclusively caused by Co atoms located at the edge surfaces. This is probably the first direct evidence for the edge location of the Co atoms in Co-Mo-S. The AEM results alone do not allow one to conclude whether the edge Co atoms are located at edge substitutional (position (d)) or edge intercalation positions (position (b)). However, in view of the earlier results discussed above, the edge intercalation position is not likely.

The Co coverage at the edges given in Table I probably corresponds close to the maximum achievable since further sulfiding at high temperatures leads to partial segregation of Co from Co-Mo-S to form Co_9S_8.

AEM studies of unsupported Ni-Mo catalysts with large crystals suggest that the Ni atoms are also located at the MoS_2 edges but the results are more qualitative than in the case of the Co-Mo catalysts.

Previous studies of MoS_2 single crystals (27) have shown that the NO adsorption occurs at the edge planes. Thus, adsorption studies using NO should provide the possibility of further elucidating the properties of Mo catalysts promoted by either Co or Ni. In the present NO adsorption studies infrared spectroscopy was used since this allowed distinction between the adsorption occurring on molybdenum and promoter atoms (see (20) for details). The Ni-Mo and Co-Mo catalysts studied were prepared by adding different amounts of promoter atoms to the same Mo/Al_2O_3 catalyst. EXAFS studies on the sulfided catalysts indicate that the size of the small MoS_2 domains is about the same for all the catalysts (14,16). Thus, the total number of Mo edge sites is also expected to be the same. Nevertheless, with increasing Co loading increasing amount of promoter atoms adsorbing NO is seen, whereas the number of Mo edge atoms available for adsorption decreases (Fig. 2a and c). This indicates that the Co or Ni promoter atoms at the MoS_2 edges partially cover the Mo atoms. Comparison with MES results (12) shows that these Co atoms form Co-Mo-S structures. From the NO adsorption (Fig. 2a and c) and activity data (Fig. 2b and d), it is clear that for the promoted catalysts the HDS and hydrogenation activities are mainly associated with the edge promoter atoms while the edge Mo sites are less important. The activity results also show that the hydrogenation is less promoted than the HDS reaction. This is in agreement with previous results in the literature (30-32).

By combining the IR and AEM results it follows that the nature of the active phases is similar in supported and unsupported HDS catalysts. This is in accordance with MES results (6, 9, 11) which show that similar Co-Mo-S type structures exist in both systems. From these results one may understand why Voorhoeve and Stuiver (33)

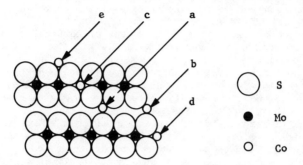

Figure 1. Schematic picture showing some of the possible locations of Co in the MoS$_2$ structure. For a description of the positions a)-e), see text.

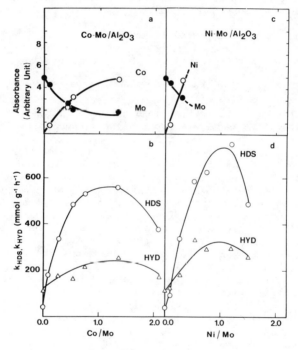

Figure 2. IR absorbances of the NO absorption bands and activi-
ty data for series of sulfided Co-Mo/Al$_2$O$_3$ ((a) and (b)) and
Ni-Mo/Al$_2$O$_3$ catalysts ((c) and (d)).
(Due to opagueness of the high Ni loading samples, the IR data
are only shown for catalysts with Ni/Mo ratios less than 0.3).
Figure adapted from Refs. (20, 32).

and several later authors (see e.g., (3, 4, 13, 17, 34)) have observed many catalytic similarities between unsupported and supported
catalysts.

Magnetic susceptibility has previously been used in many instances to study Co-Mo HDS catalysts (35-42). Most of the studies
have been devoted to the calcined state of the catalysts and only in
few cases to the sulfided state (39-42). In the first studies of
sulfided Co-Mo/Al$_2$O$_3$ catalysts, the magnetic moments were above $4\mu_B$
and were quite close to the values found for the catalysts before
sulfiding (39, 41). This was taken as an indication that the Co atoms are not sulfided or only sulfided to a very small extent. In a
more recent study (42), the magnetic moment was observed to decrease
after sulfiding to a value of about 3.2 μ_B. This decrease indicates
that a sulfided Co surface phase was formed and on the basis of various assumptions concerning the Co phase distribution, a magnetic
moment of about 1.73 μ_B (i.e. Co(II) low-spin) was estimated for the
surface Co phase.

In view of the recent Mössbauer investigation, the differences
in magnetic moments observed in previous studies are probably related to differences in the Co phase distribution which, in fact,
has been found to be sensitive to the choice of preparation parameters. To avoid such ambiguities, it is thus necessary to measure
the magnetic susceptibility on samples where the Co phase distribution is known. This will also allow one to obtain magnetic properties for the catalytically active Co-Mo-S species.

In order to avoid the contribution from Co in the alumina, we
have studied unsupported Co-Mo catalysts for which all the Co atoms
were found by MES to be present as Co-Mo-S. Measurements of the magnetic susceptibility vs. temperature show (see Fig. 3) that the effective moment per cobalt atom varies smoothly from 0.39 μ_B at 5.1 K
to 0.73 μ_B at 275 K. The very low magnetic moment indicates that Co
is in a sulfide environment. However, the results are not compatible
with Co in any magnetically isolated configuration (i.e. high or low
spin Co(II) or Co(III)) independent of the geometry and nature of
surrounding non-magnetic ligating atoms. Rather the result suggests
that extensive electron delocalization is occurring due to the interactions with the neighboring non-magnetic atoms.

The reduced low temperature moment indicates some degree of
antiferromagnetic ordering of the spin lattice. The antiferromagnetic interactions are expected to take place via sulfur atoms along
the long chains of nearest neighbor Co centers at the edges of the
MoS$_2$ particles. In view of the above findings, the neighboring Mo
atoms may also be involved in the exchange paths.

The determination of the magnetic moment of Co in Co-Mo-S for
the more important alumina supported catalysts is more difficult
since the Co atoms may be present in several different phases (e.g.,
Co-Mo-S, Co in alumina, and Co$_9$S$_8$). Therefore, in order to calculate
the magnetic moment of Co-Mo-S it is necessary to know both the
phase composition as well as the magnetic moments of Co$_9$S$_8$ and Co in
alumina. Consequently, magnetic measurements were carried out on catalysts for which the Co phase composition previously has been determined by use of MES (12). The magnetic susceptibility of Co$_9$S$_8$
has been found by several investigators (see e.g., (43, 44)) to be
very low and is not expected to give any significant contribution to

the magnetic moments. Although the Co atoms present in the alumina
are found in both octahedral and tetrahedral coordination in calcined Co-Mo catalysts (45, 46), detailed MES studies (45) indicate
that only the tetrahedrally coordinated Co atoms remain after sulfidation. The magnetic moment of such Co species is expected to be
around 4.1 μ_B (42). Thus on basis of the Co phase composition, a
moment of 4.1 μ_B for Co in the alumina and neglecting the contribution from Co_9S_8, the magnetic moment of Co in Co-Mo-S can be estimated by use of the expression:

$$\mu^2 = \sum_i a_i \mu_i^2 \tag{1}$$

where i represents Co in all different configurations. In Fig. 4 we
have shown the inverse susceptibilities of catalysts with different
Co/Mo ratios and in Table II the resulting average magnetic moments
of the catalysts are given together with the calculated moments for
Co-Mo-S.

Table II. Magnetic moments at room temperature of Co-Mo/Al_2O_3 Catalysts

Catalyst	Co phase distribution (at %)			Magnetic moments (μ_B)	
Co/Mo	Co:Al_2O_3	Co_9S_8	Co-Mo-S	μ_{total}	$\mu_{Co-Mo-S}$
0.09	11±4	0±4	89±4	2.05	1.41-1.80
0.27	15±5	0±4	85±5	1.95	0.74-1.54
1.19	8±4	73±4	20±4	1.52	1.35-2.61
2.09	-	-	-	1.12	-

For each Co:Mo ratio, an interval is given for the derived moment of the Co atoms in Co-Mo-S, as calculated from the estimated
uncertainties of the Co phase distributions. The intervals are wide
but overlapping with an average value of around 1.4 μ_B per Co atom.
It is seen that although the content of Co in the alumina for all
the catalysts is very small it gives rise to the largest contribution to the observed moment. Therefore, the uncertainties in the
magnetic moment of Co in Co-Mo-S (see Table II) are mainly determined by the uncertainty in the determination of the content of Co
in alumina.

The magnetic moment of the unsupported catalyst (pure Co-Mo-S
phase) is temperature dependent, and its room temperature value is
substantially smaller than that found in the supported catalysts.
These results suggest that the electron delocalization and exchange
effects are more important for Co-Mo-S in the unsupported catalysts
than in the supported catalysts. A possible explanation for this behavior is the presence of support interactions. Such interactions
may involve oxygen bridges between the Mo atoms in Co-Mo-S and the
alumina. The presence of such oxygen ligands in the MoS_2 structure,

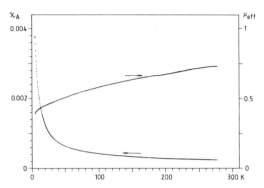

Figure 3. Magnetic susceptibility (left scale; cgs units per gram atom Co) and effective moment (right scale; Bohr magnetons) versus temperature for an unsupported Co–Mo HDS catalyst exhibiting Co–Mo–S as the only Co phase (295 data points are shown).

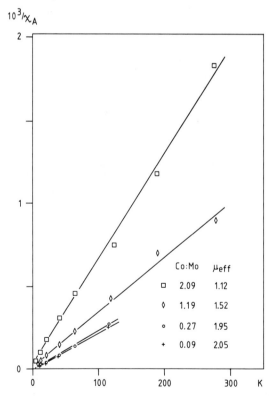

Figure 4. Inverse magnetic susceptibility (cgs units gram atom Co) versus temperature for a series of sulfided Co–Mo/Al$_2$O$_3$ catalysts with different Co/Mo ratios.

which is a semi-conductor, may very well alter its electronic and magnetic properties in a way that it becomes less conductive. This may result in less electron delocalization and thus in a higher moment of the Co atoms present in the alumina supported Co-Mo-S phase. It is also possible that in the alumina supported catalysts, which have the highest MoS_2 dispersion, Mo^{3+} or Mo^{5+} species may contribute to the observed moment.

Exposure of the catalysts to even small traces of oxygen leads to large increases in the observed magnetic moment. In agreement with previous MES (9, 11) and EXAFS studies (15, 16), this shows that the Co edge atoms easily coordinate to oxygen. This result also implies that it is important to carry out in situ studies of Co-Mo-S.

Acknowledgments

The authors are grateful to K. Reiter, J. Refslund Andersen, J.W. Ørnbo, and F. Rasmussen for technical assistance. One of us (Erik Pedersen) acknowledges support by the Danish Natural Science Research Council through grant numbers 511-742, 511-3993 and 511-10516.

Literature Cited

1. de Beer, V.H.J.; Schuit, G.C.A. In "Preparation of Catalysts"; B. Delmon, P.A. Jacobs, and G. Poncelet, Eds.; Elsevier: Amsterdam, 1976; p. 343.

2. Massoth, F.E. In "Advances in Catalysis"; D.D. Eley, H. Pines, and P.B. Weiz, Eds.; Academic Press, New York, 1978; Vol. 27, p. 265.

3. Delmon, B. In "Proceedings of the Climax Third Intern. Conf. on Chemistry and Uses of Molybdenum"; H.F. Barry and P.C.H. Mitchell, Eds.; Climax Molybdenum Co., Ann Arbor: Michigan, 1979; p. 73.

4. Grange, P. Catal. Rev.-Sci. Eng. 1980, 21, 135.

5. Ratnasamy, P.; Sivasanker, S. Catal. Rev.-Sci. Eng. 1980, 22, 371.

6. Topsøe, H.; Clausen, B.S.; Candia, R.; Wivel, C.; Mørup, S. Bull. Soc. Chim. Belg. 1981, 90, 1189.

7. Mitchell, P.C.H. In "Catalysis"; C. Kemball and D.A. Dowden, Eds.; Specialist Periodica Report, Royal Society of Chemistry: London, 1980; Vol. 4, p. 175.

8. Topsøe, H. In "Surface Properties and Catalysis by Non-Metals"; J.P. Bonnelle et al., Eds.; 1983, p. 329.

9. Clausen, B.S.; Mørup, S.; Topsøe, H.; Candia, R. J. Phys. Colloq. 1976, C6, p. C6-249.

10. Topsøe, H; Clausen, B.S.; Burriesci, N.; Candia, R.; Mørup, S. In "Preparation of Catalysts II"; B. Delmon, Grange, P., Jacobs, P.A., Eds.; Elsevier Scientific Publishing Company: Amsterdam, 1979; p. 479.

11. Topsøe, H.; Clausen, B.S.; Candia, R.; Wivel, C.; Mørup, S. J. Catal. 1981, 68, 435.

12. Wivel, C.; Candia, R.; Clausen, B.S.; Mørup, S., Topsøe, H. J. Catal. 1981, 68, 453.
13. Candia, R.; Clausen, B.S.; Topsøe, H. J. Catal. 1982, 77, 564.
14. Clausen, B.S.; Topsøe, H.; Candia, R., Villadsen, J.; Lengeler, B.; Als-Nielsen, J.; Christensen, F. J. Phys. Chem. 1981, 85, 3868.
15. Clausen, B.S.; Lengeler, B.; Candia, R., Als-Nielsen, J.; Topsøe, H. Bull. Soc. Chim. Belg. 1981, 90, 1249.
16. Clausen, B.S.; Topsøe, H.; Candia, R.; Lengeler, B. In "Catalytic Materials: Relationship Between Structure and Reactivity"; American Chemical Society: Washington, D.C. 1984; p. 71.
17. Candia, R.; Clausen, B.S.; Topsøe, H. Bull. Soc. Chim. Belg. 1981, 90, 1225.
18. Sørensen, O.; Clausen, B.S.; Candia, R.; Topsøe, H. Appl. Catal. in press.
19. Topsøe, N.-Y.; Topsøe, H. J. Catal. 1982, 75, 354.
20. Topsøe, N.-Y.; Topsøe, H. J. Catal. 1983, 84, 386.
21. Michelsen, K.; Pedersen, E. Acta Chem. Scand. A32,847 and references therein (1978)
22. Johansson, T.; Nielsen, K.G. J. Phys. E: Sci. Instrum. 1976, 9, 852.
23. Topsøe, N.-Y. J. Catal. 1980, 64, 235.
24. Topsøe, N.-Y.; Topsøe, H. Bull. Soc. Chim. Belg. 1981, 90, 1311.
25. Groszek, A.J.; Witheridge, R.E. Powder Metall. 1972, 15, 115.
26. Tanake, K.; Okuhare, T. Catal. Rev. Sci. Eng. 1982, 15, 249.
27. Suzuki, K.; Soma, M.; Onishi, T.; Tamaru, K. J. Electron Spectrosc. Relat. Phenom. 1981, 24, 283.
28. Stevens, G.C.; Edmonds, T. J. Catal. 1975, 37, 544.
29. Salmeron, M.; Somorjai, G.A.; Wold, A.; Chianelli, R.; Liang, K.S. Chem. Phys. Lettr. 1982, 90, 105.
30. Hargreaves, A.E.; Ross, J.R.H. In "Proc. 6th Int. Congr. Catal". G.V. Bond; P.B. Wells; F.C. Tompkins, Eds. Chem. Soc.: London, 2, 1977, p. 937.
31. Massoth, F.E.; Chung, K.S. In "Proc. 7th Int. Congr. Catal.", T. Seiyama; K. Tanabe, Eds. Elsevier: New York, 1980, p. 629.
32. Candia, R.; Clausen, B.S.; Bartholdy, J.; Topsøe, N.-Y.; Lengeler, B.; Topsøe, H. In "Proc. 8th Int. Congr. Catal.", Verlag Chemie: Weinheim, 1984, Vol. II, p. 375.
33. Voorhoeve, R.J.H.; Stuiver, J.C. M. J. Catal. 1971, 23, 228.
34. Furimsky, E.; Amberg, C.H. Can. J. Chem. 1975, 53, 2542.
35. Richardson, J.T.; Ind. Eng. Chem. Fundam. 1964, 3, 154.
36. Aschley, J.H.; Mitchell, P.C.H. J. Chem. Soc. A 1968, 2821.
37. Lipsch, J.M.J.G.; Schuit, G.C.A. J. Catal. 1969, 15, 163.
38. Lo Jacono, M.; Cimino, A.; Schuit, G.C.A. Gass. Chim. Ital. 1973, 103, 1281.
39. Mitchell, P.C.H.; Trifiro, F. J. Catal. 1974, 33, 350.
40. Perrichon, V.; Vialle, J.; Turlier, P.; Delvaux, G.; Grange, P.; Delmon, B. Comptes rendus, 1976, 282 série C, 85.
41. Ramaswamy, A.V.; Sivasanker, S.; Ratnasamy, P. J. Catal. 1976, 42, 107.
42. Chiplunker, P.; Martinez, N.P.; Mitchell, P.C.H. Bull. Soc. Chim. Belg. 1981, 90, 1319.

43. Townsend, M.G.; Horwood, J.L.; Tremblay, R.J.; Ripley, L.G.;
    Phys. Stat. Soc. (a) 1972, K 137.
44. Knop, O.; Huang, C.-Y.; Reid, K.I.G.; Carlow, J.S.; Woodhams,
    F.W.D. J. Solid State Chem. 1976, 16, 97.
45. Wivel, C.; Clausen, B.S.; Candia, R.; Mørup, S.; Topsøe, H. J.
    Catal. 1984, 87, 497.
46. Candia, R.; Topsøe, N.-Y.; Clausen, B.S.; Wivel, C.; Nevald,
    R.; Mørup, S.; Topsøe, H. In "Proceedings of the Climax Fourth
    Intern. Conf. on Chemistry, and uses of Molybdenum", Ann Arbor:
    Michigan, 1982; p. 374.

RECEIVED October 4, 1984

15

Preparation and Properties of Cobalt Sulfide, Nickel Sulfide, and Iron Sulfide

D. M. PASQUARIELLO, R. KERSHAW, J. D. PASSARETTI[1], K. DWIGHT, and A. WOLD[2]

Department of Chemistry, Brown University, Providence, RI 02912

Co_9S_8, Ni_3S_2, and Fe_7S_8 were prepared as single-phase polycrystalline materials by heating the appropriate metal sulfates in a controlled mixture of H_2 and H_2S at low temperature. The products were characterized by x-ray diffraction, thermogravimetric analysis, and magnetic susceptibility measurement. The x-ray diffraction pattern and field dependent magnetic susceptibility of Fe_7S_8 were affected by the thermal history of the sample. The observed differences can be related to the vacancy ordering associated with ferrimagnetic Fe_7S_8.

The transition metal sulfides Co_9S_8, Ni_3S_2 and Fe_7S_8 have been identified as possible promotors in hydrodesulfurization catalysts. However, the actual catalysts are amorphous, and discrete sulfide phases have never been observed in γ-Al_2O_3 supported systems within the composition range of commercial catalysts.

Since the preparation of Co_9S_8, Ni_3S_2, and Fe_7S_8 is difficult to achieve by direct combination of the elements, a low temperature synthesis involving the treatment of anhydrous sulfates in a controlled H_2/H_2S atmosphere was developed at Brown University (1). Since magnetic susceptibility appears to be capable of distinguishing the various members of the Fe-S, Co-S and Ni-S systems (2-4), it was decided to characterize the low temperature single-phase products by magnetic susceptibility measurements as well as x-ray diffraction analysis.

Delafosse, et al. (5), have shown that sulfides of nickel and cobalt can be prepared by heating their anhydrous sulfates in a stream of H_2/H_2S at low temperatures. However, the experimental conditions for obtaining pure Ni_3S_2 and Co_9S_8 were not specified. It has been shown (2), (6) that both Co_9S_8 and Ni_3S_2 permit little variation from ideal stoichiometry. For both compounds, there is no observable variation in the lattice parameter as determined from x-ray analyses, and magnetic measurements of Co_9S_8 have confirmed its narrow homogeneity range.

[1] Current address: Exxon Research & Engineering Company, Annandale, NJ 08801
[2] Author to whom correspondence should be directed.

0097–6156/85/0279–0247$06.00/0
© 1985 American Chemical Society

Synthetic samples of the low temperature phase of Fe_7S_8 have been prepared by Lotgering (4) and magnetic measurements confirmed the work of other investigators (7-9) that the spontaneous magnetism of Fe_7S_8 represents a ferrimagnetic structure which is based upon an ordering of iron vacancies. This can be represented by the formula $Fe_4^{\rightarrow}[Fe_3^{\leftarrow} \; \Box \;]S_8$. If this model is correct, then randomization of the vacancies should affect markedly the observed magnetic behavior.

Experimental Section

Preparation of Samples. The sulfides Co_9S_8, Ni_3S_2, and Fe_7S_8 were prepared by treating pre-dried sulfate salts of cobalt, nickel, and iron with a mixture H_2 and H_2S in a vertical reactor (Figure 1) at $325°C$ for Fe_7S_8, and $525°C$ for Co_9S_8 and Ni_3S_2. Cobalt and nickel sulfates were dried initially at $135°C$; preliminary drying of ferric sulfate was unnecessary. After placement of the sulfate in the reactor tube, the system was purged with nitrogen, and a drying step followed. After one hour of drying under a nitrogen flow, the desired flow rates for H_2 and H_2S were selected and allowed to equilibrate. At this point, the temperature was elevated to ensure complete reaction. For both Co_9S_8 and Ni_3S_2, the reactor tube was removed from the furnace at the end of the reaction and air quenched to room temperature. The quenched samples of Fe_7S_8 were prepared in a silica reactor tube (fitted with a Vycor frit) which was cooled rapidly with ice water at the end of the reaction. An annealed sample of Fe_7S_8 was prepared by heating the quenched product in a sealed evacuated silica tube for two weeks at' $300°C$. The tube was allowed to reach room temperature overnight. Slow-cooled samples of Fe_7S_8 were prepared by lowering the temperature of the reactor from $325°C$ to $175°C$ at a rate of $1°C/min$. The reactor tube was then removed from the furnace and allowed to reach room temperature. For all the syntheses, once the reactor tube reached room temperature, the system was purged with nitrogen before the samples were removed.

The experimental conditions for the preparation of Co_9S_8, Ni_3S_2, and Fe_7S_8 are given in Table I.

Characterization of Samples. Powder diffraction patterns of the samples were obtained with a Philips diffractometer using monochromated high-intensity $CuK\alpha_1$ radiation ($\lambda = 1.5405Å$). For qualitative identification of the phases present, the patterns were taken from $12° < 2\theta < 72°$ with a scan rate of $1°$ $2\theta/min$ and a chart speed of 30 in/hr. The scan rate used to obtain x-ray patterns for precision cell constant determination was $0.25°$ $2\theta/min$ with a chart speed of 30 in/hr. Cell parameters were determined by a least-squares refinement of the reflections.

The crystallite size was determined by the Scherrer method, and a shape factor of 0.9 was applied (10). A computer program was used to digitize the selected x-ray (slow scan) peaks and determine the peak width.

Thermogravimetric analysis was performed for each material using a Cahn electrobalance (Model RG). Each sulfide was first heated in a stream of oxygen and then reduced to the metal in a stream of $85\%Ar-15\%H_2$.

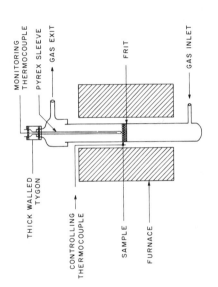

Figure 1. Reactor for the preparation of Co_9S_8, Ni_3S_2, and Fe_7S_8.

Table I. Reaction Conditions

Reagent	Preliminary Drying	Drying Temp (°C) Under N₂ (1 hr)	H₂/H₂S (v/v)	Temp (°C)	Time (Hrs)	Product
$CoSO_4 \cdot 7H_2O$ (Baker W093)	135°C/4 hrs	250	40:1	525	4	Co_9S_8
$NiSO_4 \cdot 6H_2O$ (Baker 2808)	135°C/4 hrs	250	40:1	525	2	Ni_3S_2
$Fe_2(SO_4)_3 \cdot nH_2O$ (Fisher 774179)	-----	175	10:1	325	6	Fe_7S_8

Magnetic data were obtained over the temperature range 80-300K using a Faraday balance equipped with a Cahn electrobalance (Model RG). Measurements were performed at field strengths between 6.2 and 10.4 kOe. The balance was calibrated using platinum wire (χ_g = 0.991 x 10^{-6} emu/g at 273K) as a standard; temperatures were measured with a Ga-As diode. The core diamagnetic correction was not applied to these measurements because of the large uncertainty in the magnitude of the correction relative to the susceptibility of the materials studied.

Results and Discussion

Pure Co_9S_8 is difficult to prepare by direct combination of the elements. Sulfur deficient products (less than 47.06 atomic percent sulfur) contained free cobalt, as indicated by the presence of the (111) reflection of cubic cobalt, which is in agreement with Lindquist (12). It has been shown that such samples show a magnetic susceptibility which is strongly field dependent. The magnetic data for stoichiometric Co_9S_8 showed a field-independent susceptibility consistent with Pauli paramagnetism, and the observed value of 0.85 x 10^{-6} emu/g is in reasonable agreement with the results of Knop (13).
Stoichiometric samples of Co_9S_8 can best be prepared by heating cobalt sulfate in a stream of H_2/H_2S at 525°C. Whereas it takes almost two weeks to obtain Co_9S_8 by the direct combination of the elements, pure single-phase products can be obtained from the sulfate in six hours.
    Ni_3S_2 and Fe_7S_8 are also difficult to prepare by direct combination. Kullerud and Yund (6) reacted nickel and sulfur for 168 hours at 500°C, and Lotgering (4) annealed Fe_7S_8 for one month at 270°C. The technique for heating sulfates in a controlled H_2/H_2S atmosphere is therefore a rapid method for obtaining homogeneous single-phase products of sulfides which resist preparation by other methods.
    X-ray and thermogravimetric analyses of the products are given in Table II. Cell constants of a = 9.930(2) for Co_9S_8 and a = 5.738(2), c = 7.126(2) for Ni_3S_2 correspond with those reported previously (13-14). The relative crystallite sizes are also given in Table II.

Table II. X-ray and TGA Analysis

Sulfide	Cell Parameters	Crystallite Size(A)	% Metal Obs.	% Metal Calc.
Co_9S_8	a = 9.930(2)	380	67.0(2)	67.4
Ni_3S_2	a = 5.738(2)	400	73.1(2)	73.3
	c = 7.126(2)			
(quenched)	a = 3.447(2)	290	60.4(2)	60.4
Fe_7S_8	c = 5.747(2)			

The magnetic susceptibility was found to be field independent for both Co_9S_8 and Ni_3S_2. This indicates the absence of any ferromagnetic impurity. In addition, the susceptibilities for both of these materials are temperature independent, and their respective values of 1.3×10^{-6} and 0.6×10^{-6} emu/g are consistent with Pauli paramagnetism.

The cell constants given in Table II for a quenched sample of Fe_7S_8, a = 3.447(2) and c = 5.747(2)Å correspond to the values reported by Erd, et al., (15) for the hexagonal pseudocell. The x-ray pattern of an annealed Fe_7S_8 yields d-spacings which correspond to those reported by Erd, et al., (15) and calculated from the monoclinic superlattice reported by Tokanami, et al. (16). These d-values are compared in Table III. It can be seen that annealing of Fe_7S_8 samples generates an ordered monoclinic cell. A sample of Fe_7S_8 was slow cooled at 1°C/min from 325°C to 175°C and then quenched to room temperature. The resulting x-ray diffraction pattern showed the onset of ordering, as indicated by the appearance of some of the superlattice peaks.

Magnetic susceptibility measurements were also able to follow the ordering process in Fe_7S_8. Quenched samples from 325°C showed field independent magnetic susceptibility, whereas the annealed sample indicated strong field dependency. The results of these studies are shown in Figure 2. Here the intercept gives the magnitude of the susceptibility of a sample, and the slope is proportional to its spontaneous magnetization.

Bertaut (8) discussed the ferrimagnetic behavior of naturally occurring Fe_7S_8 samples in terms of the ordering of iron vacancies as well as of spins. In this study, the quenched Fe_7S_8 shows a temperature independent susceptibility from liquid nitrogen to room temperature, which is consistent with a random distribution of iron vacancies. The observed magnitude of 25×10^{-6} emu/g for the susceptibility of the quenched sample is what would be anticipated for an anti-ferromagnet well below T_N. Observation of field dependency for annealed samples of Fe_7S_8 coincides with the appearance of superlattice lines in the x-ray diffraction patterns.

Summary

The treatment of the sulfate salts of cobalt, nickel, and iron with a controlled mixture of H_2 and H_2S at low temperatures yielded Co_9S_8, Ni_3S_2, and Fe_7S_8. The sulfides prepared here were characterized by x-ray diffraction, magnetic susceptibility, and thermogravimetric analysis. The method of preparation was found to yield single phase materials which were free of ferromagnetic impurities. Co_9S_8 and Ni_3S_2 exhibited temperature independent magnetic susceptibility which is consistent with Pauli paramagnetism.

The field dependent magnetic susceptibility measurements for Fe_7S_8 were sensitive to the thermal history of the sample. Annealed samples showed strong field dependent behavior (i.e., large spontaneous magnetization), whereas quenched samples did not. Slow-cooled samples exhibited less field dependent behavior than the annealed samples which indicated less ordering of the vacancies. These observations are consistent with the Bertaut model for vacancy ordering in ferrimagnetic Fe_7S_8.

Table III.   Diffraction Data for Fe_7S_8 Samples

Quenched	Annealed	Monoclinic[a] Superlattice	Calc. [b] Superlattice
	5.79	5.75	5.75
		5.29	5.27
		4.72	4.68
	3.11		3.13
2.98	2.97	2.97	2.97
2.87	2.85	2.84	2.85
	2.70	2.70	2.74
2.64	2.64	2.64	2.64
	2.52		2.55
	2.42		2.36
		2.27	2.25
	2.21	2.21	2.23
	2.15	2.16	2.15
2.07	2.06	2.06	2.06
		2.01	2.01
	1.971		1.973
		1.946	1.954
	1.911	1.914	1.917
1.720	1.717	1.717	1.717
	1.630	1.632	1.631
1.614	1.604	1.606	1.600
	1.561		1.565
1.491	1.498		1.496
1.478	1.487	1.488	1.488
1.442	1.442	1.439	1.438
	1.430	1.424	1.424
1.325	1.320	1.320	1.317
1.296	1.289	1.286	1.289

(a)Reference 15
(b)Reference 16
NOTE: Because of the complexity of the structure, unambiguous
       assignment of the Miller indices at high 2θ values is not
       possible from powder data.

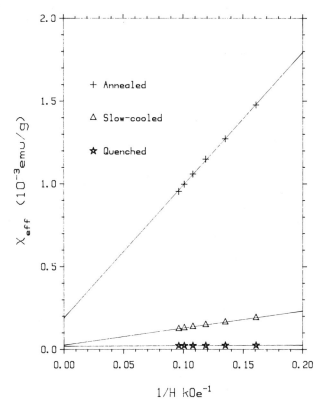

Figure 2. Magnetic susceptibility versus reciprocal field for Fe_7S_8 samples.

Acknowledgments

Acknowledgment is made to both the Exxon Laboratories, Linden, New Jersey, and to the National Science Foundation (Grant DMR79-23605) for the support of D. Pasquariello. In addition, we would like to thank the National Science Foundation (Grant DMR79-23605) for the support of K. Dwight. A. Wold would like to thank the GTE Laboratories (Waltham, Massachusetts) of the GTE Corporation for partial support during this work. Acknowledgment is also made to Brown University's Materials Research laboratory program which is funded through the National Science Foundation.

Literature Cited

1. D. M. Pasquariello, R. Kershaw, J. D. Passaretti, K. Dwight, A. Wold, Inorg. Chem., 1984.
2. K. Kim, K. Dwight, A. Wold, R. R. Chianelli, Mat. Res. Bull., 1981, 16, 1319.
3. E. H. M. Badger, R. H. Griffeth, W. B. S. Newling, Proc. Roy. Soc. A, 1949, 197A, 184.
4. F. K. Lotgering, Philips Res. Rep., 1956, 11, 190.
5. D. Delafosse, P. Barret, C. R. Acad. Sci. Paris, 1960, 251, 2964.
    D. Delafosse, M. Abon, P. Barret, Bull. Soc. Chim. France, 1961, 164, 1110.
    D. Delafosse, P. Barret, C. R. Acad. Sci. Paris, 1961, 252, 280.
    D. Delafosse, P. Barret, C. R. Acad. Sci. Paris, 1961, 252, 888.
6. G. Kullerud, R. A. Yund, J. of Petrology, 1962, 3(1), 126.
7. L. Neel, Rev. Mod. Phys., 1953, 25, 58.
8. E. F. Bertaut, Bull. Soc. France Miner. Crist., 1956, 79, 276.
9. R. Benoit, C. R. Acad. Sci. Paris, 1952, 234, 2174.
10. W. J. Croft, Annals of the N.Y. Acad. Sci., 1956, 62, 464.
11. M. Lindquist, D. Lundquist, A. Westgren, Svensk. Kem. Tidskr., 1936, 48, 156.
12. O. Knop, M. A. Ibrahim, Canad. J. Chem., 1961, 39, 297.
13. V. Rajamani, C. T. Prewitt, Can. Miner., 1975, 13, 75.
14. M. A. Peacock, Univ. Toronto Studies, 1947, 51, 59.
15. R. C. Erd, H. T. Evans, D. H. Richter, Amer. Miner., 1957, 42, 309.
16. M. Tokonami, K. Nishiguchi, N. Morimoto, Amer. Miner., 1972, 57, 1066.

RECEIVED December 27, 1984

ZEOLITES AND CLAYS

Zeolite Chemistry in Catalysis

JACQUES C. VEDRINE

Institut de Recherches sur la Catalyse, Albert Einstein, 69626 Villeurbanne, France

The importance of molecular sieves catalysts in industrial catalysis has increased significantly over the past two decades. To date, all commercial applications of zeolite catalysts have involved acidic zeolites, particularly the ultra stable rare earth Y-type zeolite used in the catalytic cracking processes. Recent investigations of zeolite chemistry have revealed some particular features of both basic and acidic zeolites and have opened the new field of improved fuel processing. The new chemical evidence has raised the possibility that physico-chemical features play a role in catalysis. A large effort has therefore been devoted to the shape-selectivity properties of such materials. Moreover, increasing interest has been focussed recently on the understanding of crystallization of zeolites during synthesis, in the synthesis of new zeolites and in the chemical modifications of zeolites with the objective of expanding the applications of such materials.

The word zeolite stems from the Greek "zeo" (boiling) and "litos" (stone) meaning a material able to eliminate large amounts of water when heated. More than forty different structures of zeolites have been identified while grea progress has been gained in synthetic zeolites, particularly by Linde and Mobil . Zeolites are usually alumino silicates with a general formula

$$M_x D_y (Al_{x+2y} Si_{n-(x+2y)} O_{2n}), mH_2O$$

where M and D designate respectively a mono-(H^+, Na^+, K^+, Cs..) or a divalent (Ba^{2+}, Mg^{2+}, Ca^{2+},...) cation. They are formed of AlO_4

0097–6156/85/0279–0257$06.00/0

and SiO$_4$ tetrahedra bonded together via the oxygen atoms and assembled in such such a way to constitute cavities, cages and/or channels leading to a regular lattice. The cations compensate the negative charge born by Al due to its fourfold coordination.

The dehydration of these materials makes void volume available to molecules whose shape and size have to be compatible with the size of the cavities and pores. The diameter of these cavities or pores depends on the type of zeolite and varies from 0.4 nm (Linde type A) to 0.75 nm (Linde-Y) which corresponds typically to the size of molecules and is expressed in the word "molecular sieve" given to such materials. It follows that diffusivities of reagents or products often play a determining role in catalysis. This holds true particularly in shape selectivity features.

In Figure 1 examples of structure of some zeolites are given to picture some typical spatial arrangements of channels, cavities, etc.

In Tables I and II are reported the dimensional properties of current zeolites with special emphasis on the tri- or mono-dimensional interconnections between cages, cavities or channels which allow more or less the traffic of reagent and/or product molecules through the zeolitic network. On the basis of pore size, zeolites can be classified in three groups : the large pore zeolites, which show twelve-membered rings (such as mordenite and faujasite), the intermediate pore zeolites with 10-membered rings (such as ZSM-5, ZSM-11, ferrierite) and the small pore zeolites with 8-membered rings (such as erionite, A, chabazite, ZK-5). The size of the pores or cavities are comparable to the critical molecular dimensions for some hydrocarbons as shown in Table III.

A zeolite framework bears a net negative charge per unit cell equal to the number of tetracoordinated Al it contains. This negative charge is compensated by cations as discussed above. These cations are exchangeable which allows one to introduce a desired cation. They are not located in the framework at lattice atom position but are located in some defined sites within the cavities or cages. Such locations could be characterized by X-ray diffraction studies and were designated for Y type zeolite as S$_I$ (centre of the hexagonal prism), S$_{I'}$ and S$_{II'}$ within the sodalite cage (near the hexagonal prism and near the window toward the supercage, res the sodalite cage (near the hexagonal prism and near the window toward the supercage, respectively) and S$_{II}$ in the large cage (<u>1</u>). It was thus interesting to determine the location of exchanged cations and it was observed that dehydration or outgassing results in a migration of the cations from S$_{II}$ or S$_{II'}$ sites towards inner sites as S$_I$ or S$_{I'}$ and vice versa (<u>2, 3</u>). The mobility of such cations under ambient atmospheres and temperatures is one of the most striking features of zeolite matrices.

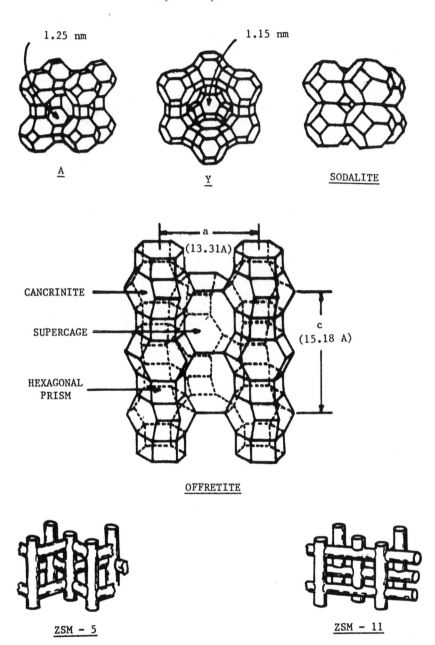

Fig.1 : Structures of some peculiar zeolitic frameworks.

Table I. Pore structure of some zeolites.

Zeolites	Size of channels (nm)	Interconnection system
Chabazite	8 0.36x0.37	3 D
Erionite	$\bar{8}$ 0.36x0.52	3 D
Faujasite (X, Y)	$\bar{1}2$ 0.74	3 D
Ferrierite	$\overline{10}$ 0.43x0.55 ↔ 8 0.34x0.48	1 D
ZK-5	$\bar{8}^-$ 0.39	3 D
Linde A	$\bar{8}$ 0.41	3 D
Linde L	$\bar{1}2$ 0.71	1 D
ZSM-11	$\overline{10}$ 0.51x0.55	3 D
ZSM-5	$\overline{10}$ 0.54x0.56 ↔ 10 0.51x0.55	3 D
Mordenite Z	$\bar{1}2$ 0.67x0.70 ↔ $\bar{8}$ 0.29x0.57	1 D
Offretite	$\bar{1}2$ 0.64 (1D) ↔ $\bar{8}$ 0.36x0.52 (2D)	
Rho	$\bar{8}^-$ 0.39x0.51	3 D

The underlined numbers correspond to the number of O in the ring opening. 1 D, 2 D, 3 D represents the dimension of the inter-connecting system.

Table II. Size of cavities of some zeolites

Zeolites	Type of cavities	free dimension (nm)
Chabazite	20-hedron	0.65 x 1.1
Erionite	23-hedron	0.63 x 1.3
Faujasite (X, Y)	Sodalite	0.66
	Supercage (26-hedron)	1.18
Offretite	14-hedron	0.60 x 0.74
Sodalite	Sodalite cage	0.66
ZK-5	18-hedron	0.66 x 1.08
	26-hedron	1.14

Table III. Critical molecular sizes

Molecules	Critical diamter (nm)	Chain length (nm)
n-paraffins		
C_2	0.47	0.53
C_3	0.47	0.65
C_4	0.47	0.78
C_6	0.47	1.03
C_{10}	0.47	1.53
iso-paraffins	0.54	-
aromatics		
benzene	0.68	0.34
toluene	0.68	0.34
p-xylene	0.68	0.34 - 0.40
o-,m-xylene	0.74	0.34 - 0.40
1,3,5-trimethyl-benzene	0.84	0.34 - 0.40
1,3,5-triethyl-benzene	0.92	0.34 - 0.40
1,3,5-triisopropyl-benzene	0.94	0.34 - 0.40
1,2,4,5-tetramethyl-benzene	0.74	0.34 - 0.40
1,2,3,5-tetramethyl-benzene	0.86	0.34 - 0.40
naphtalene	0.69	0.4

A question which obviously arises when alkali exchangeable cations from the synthesis are exchanged by other cations is to know if the latter cations are all at exchangeable locations or if a part of them is not in such locations and therefore may form an occluded compound. It was for instance shown (4) that for calcined Mg-Y and Ca-Y zeolites (T=600°C) a part of the Mg and Ca cations forms small particles of MgO and CaO. X ray diffraction was unable to evidence such small particles but ESR technique of γ-irradiated samples clearly showed their presence. Another example can be found in the Rh-exchanged zeolites by using Rh salts of different size (amines, chlorides...) for zeolites of different pore sizes. It was shown by XPS (5) (analysis of the Rh/(Si+Al) ratios vs the chemical content) that facile using Rh salts of different size (amines, chlorides...) for zeolites of different pore sizes. It was shown by XPS (5) (analysis of the Rh/(Si+Al) ratios vs the chemical content) that facile exchange occured in the large-pore zeolites Y and mordenite. Rh ions were found in the channels and on the surface of the medium-pore zeolites ZSM-34 (offretite-type), ZSM-5 and ZSM-11 suggesting that both exchange and surface hydrolysis occured. On the small pore zeolites (erionite, A, ZK-5) surface hydrolysis was mainly observed with surface Rh deposition of a Rh-oxy-hydroxy hydrate complex.

Reducible cations may be reduced for instance under H_2 atmosphere at 300°C resulting in metallic particles which may be encaged inside the zeolite cavities (usually free diameter ϕ <1nm) or may migrate at the surface of the zeolite grains or be trapped in defects or broken cavities for metallic particle sizes larger than 1 nm (6).Many studies on Pt, Pd, Rh, Ir particles (6-8) on zeolites have been carried out with the aim to prepare a narrow size distribution of very small metallic particles and to study their properties.

Chemistry of Zeolites

The intracrystalline channel cavity-pore-cage system in zeolites is surrounded by the lattice and therefore is submitted to the zeolite crystal field. This results in solvent-like and even electrolyte-type properties. One has seen above how cations could be easily exchangeable. It may also exist an interaction between any occluded ionic compound and the zeolitic framework. Salts, especially salts of univalent anions, have been shown to penetrate the zeolite structure and fill the available space even if the openings of the cavities (as the O_6-ring of 0.24 nm in size in sodalite cage of Y zeolite) is smaller than the size of the anion (ClO_3^-, NO_3^- for instance). The interesting feature is then the enhanced thermal stability of the occluded salt.

A particularly important reaction in zeolites generates acidic OH groups attached to framework cations which are directly responsible for acidic behaviour. Their acidic features include the nature of the sites, their strength and distribution in strength, their concentration and are characterized by many methods (9) such as IR spectroscopy, TPD, NMR, microcalorimetry, Hammett's indicators, etc.

Ionization effects have also been shown in the case of occlusion of alkali metals or gases. For instance alkali metals were reported (10) to reduce NaX and NaY zeolites. Sodium ions capture the electron from the occluded alkali atoms forming in the large cavities Na_6^{5+} and Na_4^{3+} ionized ensembles respectively as evidenced and characterized by ESR. Transition metal ions at exchangeable cation locations may be easily "solvated" resulting in cationic complexes. For instance, Cu^{2+} ions may react with NH_3 or pyridine giving rise to $Cu(NH_3)_4^{2+}$ or $Cu(pyr.)_4^{2+}$ complexes within the zeolite cavities (11). Certain transition metal ions may also ionize NO or NO_2 molecules, forming electron transfer complexes such as (12-13) :

$$Cu^{2+}-Y + NO \rightarrow (Cu^+-NO^+)-Y$$
$$Ni^{2+}-Y + NO \rightarrow (Ni^+-NO^+)-Y$$

All these reactions although, endothermic ($\simeq$ 500 kJ.mol^{-1}) occur at moderate temperature, demonstrating the high affinity of zeolites for ionic species.

Transition metal ion may be reduced or oxidized in zeolites to unusual oxidation state. For instance Pd^+ and Pd^{3+} could be observed in Y-type zeolite by reducing or oxidizing Pd^{2+} (14). Ni^+ ions (d^9) could also be formed for instance in Ni Ca-Y zeolite reduced at 200°C by hydrogen. Reactions of such ions with CO, O_2, NO at ambient temperature result in Ni(I) complexes characterized by ESR and UV-vis. spectroscopies (15).

Transition metal ions complexes may be entrapped into zeolitic cavities. Moreover a desired complex may be directly synthesized into the zeolite cavities. For instance Rh(III) ions in Y-zeolite are converted by CO into a Rh(I) dicarbonyl complex $\frac{CO}{}$ at 2100 and 2040 cm^{-1}) as evidenced by IR and XPS techniques (16). $Rh_6(CO)_{16}$, $Rh_4(CO)_{12}$, $Ir_6(CO)_{16}$ and $Ir_4(CO)_{12}$ are potentially active catalysts for reactions such as hydroformylation of olefins, hydrogenation of olefins and aromatics, etc. Immobilization of these clusters into a zeolite matrix is expected to increase their stability toward aggregation. When Rh exchanged Y zeolite was reacted with $CO:H_2$ equimolecular mixture at 130°C under 80 bars, $Rh_6(CO)_{16}$ was shown to be formed and entrapped within the cavities whereas sublimed $Rh_6(CO)_{16}$ was shown not to enter the same cavities but to sit on the external surface (18). These Rh and Ir carbonyl compounds entrapped within the zeolites exhibited interesting catalytic properties for instance in the hydroformylation of 1-hexene (19). To date however their catalytic behaviour is not good enough to industrially replace the well known homogeneous processes. Further work in that field remains of importance in the future, particularly when one is looking for the heterogeneization of homogeneous catalytic processes.

Zeolite Catalysis

Acidic Catalysis. The interest of zeolites in catalytic reactions has appeared in the early sixties with the observation that small quantities of zeolites incorporated in silica-alumina and silica-clay materials significantly improve the properties of petroleum cracking catalysts (19). The resulting savings to the industry amounts to over several billions dollars a year.

Another property of zeolites is the high conversion rates in the channel system. It was also observed that with different spatial configurations of channels, cavities, windows, etc, the catalytic properties are changed and the selectivity orientates toward less bulky molecules due to limitation in void volume near the active sites or to resistance to diffusivity. This feature termed shape-selectivity, was first proposed by McBain (20) demonstrated experimentaly by Weisz et al (21) and reviewed recently (22). For instance CaA zeolite was observed to give selective dehydration of n-butanol in the presence of more bulky i-butanol (23) while CaX non selective zeolite converted both alcohols. In a mixture of linear and branched paraffins, the combustion of the linear ones was selectively observed on Pt/CaA zeolite (24). Moreover, selective cracking of linear paraffins was obtained from petroleum reformate streams resulting in an improvement of the octane number known to be higher for branched paraffins and for aromatics than for linear paraffins. Shape selectivity usually combines acidic sites within

the zeolite pore structure with the spatial arrangement and
interconnection of the channels and cavities. Industrial
applications of zeolites are summarized in Table IV.

Table IV. Main commercial processes involving zeolites

Process	Catalyst	Benefit
Catalytic cracking	RE-Y, US-Y	High conversion rate
Hydrocracking	Co, Mo, W, Ni on faujasite, mordenite, erionite	High conversion rate
Toluene dispropor-tionation	ZSM-5	Increase in p-xylene yield
Alkylation of aromatics	ZSM-5	Increase in p-xylene yield. Increase in ethyl-benzene yield. Low by-product formation
Xylene isomeri-zation	ZSM-5	Increase in p-xylene yield. Low by-product yield
Dewaxing	Ni/ZSM-5	Increase in octane number
Selectoforming	Ni/erionite	Increase in octane number LPG production
Methanol to gasoline	ZSM-5	High yield of gasoline with high octane number

Basic Catalysis. The catalytic properties of alkali zeolites
free of acidic sites have been investigated for the cracking of
hexanes (25, 26). At 500°C K-Y zeolite cracks easily n-hexane and
its isomers resulting in product distributions markedly different
from those obtained over acidic zeolites or even by thermal
cracking (pyrolysis). Free radical-type mechanism predominates on
the zeolite surface. The relative rates of H atom abstraction
(bimolecular) and ß-scission (unimolecular) are greatly affected
by the zeolite matrix. Zeolites also concentrate hydrocarbon
reactants within the crystal, which enhances the rate of
bimolecular reaction step. Comparison with silicalite (Al-free
ZSM-5 zeolite) and quartz chips has been done in order to
characterize the zeolitic effect. Silicalite behaves as inert
quartz chips with no effect on the rate of H-abstraction step,
i.e. no effect on thermal cracking reaction. It follows that
alkali ions in K-Y zeolites play a role in the rate of ß-scission
while the zeolite matrix itself has only a concentration effect.

Synthesis of zeolites. The chemistry of the synthesis of such materials is very complicated and not well understood . The crystallization of such solids involves first the polymerization of silicate anions in the solution mixture which depends on the pH, on the OH^-/H_2O ratio, on the reactant relative concentration, on the stirring, on the temperature, etc. Usually synthesis occurs under pressure. The synthesis conditions (concentrations of reactants, temperature, pH, duration of synthesis ...) result in different sizes and morphologies of zeolitic grains and in homogeneous or heterogeneous Al distribution along the zeolitic framework and along the grains. It follows that catalytic properties, particularly shape selectivity, and aging by coking, are greatly influenced by such differences. For instance high Si/Al ratios and monomeric Na silicate in the mother solution result in small ZSM-5 microcrystallites while Al rich ingredients and polymeric silica yield large crystals (27). There is a competitive interaction between alkali and organic cations with the alumino-silicate polymeric anions.

The use of organic cations has opened a new field in the synthesis of new zeolites (28-30). These cations are thought to increase the solubility of the silicate and aluminate ions and to arrange the water molecules in a template-like fashion . Using tetramethylammonium cations X and Y type zeolites were synthetized. Alkylammonium (R_4N^+), alkylphosphonium (R_4P^+) and other organic complexes were also used. More recently the use of tetrapropylammonium and tetrabutylammonium cations resulted in the formation of new zeolites designated ZSM-5 (31) and ZSM-11 (32) respectively by Mobil. Their interconnecting channels (see Fig. 1) present a very interesting spatial configuration because of their structure (no cavities), the limited size of the channels ($\simeq$ 0.55 nm) and the three dimensional traffic of molecules. The limited size of the product hydrocarbons (C_n with n < 10) results in low coking and deactivation rates in methanol conversion reactions (33). This is a characteristic of selective zeolites (34).

Modification of Zeolites. The purpose of any modification is to improve catalytic properties. The most common modification is the dealumination of zeolites which has been shown to increase crystalline thermal stability. Three dealumination procedures can be used (35) : chemical dealumination, thermal or hydrothermal dealumination and combination of thermal and chemical dealuminations.

i.   The chemical dealumination (36) may occur under chemical treatment. A suitable reagent in solution (acids, salts, chelating agents) or in the vapor phase ($SiCl_4$) may be used. Hydrochloric acid, chromium chloride, ammonium fluorosilicate (36), EDTA (37), acetylacetone (38) have been employed often. The aluminum is extracted from the lattice, leaving a nest of hydroxyl groups. Further calcination at temperatures as high as 500 or 600°C eliminates these hydroxyl groups and presumably leads to silicon transport to fill the vacancies formed (39) or even leads to a reconstruction of the lattice,

Si atoms replacing the missing Al atoms (40). A novel dealumination procedure was used by Beyer et al (41). It consists on high temperature ($\sim$ 500°C) reaction of the zeolite with a volatile compound, namely $SiCl_4$ vapor. Framework substitution of Si from $SiCl_4$ for Al takes place, while the resulting $AlCl_3$ is volatilized. Mordenite was then less dealuminated (26 %) than faujasite Y ($SiO_2 : Al_2O_3$ from 5 to 46), presumably because of a limited accessibility to $SiCl_4$.

ii. **Hydrothermal Treatment** : the samples are heated at 600-700°C under N_2 flow saturated with water vapor. Al atoms are extracted from the lattice while an Al-O compound is deposited in the cavities or channels. The structure of this Al-O compound is unknown. Species such as $Al(OH)^{2+}$, Al^{3+} (42) hydroxy aluminum acid (43) etc, have been suggested.

iii. **Hydrothermal and Chemical Treatments** : very high extents of dealumination can be reached by carrying out successive hydrothermal and chemical treatments. The chemical reagent is often hydrochloric acid which dissolves the Al-O compound formed using hydrothermal treatment. Operating several times in such a cyclic way, allows one to obtain highly dealuminated and highly stable mordenites (44,45) and faujasites (46).

These dealumination procedures result in Al deficient zeolites of the high thermal stability. In the case of faujasite Y, the so called ultra-stable Y zeolite (US-Y) is used as a catalyst for cracking catalysis (47). It was reported that the total acidity of (48) Al-deficient zeolites was less than that of the parent zeolite but with stronger acidic sites (49). For mordenite, this acidity decreases linearly with Al content (50, 51). However calorimetric measurement of the NH_3 heat of adsorption has shown that when the total number of acidic sites decreases regularly with dealumination, as could be reasonably expected, the strength of the strongest acid sites is enhanced (52).

It could also happen that a mild acid dealumination results in a more active material than the parent zeolite. It was then suggested that the acid treatment has either removed amorphous materials within the channels lowering the diffusion resistance or generated stronger acid sites, which enhanced the catalytic activity or even both. It is interesting to note that a strong dealumination of a mordenite from La Grande Paroisse using the above cyclic procedure results in a material capable of preferentially converting methanol into propene with a high catalyst life time and limited coking (45). It is suggested that the low acidity of the material prevents polycondensation of propene, particularly to alkylated aromatics, which are precursors of coke formation.

Thermal dealumination also tends to decrease the effective pore diameter either by shrinkage of the unit cell and/or by generating amorphous materials in the zeolite channels. This results in an increase of the resistance to diffusivity and an increased shape selectivity. This has opened several new potential applications for zeolite matrices with controlled selectivity.

Chemical Modifications of Zeolites.   Some chemical treatments may
modify the zeolite material without dealumination. The purpose of
such treatments is either to dissolve some amorphous materials
located within the channels or cavities as discussed above or to
incorporate some chemical compound onto active sites within the
channels or cavities or even to artificially introduce amorphous
compound or bulky cations into the zeolite pores or channels. In the
latter two cases it is purposedly desired to reduce the pore volume
or the pore mouth resulting in larger diffusivity resistance and,
subsequently, produce different catalytic properties. In the latter
case the active sites may or may not be modified but the shape
selectivity is expected to be enhanced by coating the inner walls of
the pore and thereby increase resistance to diffusivity. For
instance in the case of ZSM-5 type zeolite many compounds of P (53,
55), Mg (53), B (56), Zn, Sb, Si ... (57) have been introduced,
resulting in an improvement of selective para-xylene formation by
toluene alkylation with methanol. When a phosphorus compound is
introduced using several P compounds e.g. trimethylphosphite, it was
shown by XPS, microcalorimetry and ir techniques (55) that, firstly,
the P compound entered the channels and did not plug or deposit only
on the outer surface of the zeolite grains and secondly the P
compound partially neutralized the catalytically active acidic
hydroxyl groups. The P oxygenated compound was suggested (55) to be
attached to lattice oxygen and to lay within the zeolite channels.
This resulted in an enhanced resistance to diffusivity and a better
yield in para-xylene with respect to the other isomers in toluene
alkylation reaction with methanol (55).

Effect of the Size of the Zeolite Grains : Effect on Shape
Selectivity Properties

When refering to shape selectivity properties related to
diffusivity, it seems obvious that the larger the zeolite grain,
the higher will be the volume/surface ratios and the shape
selectivity, since the reaction will be more diffusion controlled.
The external surface area represents different percents of the total
zeolite area depending on the size of the grains which could be
important if the active sites at the external surface also play a
role in the selectivity. For instance in the case of toluene
alkylation by methanol, the external surface acid sites will favor
the thermodynamical equilibrium due to isomerization reactions
(o:m:p-xylene ≃ 25:50:25 at 400°C) while diffusivity resistance will
favor the less bulky isomer namely the para-xylene. It may therefore
be useful to neutralize the external surface acidity either by some
bulky basic molecules or by terminating the synthesis with some Al
free layers of siliceous zeolite.
     The concept of molecular shape-selective catalysis is based on
the action of catalytically active sites internal to the zeolitic
framework, to diffusivity resistance either to reactant molecules or
to product molecules or to both and to void limitation to reaction
intermediates.This implies an intimate interaction between the
shape, size and configuration of the molecules and the dimension,
geometry and tortuosity of the channels and cages of the zeolite.
Several types of effects exist :

i.   **Reactant Selectivity** occurs if only a part of the reactant
molecules is able to enter and to diffuse freely within the
zeolite structure and therefore to react on the internal active
sites while more bulky molecules do not enter and do not react.
This property is used for shape-selective catalytic cracking,
hydrocracking and selectoforming since linear molecules will be
preferentially cracked with respect to branched molecules.

ii. **Product Selectivity** occurs if reactant molecules can enter
freely into the zeolite framework but from the product
molecules the less bulky ones diffuse much more freely than the
other molecules. This aspect is important for the production of
para-aromatic compounds and for the resistance to deactivation
by coking of zeolites.

iii. **Restricted Transition-State Molecular Shape Selectivity** is
observed when local configuration constraints at the active
sites prevent or decrease the occurence probability of a given
transition state (e.g. bimolecular) of reaction intermediate
and therefore modify the usual reaction mechanism. This holds
true particularly for acid type reactions since tertiary
carbonium ion are more easily formed than secondary and much
more than primary ions. Shape selectivity may then completely
change this chemical behavior and even reverse the order.

This effect is presumably responsible of the low yield in
by-products in xylene isomerization (near-absence of
transalkylation) and is rather beneficial in industrial processes.
The effect of the crystal size on shape selectivity properties
of zeolites has been nicely demonstrated by researchers at Mobil for
ZSM-5 zeolite (53, 55-59). Many examples can be found in ref. 21,
22, 60, 61. Shape selectivity obviously increases with the size of
the grains since the path length of the molecules increases.

Table V. Toluene alkylation by methanol at 400°C. (62)

Zeolite	H-ZSM-5				H-ZSM-11		
grain size (μm)	0.3	0.5-1	1-3	0.4-0.8	0.6-1.2	4-8	1-3[**]
Toluene conversion[*]	19	16	21	26	25	23	20
% xylenes	80	88	86	80	81	81	80
para-xylene	32	52	62	32	25	38	58
meta-xylene	49	37	27	43	53	43	30
ortho-xylene	19	11	11	25	22	19	12

WHSV ≈ 5 h^{-1}

[*] CH_3OH : Toluene = 1:4 molar leading to a maximum conversion of
toluene of 25 %. Methanol is totally converted.

[**] well crystallized twinned crystals kindly supplied by P. Jacobs
(Leuven).

The effect of grain size is not always observed experimentally. Table V reports some data about the alkylation of toluene with methanol as a function of the grain size of ZSM-5 and ZSM-11 zeolite samples. It clearly appears that, as expected, a higher selectivity to para-xylene is obtained when the grain size of the ZSM-5 zeolite increases. This does not seem to hold true for ZSM-11, samples prepared in our laboratory. However, a detailed high resolution electron microscopy (HREM) study of the morphology of the grains has shown (63) that :

i. our ZSM-5 samples are mainly single crystal type or twinned crystals,
ii. our ZSM-11 samples are indeed formed of aggregates of tiny particles (10-50 nm in size) as shown by careful HREM analysis of thin sections of the grains.

It follows that our ZSM-11 samples present actually a much higher actual surface to bulk ratio of zeolite material than ZSM-5 samples for the same "apparent" grain size. This result is supported by an intense IR band at 3740 cm^{-1} due to terminal silanol groups for ZSM-11 samples at the surface of the tiny particles at variance with ZSM-5 crystals which present only a weak band. The low shape selectivity of ZSM-11 samples for para-xylene formation is therefore due to the morphology of the grains (aggregates of tiny crystals and therefore smaller channel length) rather than to the difference in channel tortuosity between ZSM-11 and ZSM-5 zeolites (Fig. 1) as suggested previously (64).

Synthesis of Novel Zeolites

In an attempt to synthetize novel zeolite materials with different catalytic or thermal properties, zeolitic structures related to ZSM-5 zeolite have been obtained. The purpose was either to get better catalytic performances in shape selectivity and resistance to aging or to overcome Mobil patents. In such a way Al free ZSM-5 (designated silicalite) (67), B-ZSM-5 (Al replaced by B in boralite (60) or borosilicates (61)), Fe-ZSM-5 (Al replaced by Fe) (68) zeolites have been prepared. Also, Cr, V, Ge, Zr, Ti, ... were reported to have been used to replace Al or Si in zeolites. Some interesting improvements in catalytic properties have then been claimed. The key question is whether or not the substituted element is the active site or if residual Al is doing all the work !
For obvious reasons much effort has been directed toward introducing elements other than Si and Al into zeolites such as A, X, Y, mordenite. One may expect differences in pore channel and cage systems and in the active site properties. Since zeolites are often considered as a "solvent" for reactants because of electric field features, one may expect some modifications in such "solvent" properties. At the present time the boralites seem to have interesting new properties. A decrease in unit cell dimensions have been observed as a function of B incorporation (66-67) indicating that B is actually incorporated into the framework, which was supported by MASNMR studies of ^{11}B.

Aluminophosphate zeolites with interesting molecular sieve properties have also been synthesized (69) but the pentavalency and trivalency of P and Al respectively do not result in a supplementary negative charge as observed in silico-aluminate zeolite, and subsequently no acidity is expected. However, alumino-silico-phosphate zeolites with variable Al and P contents may well be of interest in the future.

Conclusions and future in catalysis

Zeolite matrices have appeared recently to be of fascinating interest both for industrial and universitary purposes. For the industry many applications, particularly for acidic zeolites, have been found saving billions of dollars each year. For university or basic research the well defined structure of the zeolite materials constitutes a relatively simple model. The main properties of the zeolites in catalysis arise from (70)
- the acidity of Brönsted or Lewis type
- the exchangeable cations (nature, location, accessibility of catalytically active cations)
- the metallic particles entrapped into the zeolitic cavities
- the size of the channel, openings and cavities, which regulates the diffusivity of reactants or products and the size of the intermediate reaction complexes, resulting in "shape selective" properties
The zeolitic framework involving channels, cavities, cages induces very peculiar properties. Due to local crystal and electrical field particular ionization properties are observed with a solvent-like nature.
The commercial use of zeolites depends on their properties controlled by their structural chemistry (their uses as adsorbents and not as catalytic materials are not discussed in this paper), their availability and their cost. The large geological exploration efforts for zeolite deposits throughout the world have probably resulted in the identification of all zeolitic material of commercial significance. The great improvement in synthesis methods has opened a very large field for commercial production of synthetic zeolites. The emergence of new zeolites, of new chemical or physical modifications of "old" zeolites and of zeolite ion exchange processes and applications is promising for the future. Note also that the hydrophobic or hydrophylic properties of zeolites depending on their Al content is important in catalysis for instance in methanol conversion reaction (hydrophobicity increases with dealumination).
The wide possibilities to synthesize and/or to modify all kinds of zeolites such as they present channel networks which differ by their spatial distribution (allowing more or less the circulation of hydrocarbons) and by their size, such as they may bear active sites at different locations (acidic sites, catalytically active transition metal ions, entrapped metallic particles or clusters...) make the zeolite future look bright. At the present time it seems that the zeolites constitute an excellent and promising field of research, where the imagination and the ideas of the chemists may

fully find an area for accomplishment in new zeolites, in new
catalytic reactions and in improved catalytic processes.

Literature Cited

1. Gates, B.C.; Katzer, J.R. and Schuit G.C.A., in " Chemistry of
   Catalytic Processes", Mc Graw Hill, New York, 1979, p. 30.
2. Gallezot, P.; Ben Taârit, Y. and Imelik B., J. Catal., 1972,
   26, 295 and J. Phys. Chem., 1973, 77, 2556.
3. Tikhomirova N.N.; Nikolaeva, I.V.; Demkin, V.V.;
   Rosolovskaya, E.N. and
   Topchieva, K.V., J. Catal. 1973, 29, 105 and 1973, 29, 500.
4. Abou kaïs A., Mirodatos C., Massardier J., Barthomeuf D. and Vedrine
   J.C., J. Phys. Chem., 1977, 81, 397.
5. Shannon R.D., Vedrine J.C., Naccache C. and Lefebvre F., J. Catal.,1984,
   88, 431.
6. Gallezot, P.; Mutin, I.; Dalmai-Imelik, G. and Imelik, B.,
   J. Microsc. Spectr. Electron, 1976, 1, 1.
7. Primet, M. and Ben Taârit, Y., J. Phys. Chem., 1977, 81, 1317.
   Gallezot, P. and Imelik, B., Adv. Chem. Ser., 1971, 121, 66.
8. Kaufherr, N.; Primet, M., Dufaux, M. and Naccache, C., C.R.
   Acad. Sci. Paris, Ser. C, 1978, 286, 131.
   Primet, M., J.C.S. Faraday Trans. I, 1978, 74, 2570.
9. Vedrine, J.C.; Auroux, A. and Coudurier, G., in "Catalytic
   Materials", Whyte T.E. et al, eds., ACS Sympos. Ser., 1984,
   248, 253.
10. Kasai, P.H. and Bishop, R.J., in"Zeolite Chemistry and
    Catalysis", ACS Monograph, Rabo, J.A. ed., 1971, 171, 350.
11. Naccache, C. and Ben Taârit, Y., Chem. Phys. Letters, 1971,
    11, 11.
    Turkevich, J.; Ono, Y. and Soria, J., J. Catal., 1972, 25, 44.
    Vedrine, J.C.; Derouane, E.G. and Ben Taârit, Y., J. Phys.
    Chem., 1974, 78, 531.
12. Rabo, J.A.; Angell, C.L.; Kasai, P.H. and Schomeker, V.,
    Discussions Faraday 1966, 41, 326.
13. Naccache, C., Che, M. and Ben Taârit, Y., Chem. Phys. Letters,
    1972, 13, 109.
    Naccache, C. and Ben Taârit, Y., JCS Faraday Trans. I, 1973,
    69, 1475.
14. Naccache, C.; Primet, M. and Mathieu, M.V., ACS Sympos. Ser.,
    1973, 121, 266.
    Che, M.; Dutel, J.F.; Gallezot, P. and Primet, M., J. Phys.
    Chem., 1976, 80, 2371.
15. Garbowski, E. and Vedrine, J.C., Chem. Phys. Letters, 1977,
    48, 550.
16. Primet M., Vedrine J.C. and Naccache C., J. Mol. Catal., 1978,
    4, 411.
17. Gelin P., Ben Taarit Y. and Naccache C., J. Catal., 1979,
    59, 537.
18. Mantovani E., Palladino N., and Zandri A. J. Mol. Catal.,
    1977, 3, 285.

19. Planck, C.J. and Rosinski, E.J., US patent 3,140,249 (1964) ; 3,140,253 (1964), 3,210,267 (1965), 3,271,418 (1966) and Chem. Eng. Prog. Symp. Ser., 1967, 73, 2.
20. McBain, J.W., The sorption of gases and vapors by solids, Rutledge, London 1952.
21. Weisz, P.B.; Frilette, V.J. and Golden, R.L., J. Catal., 1962, 1, 301.
22. Weisz, P.B., Proceed. 7th Intern. Cong. on Catal., Tokyo, Seiyama, T. and Tanabe, K., eds, Elsevier Scient. Pub. Co, Amsterdam, 1981, 7, 1.
23. Weisz, P.B.; Frilette, V.J.; Maatman, R.W. and Mower, E.B., J. Catal., 1962, 1, 307.
24. Weisz, P.B.; Erdol und Köhle, 1965, 18, 527.
25. Rabo, J.A. and Poutsma, M.L., Adv. Chem. Ser., 1971, 102, 284.
26. Poutsma, M.L. and Schaffer, S.R., J. Phys. Chem., 1973, 77, 158.
27. Gabelica, Z.; Derouane, E.G. and Biom, N., in "Catalytic Materials", Whyte T.E. et al, eds., ACS Sympos. Ser., 1984, 248, 219.
28. Barrer, R.M. and Denny, P.J., J. Chem. Soc., 1961, p. 971. Barrer, R.M.; Denny, P.J. and Flanigen, E.M., US Patent 3,306,922 (1962).
29. Kerr, G.T., J. Inorg. Chem., 1966, 5, 1537 and 1539, US Patent 3,247,195 (1966).
30. Breck, D.W. in "Zeolite Molecular Sieves", Wiley and sons, New York, 1974, p. 305.
31. Argauer, R.J. and Landolt, G.R., US Patent 3,702,886 (1972). Kokotailo, G.T.; Lawton, S.L.; Olson, D.H. and Meier, W.M., Nature, 1978, 272, 437.
32. Chu, P., US Patent 3,709,979 (1972). Kokotailo, G.T.; Chu, P.; Lawton, S.L. and Meier, W.M., Nature, 1978, 275, 119.
33. Dejaifve, P.; Auroux, A.; Gravelle, P.C.; Vedrine, J.C.; Gabelica, Z. and Derouane, E.G., J. Catal., 1981, 70, 123.
34. Derouane, E.G. and Vedrine, J.C., J. Molec. Catal., 1980, 8, 479.
35. Scherzer, J., in "Catalytic Materials", Whyte T.E. et al, eds., ACS Sympos. Ser., 1984, 248, 157.
36. Barrer, R.M. and Makki, M.B., Can. J. Chem., 1964, 42, 1481.
37. Kerr, G.T., J. Phys. Chem., 1968, 72, 2594 and 1969, 73, 2780.
38. Beaumont, R. and Barthomeuf, D., J. Catal., 1972, 26, 218 ; 1973, 27, 45 and 1973, 30, 288.
39. Maher, P.K.; Hunter, F.D. and Scherzer, J., Adv. Chem. Ser., 1971, 101, 266. Gallezot, P.; Beaumont, R. and Barthomeuf, D., J. Phys. Chem., 1974, 78, 1550.
40. Gross, Th.; Lohse, V.; Engelhardt, G.; Richter, K.H. and Patzelova, V., Zeolites, 1984, 4, 25.
41. Beyer, H.K. and Belenykaya, I., in "Catalysis by Zeolites", Imelik, B. et al, eds., Studies in Surf. Sci. and Catal., Elsevier Scient. Publ. Co, Amsterdam, 1980, 5, 203.
42. Kerr, G.T., J. Phys. Chem., 1967, 71, 4155 and J. Catal., 1969, 15, 200.

43. Vedrine, J.C.; Abou-Kaïs, A.; Massardier, J. and
    Dalmai-Imelik, G., J. Catal., 1973, 29, 120.
44. Chen, N.Y. and Smith, F.A., Inorg. Chem., 1976, 15, 295.
45. Bandiera, J.; Hamon, C. and Naccache, C., Proceed. 6th Intern.
    Conf. on Zeolites, Reno, july 1983, in press.
46. Scherzer, J., J. Catal., 1978, 54, 285.
47. McDaniel, C.F. and Maher, P.K., Proceed. Confer. Molec.
    Sieves, Soc. Chem. Ind., London, 1968, p. 186.
48. Mirodatos. C., Ha, B.H.; Otsuka, K. and Barthomeuf, D.,
    Proceed. 5th Intern. Conf. on Zeolites, Napoli, june 1980,
    Rees, L.V.C. ed., Heyden, London, 1980, p. 138.
49. Barthomeuf, D., Molecular Sieves II, ACS Sympos. Ser., 1977,
    40, 453.
50. Eberly, P.E.; Kimberlin, C.N. and Voorhies, J., J. Catal.,
    1971, 22, 419.
51. Vedrine, J.C.; Auroux, A.; Coudurier, G.; Engelhard,
    P.; Gallez, J.P. and Szabo, G., Proceed. 6th Intern. Conf. on
    Zeolites, Reno, july 1983, in press.
52. Auroux, A.; Gravelle, P.C.; Vedrine, J.C. and Rekas, M.,
    Proceed. 5th Intern. Conf. on Zeolites, Napoli, june 1980,
    Rees, L.V.C. ed., Heyden, London, 1980, p. 433.
53. Chen, N.Y.; Kaeding, W.W. and Dwyer, F.G., J. Amer. Chem.
    Soc., 1979, 101, 6783.
54. Kaeding, W.W.; Chu, C.; YOung, L.B.; Winstein, B. and
    Butter, S.A., J. Catal., 1981, 67, 159.
55. Vedrine, J.C.; Auroux, A.; Dejaifve, P.; Ducarme, V.;
    Hoser, H. and Zhou, S.B., J. Catal., 1982, 73, 147.
56. Kaeding, W.W.; Chu, C.; Young, L.B. and Butter, S.A.,
    J. Catal., 1981, 69, 392.
57. Young, L.B.; Butter, S.A. and Kaeding, W.W., J. Catal., 1982,
    76, 418.
58. Haag, W.O.; Lago, R.M. and Weisz, P.B., Faraday Discussion,
    1982, 72, 317.
59. Olson, D.H. and Haag, W.O., in "Catalytic Materials", Whyte
    T.E. et al, eds., ACS Sympos. Ser., 1984, 248, 275 and
    references herein.
60. Csicsery, S.M., in "Zeolite Chemistry and Catalysis",
    Rabo, J.A. ed., ACS Monograph, 1976, 171, 680.
61. Derouane, E.G., in "Intercalation Chemistry", Academic Press
    Inc., New York, 1982, p. 101 and in "Catalysis by Zeolites",
    Imelik, B. et al, eds., Stud. Surf. Sci. Catal., Elsevier
    Scient. Publ. Co, Amsterdam, 1980, 5, 45.
62. Ducarme, V., and Védrine, J.C., submitted to Appl. Catal.
63. Auroux, A.; Dexpert, H.; Leclercq, C. and Vedrine, J.C., Appl.
    Catal., 1983, 6, 95.
64. Derouane, E.G., Dejaifve P., Gabelica Z. and Vedrine J.C.,
    Faraday Discussion 1982, 72, 331.
65. Flanigen, E.M.; Bennett, J.M.; Grose, R.W.; Cohen, J.P.;
    Patton, R.L.; Kirchner, R.L. and Smith, J.V., Nature, 1978,
    271, 512.
    Taramasso, M.; Perego, G. and Notari, B., Proceed. 5th
    Intern. Conf. Zeolites, june, Napoli, Rees, L.V.C. ed.,
    Heyden, London, 1980, p. 40.
66. Taramasso, M.; Manara, G.; Fattore, V. and Notari, B., French
    Patent 2,429,182 (1980).
67. Klotz, M.R., US Patent 4,268,420 ; 4,269,813 (1981).
    Klotz, M.R. and Ely, S.R., US Patent 4,285,919 (1981).

274   SOLID STATE CHEMISTRY IN CATALYSIS

68. Ball, W.J., Palmer, K.W. and Stewart, D.G., European Patent
    Appl., 2899 and 2900 (1979).
69. Wilson, S.T.; Lok, B.M.; Messina, C.A.; Connan, T.R. and
    Flanigen, E.M., J. Amer. Chem. Soc., 1982, 104, 1146 and in
    "Intrazeolite Chemistry", Stucky G.D. and Dwyer, F.G. eds.,
    ACS Sympos. Ser., 1983, 218, 79.
70. Naccache, C. and Ben Taarit Y., Pure and Appl. Chem. 1980,
    52, 2175.

RECEIVED January 4, 1985

The Hydroisomerization Activity of Nickel-Substituted Mica Montmorillonite Clay

R. A. VAN SANTEN, K.-H. W. RÖBSCHLÄGER, and C. A. EMEIS

Koninklijke-Shell-Laboratorium, Amsterdam, Shell Research B.V., Badhuisweg 3, Amsterdam-N, The Netherlands

Three-layer sheet aluminosilicates, when exchanged into the acidic form, are far less active as hydroisomerization catalysts than zeolites having a comparable surface proton density. However, introducing Ni^{2+} or Co^{2+} into the octahedral positions of the Al^{3+} layer in synthetic beidellite results in hydroisomerization catalysts of an activity similar to that of a zeolite. From pyridine poisoning experiments and FT/IR measurements it can be concluded that this increased activity stems from the increased acidity of the resulting NiSMM (Ni-synthetic mica montmorillonite) clay, due to reduction of Ni.
By intercalating the clay with aluminium oligomers "pillared" clays of enhanced surface area have been synthesized. The increase in catalytic activity with enhanced basal surface area indicates that the acidic sites are located in the basal plane of the synthetic clay particles.

There has been renewed interest in catalytically active clays since the report by Swift and Black (1) to the effect that replacement of octahedrally coordinated aluminium ions by nickel or cobalt in synthetic smectite clays, as done by Granquist (2), results in a new type of catalyst, called nickel- (or cobalt-) substituted mica montmorillonite (Ni(Co)SMM), which is very active in the isomerization and cracking of hydrocarbons.
Its activity is comparable to that of zeolites, but - because of its layered structure - it does not contain the small micropores of the zeolite. Thus this catalyst is a potential alternative to zeolites in cases where effects due to pore diffusion have to be avoided.
Earlier novel clay-like systems consisting of silica-alumina-silica layers (2:1 layers) with unit cell composition (3):
$$[(Al_4)octa\ (Al_x\ Si_{8-x})tetra\ O_{20}\ (OH,F)_4]^{x-}\ x\ NH_4^+.\bar{H}_2O$$
with x about 1.5 were synthesized, which had a much lower activity.

0097–6156/85/0279–0275$06.00/0

The alumina layer, in which the aluminium ions are in octahedral positions, is sandwiched between two silica layers with the tetrahedral silicon ions partly replaced by aluminium ions, giving a net negative charge to the 2:1 layers. Since these systems contain both mica-like (non-waterswellable) and montmorillonite-like layers, they have been called synthetic mica montmorillonite, SMM (4). After deammoniation the resulting proton gives SMM its acidic properties.

Much attention has been given to the highly active Ni-substituted mica montmorillonite (Ni-SMM) clays.

The activities of noble-metal-impregnated synthetic clays for the isomerization of pentane are compared in Table I. Pd-NiSMM in its protonic form has been pre-reduced at 350 °C.

Table I. Comparison of the activities of some clays

Clay[8]	Temp. (°C)	Conversion	k^{350} (°C)[c] $g.g^{-1}.h^{-1}$
Hectorite-H[a]	403	2	2×10^{-4}
Beidellite-H[a]	285	2	8×10^{-3}
Ni SMM-H[a]	250	55	2.133
Zeolite	-	-	-
Mordenite-H[7] [b]	260	55	0.73

Stoichiometry: Hectorite-H
$$(Si_8)(Mg_{5.22}Li_{0.48})O_{20}OH_{2.5}F_{1.5}$$
Stoichiometry: Beidellite-H
$$(Si_{5.78}Al_{2.22})Al_4O_{20}OH_{2.5}F_{1.5}$$
Stoichiometry: Ni-SMM-H
$$(Si_{5.26}Al_{2.74})(Al_{1.32}Ni_{4.02})O_{20}OH_{2.5}F_{1.5}$$
Stoichiometry: Mordenite-H
$$(SiO_2/Al_2O_3 = 17)$$
[a] %w Pd: 0.7; surface area 150 m^2/g; WHSV = 2 $g.g^{-1}.h^{-1}$; H_2/C_5 = 1.25.
[b] %w Pd: 0.5; LHSV = 1 $ml.ml^{-1}.h^{-1}$; H_2/C_5 = 1.81.
[c] Calculated rate of C_5 conversion at 250 °C, E_{act} = 25 kcal/mol.

The clays used all belong to the class of smectites. Except for the NiSMM clay their activities are appreciably smaller than those of the zeolites.

The nickel replacement of Al^{3+} ions in the octahedral layers occurs by hydrothermal treatment of a clay reaction mixture in which part of the aluminium ions are replaced by Ni^{2+} ions. It has been found that the nickel ions in the resulting clay occupy octahedral positions, two aluminium ions being replaced by three nickel ions, one nickel ion occupying an originally empty octahedral hole.

Therefore, this new material is a mixed dioctahedral-trioctahedral synthetic clay, in which the relative amount of trioctahedral layers is directly related to the percentage nickel in the finished material.

So far, no satisfactory explanation of its high catalytic activity has been given.   An experiment with Ni^{2+} exchanged on SMM resulted in a very poor hydroisomerization catalyst, which indicated that the high catalytic activity is due to the presence of Ni^{2+} in the clay lattice.   However, it was found that the catalyst is only active after reduction of part of its nickel.

XRD experiments indicate weak lines due to reduced metallic nickel.   The catalytic activity is enhanced by impregnation of the catalyst with a noble metal which has been found to catalyze the reduction of lattice nickel ions (5).

Work by Burch (6) and Sohn and Ozaki (7, 8) indicates that higher dispersed nickel may enter into a special kind of interaction with silica, leading to a high isomerization activity.

Heinerman et al. (5) found a relation between the amount of reduced nickel and the pentane isomerization rate, which suggests a metal-catalyzed reaction.

However, ammonia adsorption experiments in our laboratories (5) demonstrated an enhanced acidity after reduction (Figure 1).   So a dual function mechanism (9, 10), in which metal sites are responsible for the (de)hydrogenation of (alkanes) alkenes and acid sites isomerize the alkenes via a carbocation mechanism, may also explain the high isomerization activity.

With the aid of selective pyridine-poisoning experiments, we will show that isomerization of alkanes over NiSMM is a bifunctionally catalyzed reaction.

The formation of acidic and metallic sites has been investigated in detail at pressures of 2 to 5 Torr (0.3-0.6 kPa) after in-situ reduction of catalyst samples and by FT-IR investigation of pyridine adsorption as a function of temperature.   The amounts of hydrogen chemisorbed proved to be pressure-independent under these conditions.

The paper will be concluded with a discussion of crosslinking experiments of NiSMM samples with alumina oligomers.

Experimental

Catalysts

The NiSMM catalysts used were of the following global composition, as found from elemental analysis:

$(Al_{1.1}Ni_{4.4}octa$ $(Si_{6.6}Al_{1.4})tetra$ $O_{20}(OH)_{3.2}F_{0.8}(NH_4)_{1.4}$.
We used the same synthesis procedure as detailed by Heinerman et al. (5).

A typical reaction mixture used for the preparation of SMM consisted of 88 g SiO_2/Al_2O_3 (25 % Al_2O_3), 67.2 g $Al(isoprop)_3$, 9.1 g NH_4F, and some water.   About 120 g of white product was obtained after hydrothermal treatment (16 h at 300 °C) and XRD confirmed that the product thus formed was pure synthetic beidellite.   Since beidellite is NiSMM without Ni^{2+}, it can also be described as synthetic mica montmorillonite (SMM).   The composition of the clay was:

$(Al_4)octa$ $(Si_{7.3}Al_{0.7})tetra$ $O_{20}(OH)_{3.2}F_{0.9}(NH_4)_{0.7}$
By replacing part of the Al^{3+} by Ni^{2+} in the synthesis mixture and following the same procedure a NiSMM clay was obtained containing Ni in the octahedral layer.   Pt or Pd was exchanged onto the catalysts, using the tetraamine chloride complexes.

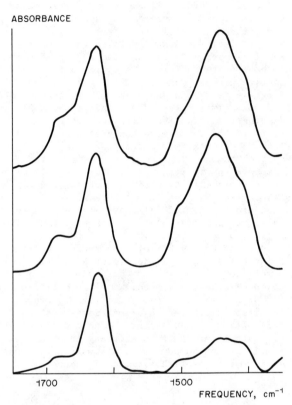

Figure 1.   FT/IR spectra of NH$_3$ adsorbed onto NiSMM
bottom: unreduced
middle: reduced
top   : after removal of zero-valent nickel
. NiSMM was outgassed at 540 $^{\circ}$C. Ammonia was adsorbed at 170 $^{\circ}$C,
  followed by evacuation at 180 $^{\circ}$C (bottom).
. Then ammonia was removed by outgassing at 540 $^{\circ}$C. After reduction in flowing hydrogen at 440 $^{\circ}$C, ammonia was adsorbed as
  described above (middle).
. Then the NiSMM pellet was outgassed at 540 $^{\circ}$C and the reduced
  nickel removed by reaction with CO at about 120 $^{\circ}$C. After outgassing at 540 $^{\circ}$C ammonia was adsorbed as described above (top).

The samples were dried at 120 °C and calcined for 0.5 h at 540 °C, prior to reduction in flowing hydrogen at temperatures from 350 to 450 °C.

The XRD pattern displayed in Figure 2 clearly demonstrates the layered structure of the material obtained. This is confirmed by transmission electronmicroscopic studies (21).

Isomerization Experiments

The catalytic experiments were performed in a conventional microflow reactor, using a few grams of 30-80 mesh catalyst particles. The reaction products were analyzed on-line by GLC with a 100-m squalane capillary column.

Infrared Experiments

Infrared spectra of pyridine adsorbed onto Pd-NiSMM and Pd-SMM were recorded at room temperature with a Digilab FTS 15 C Fourier transform infrared spectrometer.

Hydrogen Chemisorption

The chemisorption of hydrogen was studied in a volumetric apparatus at room temperature and pressures of 2 to 5 Torr (0.3-0.6 kPa) after in-situ reduction of the catalyst samples. The amounts of hydrogen chemisorbed proved to be pressure-independent under these conditions.

Results

Hydroisomerization of Paraffins

In this section we present experimental evidence for a bifunctional alkane isomerization mechanism obtained by selective poisoning of the acidic sites of Pd-NiSMM with pyridine, which was pulse-injected into the liquid hydrocarbon feed stream. The possibility of additional poisoning of the metallic sites was checked by studying the hydrogenation of benzene and the isomerization and ring opening of methylcyclopentane (MCP).

The isomerization of n-hexane at 250 °C, 26 bar, a H_2/hydrocarbon (HC) molar ratio of 22.4, and a WHSV of 3.3 $g.g^{-1}.h^{-1}$ over fresh Pd-NiSMM led to a conversion of 56 %, almost without cracking. This isomerization activity was totally and irreversibly destroyed after injection of about 10^{21} molecules pyridine per g Pd-NiSMM. Benzene hydrogenation over the poisoned catalyst (260 °C, 26 bar, H_2/HC = 25) showed that the hydrogenation function of the catalyst was still active enough to hydrogenate benzene totally to cyclohexane, indicating that the metallic sites had only been partially poisoned, it at all.

Since benzene hydrogenation was still in equilibrium over pyridine-poisoned NiSMM, we studied the conversion of MCP to determine the effect of pyridine on the acidic and metal sites. Under the conditions chosen, the main reaction of MCP is isomerization to cyclohexane (CH). In addition, MCP can undergo ring opening, which

may proceed either acid-catalyzed or metal-catalyzed. A distinction
between these mechanisms can easily be made, since acid-catalyzed
ring opening of MCP leads to the selective formation of n-hexane
(11), while metal-catalyzed ring opening of MCP leads either to the
formation of mainly methylpentanes (MP) or to a statistical cleavage
of all ring bonds, depending on the metal and the state of the metal
(12).
        The reactions of MCP at about 300 °C are displayed in Figure 3.
The initial CH/(CH-MCP) ratio shown indicates that equilibrium isomerization conversion is obtained. The second class of products
found are about 4 % hexanes, resulting from ring opening. The
n-hexane/iso-hexanes ratio found is 0.4. Since this agrees neither
with acid-catalyzed nor with metal-catalyzed ring opening, it indicates a secondary isomerization of the primarily formed ring opening
products or a contribution of both mechanisms. Isomerization is
totally and irreversibly poisoned by means of several pyridine
pulses, while ring opening goes through a minimum and reaches the
former level again when poisoning is discontinued. The ratio of
2-MP/3-MP/n-hexane is now 10:8:1, which is similar to the ratio
reported in the literature for metal-catalyzed ring opening of MCP
over Ni (13, 14). Therefore, it is concluded that MCP ring opening
over NiSMM is catalyzed by the metal sites, which are reversibly
poisoned by pyridine, while isomerization is catalyzed by the acidic
sites, which are irreversibly poisoned by pyridine. Figure 4 neatly
shows the reproducibility of the reversible poisoning of the metal-
catalyzed MCP ring opening over NiSMM. Having shown the importance
of the acidic sites of NiSMM for isomerization and in view of the
previous finding that metal sites are also required for alkane isomerization over NiSMM (5), we conclude that this reaction follows a
bifunctional mechanism.

The Acidic Sites

FT/IR measurements of adsorbed NH_3 on NiSMM catalysts have shown
that the number of sites, most probably Brønsted sites, increases
due to Ni reduction (5). This suggests that the newly formed acidic
sites are responsible for the increased activity of NiSMM compared
with e.g., beidellite, in which no Ni^{2+} is substituted for Al^{3+} in
the octahedral sites. The newly formed sites should then be highly
acidic. In order to verify this statement we have studied the thermal desorption of pyridine from Pd-NiSMM before and after reduction,
by means of FT/IR spectroscopy. For comparison, the same study was
made with Pd-exchanged synthetic beidellite.
        After admission of pyridine and pumping at 20 °C, the catalysts
contained pyridine bound to Brønsted sites (NH bending mode at 1550
cm^{-1}) and to Lewis sites (ring vibrations at 1613 cm^{-1}); see Figure 5. The relative amounts of pyridine desorbed by pumping at 150,
250, and 400 °C are given in Figure 6. The desorption was almost
complete at 400 °C. The desorption temperature of pyridine can be
regarded as a measure of the acid strength of the clay.
        In-situ reduction (16 h at 350 °C) of the sample increases the
number of free (3750 cm^{-1}) and associated (around 3550 cm^{-1}) hydroxyl groups. Pyridine adsorption gives a spectrum (see Figure 5),
which shows five times (case B) as many Brønsted sites as before

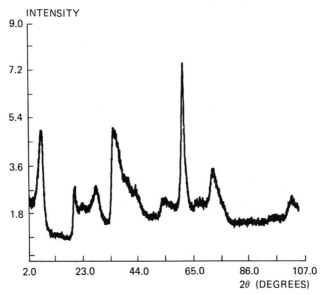

Figure 2. XRD of nickel-substituted mica montmorillonite (NiSMM) prepared according to the procedure detailed in ref. 2.

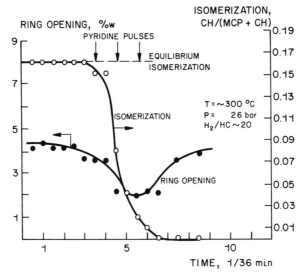

Figure 3. Effect of pyridine on the conversion of MCP. WHSV: 2 g/l.h.

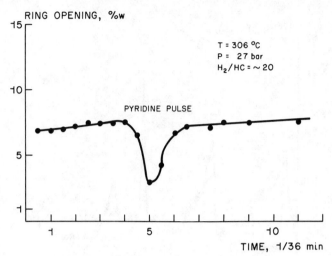

Figure 4. Effect of pyridine on the conversion of MCP, second pyridine pulse.

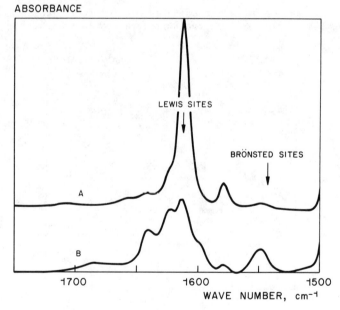

Figure 5. Fourier transform infrared spectra of pyridine adsorbed before (B) and after (A) reduction of Pd-NiSMM.

reduction (case A). The number of Lewis sites does not change much. A shift of a part of the 1613 cm^{-1} band to higher wave numbers may indicate an increase of the strength of the Lewis sites. The desorption experiments show that the newly formed Brønsted sites are indeed strongly acidic, since they adsorb pyridine at higher temperatures (see Figure 6).

The Lewis and Brønsted acidities of Pd-beidellite do not change at all during the reduction treatment. This means that (a) the water formed during Pd reduction does not disturb the balance between Lewis and Brønsted sites, and (b) no new Brønsted sites are formed during reduction of Pd. This implies that after calcination Pd is no longer located in exchange positions and is already metallic.

The number and the average strength of the Brønsted sites are lower for Pd-beidellite than for reduced Pd-NiSMM (Figure 7). Pd-NiSMM contains 15 times as many strong Brønsted sites (defined as sites which adsorb pyridine above 250 °C) as Pd-beidellite. We have measured the activity of Pd-beidellite, too. At 250 °C and 30 bar pressure (H_2/HC = 1.25; WHSV = 2 g.1^{-1}.h^{-1}) we obtained a n-pentane conversion of 1.8 %; 0.9 % was converted to iso-pentane. This corresponds to a k_{isom} of about 1 g.g^{-1}.h^{-1}. Thus the measured isomerization activities qualitatively in agreement with the number of Brønsted sites which adsorb pyridine above 250 °C.

We conclude that most of the highly acidic Brønsted sites which are responsible for the high isomerization activity of NiSMM are formed during the reduction of Ni^{2+} in the octahedral layer of the NiSMM clay.

The Metal Function

The metallic sites of NiSMM are formed by reduction of octahedral lattice Ni^{2+} and of additional Pt or Pd exchanged onto or impregnated on the catalyst prior to calcination and reduction. Pt and Pd are known to catalyze the reduction of Ni (1). Without Pt or Pd, Ni^{2+} can be reduced above 380 °C. The amounts of reduced Ni have been assessed by X-ray diffraction (5). For metal- and bifunctionally catalyzed reactions it is essential to know the dispersion and surface area of the metal.

TEM shows that after 16 h reduction at 440 °C Ni crystallites of sizes from 5 to 15 mm are formed. Hydrogen chemisorption reveals that 0.14 % Ni^0 per gram catalyst adsorbs H_2.

With Pd- or Pt-containing catalysts the problem arises how to discriminate between reduced Ni and the reduced metal. Temperature-programmed reduction experiments (5) have shown that Pd is reduced arond 80 °C. Reduction of Ni starts at 200 to 300 °C. Reoxidation and rereduction point to a possible Pd-Ni alloy formation. We have studied Pd-NiSMM and Pt-NiSMM samples after reduction at 350 and 450 °C by TEM combined with electron microprobe analysis. Metal crystallites with a maximum diameter of 20 nm are formed. Part of them contain Pd and Pt, respectively. Because of the background of lattice Ni^{2+}, reduced Ni is difficult to distinguish by this technique. Since, moreover, large metal crystallites are observed from which no Pd or Pt signal is obtained at all, it seems reasonable to assume that these crystallites are reduced Ni. The presence of alloys cannot be ruled out.

We have also tried to discriminate between reduced Ni and Pd or

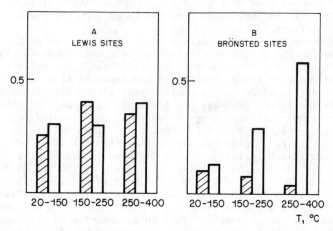

Figure 6. Amount of pyridine desorbed as a function of temperature for reduced (shaded bars) and unreduced Pd-NiSMM (nonshaded bars).

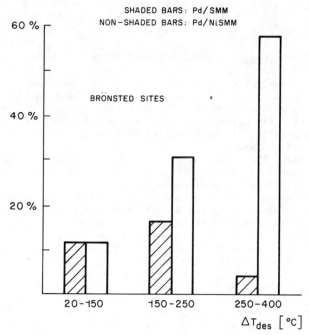

Figure 7. Amount of pyridine desorbed from reduced Pd-NiSMM catalysts relative to Pd-SMM.

Pt by hydrogen chemisorption. Since Ni reduction is slow compared with the reduction of Pd or Pt, we related the amount of H_2 chemisorbed after 1 h reduction at 350 °C to the H/Pt or H/Pd molar ratio. Pd-beidellite as reference shows that Pd reduction is then complete. The measured H/Pt and H/Pd ratios are rather low and range from 0.13 to 0.5, depending on the history of the catalyst (Table II). After prolonged reduction the H_2 chemisorption capacity of the catalysts increases due to Ni reduction. After 16 h reduction at 350 °C, the standard reduction procedure, a substantial part of the metal surface area consists of reduced Ni. The relative contribution of Ni and Pt or Pd to the metal activity of the catalysts further depends on the intrinsic activity of the various metals under the given reaction conditions.

Table II. H_2 chemisorption

Catalyst	Pd-NiSMM (B)[a]	Pd-NiSMM (A)[b]	Pd-SMM (B)[a]	Pd-SMM (A)[b]	Pt-NiSMM (B)[a]	Pt-NiSMM (B)[b]
%w Pd/Pt	0.7	0.7	0.7	0.7	0.73	0.73
Reduction treatment			H_2 chemisorption (mol/g cat.)			
1 h 120 °C	4.8		1.5			
1 h 350 °C		4.4	17.1c	5.3	8.9	11.0
16 h 350 °C	24.2	10.2			15.6	22.9
+2 h 450 °C		15.9				
H/Me ratio[d]		0.13	0.5	0.16	0.5	0.6

[a] Exchange onto dried catalyst.
[b] Exchange onto calcined catalyst.
[c] 0.5 h: 16.9 μmol/g cat.
[d] See text.

Balance between Metal and Acid Functions

In the section on hydroisomerization we have shown that paraffin isomerization over Pd-NiSMM is a bifunctionally catalyzed reaction. The metal and acid functions of the catalyst were characterized in the subsequent sections. For optimization of a bifunctional catalyst it is necessary to know whether the activity is limited by one of the catalytic functions.

It has been shown that over NiSMM without Pd or Pt the rate of n-pentane isomerization depends on the metal function (5). Pt-NiSMM and Pd-NiSMM catalysts are only about 2-5 times more active than pure reduced NiSMM (5), so it is questionable whether the mere addition of Pt or Pd is sufficient to optimize the acidic properties of NiSMM.

Since metallic and acidic sites are both created during reduction of NiSMM, it is very difficult to measure exclusively the influence of the metal function on the bifunctional activity of the catalyst.

For the catalysts listed in Table II we found a correlation

between the H_2 adsorption capacity after 16 h reduction at 350 °C and the activity for n-pentane isomerization after the same reduction procedure (Figure 8). On the basis of this observation one cannot decide whether this increase in activity is due to the increased number of metal sites available or to an increased number of strongly acidic sites formed after prolonged Ni reduction, or both. Therefore it is necessary to have an independent measure of the balance between the metal and the acid function of a bifunctional catalyst.

Such an independent measure is the selectivity of bifunctionally catalyzed consecutive reactions, e.g., the isomerization followed by hydrocracking of n-decane. By model calculations one can show that for a bifunctional catalyst limited in the acid function, the selectivity for intermediate product is high and does not change upon variation of the activity of the metal function. Large changes in selectivity because of such variations are expected for catalysts where neither the metal function nor the acidic function is rate-limiting. If a bifunctional catalyst is really limited in the metal function, this limitation is accompanied by a very low selectivity for intermediate products.

In Figure 9 the yield of iso-decane is plotted against the conversion of n-decane over NiSMM reduced at 450 °C (Ni_s^0 = 0.13 %), a 0.7 %w Pd-NiSMM and a 0.7 %w Pt-NiSMM reduced at 400 °C. The maximum yield of isodecanes is 45 % over NiSMM 60 % over Pd-NiSMM, and reaches the very high value of 80 % over Pt-NiSMM. The large differences in selectivity show that on these bifunctional catalysts the balance between metal and acidic activity is changed and no rate-limiting step exists (15). Perhaps the specific Pt-NiSMM sample tested is close to an ideal bifunctional catalyst with sufficient metal activity to balance the high acidic activity of NiSMM.

These results imply that improvement of both the metal function and the acidic function may lead to the formation of more active catalysts.

Crosslinking experiments with NiSMM clays

Swelling clays can be crosslinked (16-20) with inorganic metal hydroxide oligomers to yield thermally stable, pillared clays with properties reminiscent of zeolites.

We applied this technique to the pillaring of NiSMM, in order to determine the location of the catalytically active sites. If these are located in the basal plane, an enhanced hydroisomerization activity is expected. The NiSMM material prepared according to Swift and Black's (1) procedures turns out not to be water-swellable. An analysis of the energetics of swelling (Figure 10) shows two regions where clays are expected to be non-swellable. One region is found at high Al_{tetra}/Si_{tetra} ratios, where the negative charges on the layers stabilize the cations between the layers better than the the solvation energy of water, the other is found at low Al_{tetra}/Si_{tetra} ratios where the number of cations per layer area is so low that the attractive Van de Waals energy per unit layer area is larger than the gain of solvation energy.

We prepared water-swellable NiSMM by choosing the Si/Al/Ni ratio in the synthesis mixture such that enough, but not too much Al

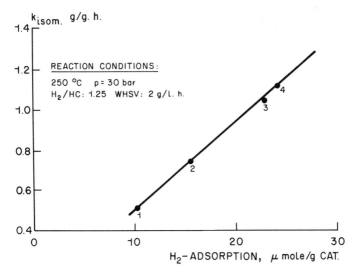

Figure 8. k_{isom} as a function of H_2 adsorption.

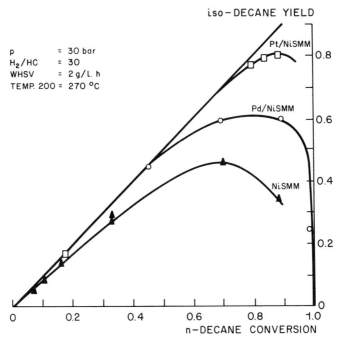

Figure 9.   Yield of iso-decanes versus n-decane conversion for various NiSMM catalysts.

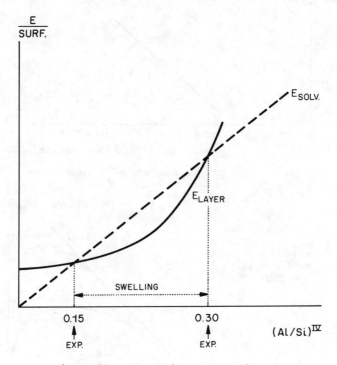

Figure 10.   Energetics of swelling.

is in a tetrahedral coordination. Good results are found for
Al_{tetra}/Si_{tetra} ratios between 0.15 and 0.3.
Crosslinking of such a NiSMM clay with a solution of a hydroxy-
alumina oligomer in H_2O leads to considerable catalytic enhancement
(sample 1, Table III). The table shows the first-order rate constant, k, for catalysis, before and after crosslinking. For sample
1 we used as crosslinking agent a hydroxy-aluminium oligomeric solution, made by refluxing metallic aluminium in 1 m HCl for 8 h and
ageing this solution for at least 10 days. Crosslinking took place
by stirring a slurry of the clay with this solution at 70 °C for
20 h. We used a ratio of 6 mmol Al/g clay in the mixture. The
Brunauer-Emmett-Teller surface area of the clay after calcination at
350 °C had increased from 170 to 230 m^2/g.

Table III.   Hydroisomerization of n-pentane at 250 °C,
using samples reduced at 343 °C[a]

Pd-Ni-SMM sample[b]	X.R.D.[c]		k $(g.g^{-1}.h^{-1})$
Before cross-linking	1.26		2.5
(1)[d]	1.73		3.6
(2)[e]	1.26	1.96	0.3

[a] weighted hourly space velocity = 2 $g.g^{-1}.h^{-1}$;
H_2/feed = 1.25 mol/mol;
$p(H_2)$ = 30 bar;
k = first-order rate constant
[b] stoichiometry of Pd-Ni-SMM: tetrahedral $Si_{6.72} Al_{1.28}$
octahedral $Al_{1.62} Ni_{3.57}$
[c] 001 reflection, nm, after drying at 110 °C
[d] crosslinked with Al oligomer
[e] crosslinked with Si-Al oligomer

The X-ray diffraction (X.R.D.) peak corresponding to a repeat
distance of 1.26 nm for the non-crosslinked sample had been completely replaced by the peak corresponding to the expanded lattice with a
repeat distance of 1.73 nm.
Completely different results were obtained with a NiSMM sample
treated with a solution containing a silica-alumina oligomer. Crosslinking was carried out at a pH of 4.8, using a solution prepared by
refluxing a mixture of chlorohydrol and a sodium silicate solution
for 24 h.
With X.R.D., only a weak signal of the basal spacing of the
expanded structure was detected, whereas the original 001 reflection
at 1.26 nm was still present. The surface area of the sample
treated as such had now decreased to 125 m^2/g.
Transmission electron micrographs (21) show a significant
amount of agglomeration in the basal direction in the product crosslinked with the silica oligomer solution. This probably is the
cause of the decrease in total surface area observed after treatment
with the silica-alumina oligomer solution. The much larger decrease
in catalytic activity (sample 2, Table III) indicates that the major-

ity of the catalytically active sites are located in the lateral
layers or the edges of lateral and basal planes.
The cause of agglomeration is basically the same as that which
induces pillaring of the clays. The positively charged oligomers
exchange with the surface cations. In this particular case the
charge of the oligomers is such that it is not only compensated by
the negative charge in the surface layers of one crystallite, but
can also exchange with charge of another surface.

Discussion and Conclusions

The isomerization of n-alkanes over NiSMM is a bifunctional reaction,
i.e., both the metal sites and the acid sites are involved in the
reaction mechanism.
The high activity of this catalyst can be ascribed to Brønsted
sites of high acidity, which are mainly formed during reduction of
lattice nickel. The acidic activity of NiSMM is so strong that,
even with 0.7 %w Pd or Pt on the catalyst, effects due to too low a
metal activity on the bifunctional activity and selectivity cannot
be excluded.
Since fluorine is contained in NiSMM prepared by the conventional procedure, one may suspect it to be responsible for the
enhanced acidity.
However, an experiment with NiSMM prepared without fluorine
gave the same enhancement after nickel reduction, so that fluorine
cannot be responsible for the increased acidity.
Elsewhere (22) a model for the highly acidic sites is discussed
that is supported by electrostatic potential calculations. According to that model the activity is due to generation of protons coordinated to oxygen ions that connect the silicon-containing tetrahedra with aluminium-containing octrahedra. Such sites, however,
can only contribute to catalysis at lateral planes, at lattice dislocations in the basal plane or at the edges of the lateral and
basal planes.
The dislocations have been observed by TEM and are partially
generated because of nickel reduction. The presence of dislocations
in the basal planes may induce some dependence of the catalytic
activity on the basal-plane surface area.
At present, rearrangement of silica tetrahedra and alumina
octahedra into zeolite-type tetrahedra in the basal plane cannot be
excluded.

Literature Cited

1. Swift, H.E. and Black, E.R., Ind. Eng. Chem., Prod. Res. Dev.
   (1974), 13, 106.
2. Granquist, W.T., U.S. Patents 3 852 405, 1974; 3 929 622, 1975,
   and 3 976 744, 1976.
3. Granquist, W.T., U.S. Patent 3 252 757, 1966.
4. Granquist, W.T., Hoffmann, G.W. and Boteler, R.C., Clays Clay
   Miner. 1972, 20, 323.
5. Heinerman, J.J.L., Freriks, I.L.C., Gaaf, J., Pott, G.T. and
   Coolegem, J.G.F., J. Catal. 1983, 80, 145.
6. Burch, R., J. Catal. 1979, 58, 220.

7. Sohn, J.R. and Osaki, A., J. Catal. 1979, 59, 303.
8. Sohn, J.R. and Osaki, A., J. Catal. 1980, 61, 29.
9. Comradt, H.L. and Garwood, W.E., Ind. Eng. Chem., Prod. Res. Dev. 1964, 308.
10. Weitkamp, J., Erdöl-Kohle-Erdgas, Petrochem.-Brennstoff Chem. 1972, 25, 494.
11. Brouwer, D., in "Chemistry and Chemical Engineering of Catalytic Processes" (Prins, R. and Schmit, G.C.A., Eds), p. 137, Sijthoff-Noordhof, 1980.
12. See, e.g., Maine., G., Plovidy, G., Prudhomme, J. and Gault, F.G., J. Catal. 1965, 4, 556.
13. Miki, Y., Yamadaya, S. and Oba, M., J. Catal. 1977, 47, 278.
14. Roberti, A., Ponec, V. and Sachtler, W.M.H., J. Catal. 1973, 28, 281
15. See Weitkamp, J., Ind. Eng. Chem., Prod. Res. Dev. 1982, 21, 552.
16. Lahav, N., Shani, U. and Shabtai, J., Clays Clay Miner, 1978, 26, 107.
17. Shabtai, J., Chim. Ind. (Milan) 1979.61, 734.
18.. Brindley, G.W. and Yamanaka, S., Am. Mineral. 1979, 64, 830.
19. Endo, T., Mortland, M.M. and Pinnavaia, T.J., Clays Clay Miner. 1981, 29, 153.
20. Vaughan, D.E.W. and Lussier, R.J., in "Proc. for the fifth International Conference on zeolites", Ed. Rees, L.V., Naples, 1980; Heyden, London, Philadelphia, Rheine.
21. Gaaf, J., Van Santen, R.A., Knoester, A. and Van Wingerden, B., J. Chem. Soc., Chem. Commun. 1983, 655.
22. Van Santen, R.A., Recl. Trav. Chim. Pays-Bas 1982, 101, 157.

RECEIVED October 26, 1984

C₁ CATALYSIS

Alkali-Promoted Copper–Zinc Oxide Catalysts for Low Alcohol Synthesis

G. A. VEDAGE, P. B. HIMELFARB, G. W. SIMMONS, and K. KLIER

Center for Surface and Coatings Research and Department of Chemistry, Lehigh University, Bethlehem, PA 18015

Alkali and barium hydroxide promotion of the Cu/ZnO methanol synthesis catalysts has been investigated with the following results: (i) at low temperatures, high H_2/CO ratios, water-free and CO_2-free synthesis gas, the alkali promote methanol synthesis rates in the order Cs>Rb>K>Na>Li; (ii) the alkali hydroxides are highly dispersed on the catalyst surface; (iii) Ba hydroxide agglomerates and shows little effects; (iv) at temperatures above 280°C and low H_2/CO ratios, the synthesis of higher alcohol is promoted by the alkali via a mechanism in which the addition of C_1 intermediate on the beta carbon of the growing alcohol chain dominates; and (v) methyl esters are side products of the higher alcohol synthesis. A method is given for a quantitative treatment of X-ray photoelectron spectra for the determination of the surface concentration of the dopant.

It has been known since the 1930's that the addition of alkali to high temperature (400–450°C), high pressure (200–250 atm) methanol synthesis catalysts promotes the synthesis of higher alcohols and oxygenates from carbon monoxide and hydrogen. One of the earliest studies was conducted by Morgan et al. (1) using Cr_2O_3/MnO catalysts. These investigators concluded that the most effective promoters for higher oxygenates were K, Rb and Cs with the latter being the most active. They postulated the formation of higher alcohols to occur by successive processes of aldolation of the C_1 intermediate with other aldehydic intermediates, followed by partial dehydration and hydrogenation. Graves (2), on the other hand, proposed that higher alcohols originate by a direct dehydrocondensation of two lower alcohol molecules. Natta et al. (3) investigated the promotion effect of alkali on higher alcohol synthesis using several catalysts derived from calcined smithsonite, $ZnCO_3$. These catalysts were found to exhibit poor stability, particularly those containing copper.

The high temperature, high pressure catalysts are of little practical interest today because of the extreme conditions required for

0097–6156/85/0279–0295$06.00/0
© 1985 American Chemical Society

their use. The more recently developed methanol synthesis catalysts
operate under milder conditions at temperatures below 300°C and
pressure below 100 atm. The effects of alkali promoters on the
synthesis of higher alcohols over the low pressure-low temperature
methanol synthesis catalysts have not been studied extensively.
Only recently, Smith and Anderson (4) have systematically studied
the effect of K_2CO_3 on the $Cu/ZnO/Al_2O_3$ catalyst in the synthesis of
higher alcohols. This potassium-promoted catalyst produced higher
alcohols and hydrocarbons with a distribution significantly different
from that of Fischer-Tropsch synthesis. Smith and Anderson developed
a mathematical model of the alcohol chain growth including branching
which did not require a detailed knowledge of the mechanism.
Although such models have proven useful in characterizing and in predicting the product composition over a single catalyst, it is still
desirable to determine the chemical mechanistic features in order to
furnish the catalyst with surface dopants that perform a specific
mechanistic function at an optimum rate.

The principal objective of this study was to address the mechanistic questions including the role of the distribution and chemical
state of the alkali by investigating the influence of different
alkali hydroxides on the activity and selectivity of a low-temperature, low-pressure methanol catalyst. The 30/70 Cu/ZnO methanol
synthesis catalyst that has been previously characterized in terms
of activity, stability, composition, morphology and structure (5,6)
was chosen for this investigation. The product distributions were
determined for each of the alkali-doped catalysts at 250°C and 288°C
with different ratios of H_2/CO and were compared with the product
distribution obtained with undoped catalyst under the same testing
conditions. To identify the precursors that may be involved in the
chain growth to higher alcohols, different alcohols were added to the
H_2/CO synthesis gas. The yield of the alcohols for which the added
alcohol is a precursor is expected to increase. The effects of CO_2
in the syngas on the activity and selectivity of the most active
alkali-doped catalyst were also determined.

The activities and selectivities of the catalysts were correlated
with the composition and microstructure of the catalysts as determined by X-ray powder diffraction, analytical scanning transmission
electron microscopy and X-ray photoelectron spectroscopy.

Results and Discussion

The surface doping of the 30/70 Cu/ZnO catalyst was performed after
the catalyst was reduced in 2% H_2/N_2 at 250°C by an addition and total
evaporation to dryness under flowing nitrogen of an alkali or an
alkaline earth hydroxide solution. In this manner, specimens of the
30/70 Cu/ZnO catalyst were doped with 0.4 atomic % of LiOH, NaOH,
KOH, RbOH, CsOH and Ba(OH)$_2$, and one specimen was doped with 0.2 at.
% Ba(OH)$_2$. These samples were then charged into the reactor under
nitrogen, tested, removed from reactor under nitrogen and subject to
X-ray powder diffraction analysis. The XPS and electron microscopic
analyses were carried out on air-exposed samples. Similar analyses
were carried out on untested catalysts.

An increase in the zinc oxide and copper particle sizes due to
the doping by alkali was apparent from the X-ray powder diffraction

results in Table I. Further evidence for change in the particle sizes
and morphologies of zinc oxide and copper was found by transmission
electron microscopy. For example, comparison of the dark field electron micrograph of ZnO in an undoped catalyst in Figure 1a, with a
similar micrograph of ZnO in a KOH-doped catalyst in Figure 1b, shows
that the dopant did induce growth of ZnO particles. These results
were further corroborated by the BET surface area measurements which
showed that the tested alkali-doped catalysts, except LiOH, 32.6 m^2/g,
had surface areas between 20 and 23 m^2/g, compared to the $Ba(OH)_2$-
doped catalysts with 28-29 m^2/g and the undoped tested catalyst with
36.5 m^2/g. The particle size measurements provide a qualitative evidence that *both* the zinc oxide and the copper particles have grown as
a consequence of doping and testing.

Table I.

The Particle Dimensions of ZnO and Cu as Determined
from X-Ray Diffraction Line Broadening[a]

Tested Catalyst[b]	Dimension (nm)			
	ZnO<10$\bar{1}$0>	ZnO<0002>	ZnO<10$\bar{1}$1>	Cu<111>
Cu/ZnO (30/70)	13.4	15.9	13.0	9.5
LiOH/Cu/ZnO (30/70/0.4)	16.9	19.8	15.2	14.0
Cu/ZnO/NaOH (30/70/0.4)	23.2	20.5	17.4	17.6
Cu/ZnO/KOH (30/70/0.4)	23.1	24.8	20.8	13.7
Cu/ZnO/RbOH (30/70/0.4)	25.2	26.5	22.0	13.7
Cu/ZnO/CsOH (30/70/0.4)	17.4	20.5	15.8	11.6
Cu/ZnO/Ba(OH)$_2$ (30/70/0.4)	14.9	18.5	14.7	13.0
Cu/ZnO/Ba(OH)$_2$ (30/70/0.2)	15.1	18.2	13.8	13.0

[a] The particle sizes were calculated using the Scherrer equation in
the form $t = 0.89\lambda/\cos\theta/(w_x^2 - w_o^2)^{\frac{1}{2}}$ where θ is the Bragg angle, w_x
the half-width of the measured reflection and w_o the instrumental
half-width in radians, cf. J. R. Anderson, "Structure of Metallic
Catalysts," AP 1975, p 366.

[b] Concentrations in terms of overall atomic % (Cu/ZnO/alkali) are
given in parentheses.

The surface analyses of the catalysts were performed in the
Physical Electronics Model 548 X-Ray Photoelectron Spectrometer with
a 400W Mg X-ray source in the required 50eV range with 100 eV pass

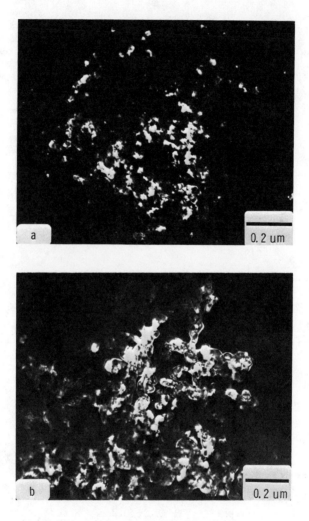

Figure 1. Dark field electron micrographs of ZnO in (a) a tested
Cu/ZnO (30/70 mol%) catalyst and (b) a tested Cu/ZnO/KOH (30/70/
0.4 mol%) catalyst. The dark field images were obtained using the
ZnO(0002) reflection (6).

energy at a scan rate of 0.2eV/s. Multiple scans were averaged by a Nicolet Model 1072 multichannel analyzer and integrated intensities were obtained with the use of a Zeiss MOP-3 automatic integrator. The measured integrated intensities were converted to surface concentration ratios by a method outlined in the Appendix. The results of the XPS analyses are summarized in Table II.

Table II.

X-Ray Photoelectron Data and Analyses

Catalyst[a]	(I_A/I_{Zn})	(X_A^S/X_{Zn}^S)[b]
Tested:		
Cu/ZnO	0.0	0
Cu/ZnO/LiOH	0.00[c]	0
Cu/ZnO/NaOH	0.0640	0.22
Cu/ZnO/KOH	0.0045	0.24
Cu/ZnO/RbOH	0.0084	0.29
Cu/ZnO/CsOH	0.0662	0.22
Cu/ZnOBa(OH)$_2$	0.0145	0.05
Cu/ZnO/Ba(OH)$_2$(0.2)	0.0151	0.05
Untested:		
Cu/ZnO	0.0	0
Cu/ZnO/NaOH	0.0247	0.09
Cu/ZnO/RbOH	0.0023	0.09
Cu/ZnO/CsOH	0.0138	0.05
Cu/ZnO/Ba(OH)$_2$	0.0123	0.04

[a] The molar ratios in all catalysts were Cu/ZnO/MOH = 30/70/0.4 except for the undoped Cu/ZnO sample and the Ba(OH)$_2$-doped sample labeled (0.2) which had the molar composition 30/70/0.2.

[b] Calculated as outlined in Appendix using Equation (A-5).

[c] Lithium was not detected because of the very low photoionization cross section, cf. Table A-I.

The atomic surface concentration ratios (X_A^S/X_{Zn}^S) show that the alkali are accumulated on the surface in the tested Cu/ZnO/MOH (M = Na,K,Rb,Cs) catalysts in approximately equal concentrations. These concentrations are near the expected ones if all the alkali ions are uniformly spread into a submonolayer as is evident from the following example: the atomic concentration of Cs, $X_{Cs}^S = X_{Zn}^S(X_{Cs}^S/X_{Zn}^S)$, where $X_{Zn}^S = 1.1744 \times 10^{19}$ S_{Zn}. Here (X_{Cs}^S/X_{Zn}^S) is the ratio of sur-

face concentrations from Table 2,1.1744×10^{19} atoms/m^2 is the surface density of zinc atoms in the $(10\overline{1}0)$ ZnO face and S_{Zn} is the area
of zinc oxide per gram of catalyst, $S_{Zn} = 10$ m^2/g of catalyst calculated from the total measured BET area 23 m^2/g of the tested Cu/ZnO/
CsOH catalyst and the area of copper, $S_{Cu} = 13$ m^2/g of catalyst obtained from the particle size in Table 1 assuming hemispherical particle shape. The particle shape of the ZnO crystallites is complex
(6) and for this reason was not used in the estimate of S_{Zn}. With
$X_{Zn}^S = 1.1744 \times 10^{20}$ atoms/g of catalyst and $(X_{Cs}^S/X_{Zn}^S) = 0.22$ from
Table II, $X_{Cs}^S = 2.58 \times 10^{19}$ surface Cs atoms per gram of catalyst.
This value is close to the total amount of cesium doped onto the
catalyst, 3.17×10^{19} atoms per gram of catalyst, and shows that over
80% of the cesium dopant is found on the surface. Nearly equal surface concentrations were found on the Na,K,Rb, and Cs- doped *tested*
catalysts. On the contrary, the surface concentrations of these elements on *untested* catalysts were found to be much smaller, indicating
that the alkali hydroxides were first occluded or agglomerated immediately after the doping and then spread into a nearly atomically
dispersed surface overlayer during testing. The barium hydroxide-
doped catalysts showed much lower surface concentration of Ba, however, even after testing, and agglomeration of $Ba(OH)_2$ into large
particles was confirmed by electron microscopic evidence in Figure 2.
The alkali, on the other hand, were dispersed on a scale below the
electron microscope resolution limit of ca. 5nm. The tendency of
the univalent ion hydroxides to spread can be understood in terms of
their low lattice energy compared to that of divalent ion hydroxides,
and it is concluded that more effective catalytic effects will be
generally achieved with alkali rather than alkaline earth compound
doping, primarily because of the agglomeration of the latter.

Although the quantitative XPS analyses may be burdened by a considerable systematic error (8), the results summarized in Table II
demonstrate convincingly that nearly equal surface concentrations of
Na,K,Rb and Cs were achieved by the employed preparation method. The
differences in surface catalytic effects summarized below are therefore caused by the different nature of the alkali ions and not by
their different surface concentrations.

Although methanol was the dominant product over all the presently
studied alkali and alkaline earth-doped Cu/ZnO catalysts as shown in
Figures 3-5, the results demonstrate three effects of the alkali promoters, namely: (i) increase of the activity of the Cu/ZnO catalyst
for methanol production; (ii) change of the CO_2-dependence of the
methanol conversion; and (iii) change of selectivity in favor of the
C_2-C_4 alcohols and esters. The most significant effects were displayed by the CsOH-doped catalyst, indicating that base-catalyzed
reactions are being promoted in the system. The promotion effects
in the $Ba(OH)_2$-doped catalysts were small, most likely due to the
agglomeration of $Ba(OH)_2$ particles shown in Figure 2, which resulted
in a low surface concentration of barium as indicated in Table II.

One of the most striking effects of the alkali hydroxides was the
change of methanol conversion at 250°C in carbon dioxide-free synthesis gas summarized in Figure 3 and in Table III. The CsOH doping had
the greatest effect, with a two-fold enhancement of the synthesis
rate. The promotion effects in the synthesis gas $CO_2/CO/H_2 = 0/30/70$
followed the order

Figure 2. Electron micrograph of a $Ba(OH)_2$ particle in the tested $Cu/ZnO/Ba(OH)_2$ (30/70/0.4 mol%) catalyst. The inset in the lower left corner is the convergent beam (40 nm diameter) diffraction pattern from an area in the $Ba(OH)_2$ crystal which is consistent with that of α-$Ba(OH)_2$ (7).

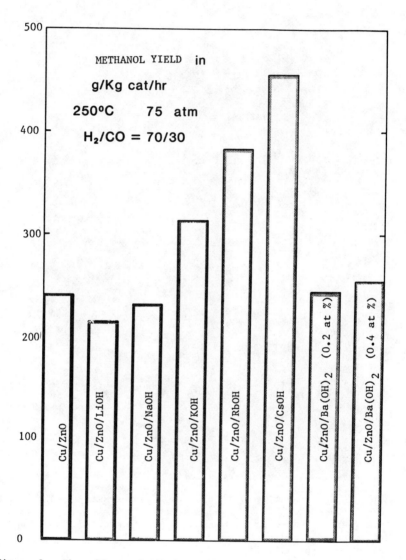

Figure 3.   The effect of alkali on the initial methanol yields at
250°C, 75 atm and a total gas flow of 15.0 l(STP)/hr over 2.45 g
of the impregnated Cu/ZnO = 30/70 mol% reduced catalyst.

Cs > Rb > K > Ba ≃ undoped > Na > Li

with equimolar amounts of the alkali or barium hydroxides added to
the Cu/ZnO catalyst. Further, at 250°C and 75 atm the alkali hydroxides did not significantly alter the high selectivity to methanol of
the Cu/ZnO catalyst.

Table III.

The Effect of CO_2 on Carbon Conversion to Methanol for
the Cu/ZnO/CsOH and Cu/ZnO Catalysts

Temperature 235°C; Pressure 75 atm; Total gas flow =
= 15 liters (STP)/hr over 2.45 g of reduced catalyst.

	% Carbon Conversion to CH_3OH	
Gas Composition	Cu/ZnO/CsOH	Cu/ZnO
$CO_2/CO/H_2$ = 0/30/70	13.7	9.0
$CO_2/CO/H_2$ = 2/28/70	22.4	51.0
$CO_2/CO/H_2$ = 6/24/70	17.0	37.0

Only 2% side products, in decreasing order methyl formate, methane and ethanol, appeared over the CsOH-doped catalyst despite the
two-fold increase in activity, and no side products were observed
over the KOH doped catalyst despite the 30% enhancement of the methanol activity compared to the undoped catalyst. When CO_2 was added
to the synthesis gas, the increase of methanol yield was smaller over
the CsOH/Cu/ZnO catalyst(Table III)than over the undoped Cu/ZnO catalyst (10), resulting in a reversal of the order of activities to
undoped Cu/ZnO>(greater than) Cu/ZnO/CsOH in the $CO_2/CO/H_2$ = 2/28/70
synthesis gas.

The enhancement of the methanol synthesis rate in the CO_2-free
gas and the reversal of this trend in the CO_2-containing synthesis
gas for the CsOH-promoted catalyst is suggested to occur in the following manner. CsOH reacts with carbon monoxide to produce a surface
formate, HCOOCs, which is then hydrogenated to methanol by hydrogen
activated on the Cu/ZnO surface in the intimate neighborhood of the
CsOH sites. This reaction accounts for the methanol synthesis rate
enhancement at low concentrations of CO_2. As the CO_2 concentration
is increased, the normal mechanism involving the whole active surface
of the Cu/ZnO components takes over with the possible withdrawal of
cesium from the action by the formation of surface carbonate. Because the surface area of the Cs-doped catalyst is only a fraction
of that of the undoped catalyst, its overall activity is lower at
intermediate CO_2 concentrations in the synthesis gas. The CsOH surface moiety is more reactive than the other alkali hydroxides because
the large ion radius of Cs^+ makes the OH^- groups the most accessible
as well as the most basic.

The conditions employed in this work at which the synthesis of

higher oxygenates, i.e. alcohols and esters was favored, were temperatures above 280°C, low H_2/CO ratios, and gas hourly space velocities
2600–5000. The results are summarized in Figures 4 and 5. Methyl
esters which were also produced are not shown in these figures. A
formal classification of the points of attack involved in the synthesis of the C_2^+ alcohols and esters is introduced below. The injection of alcohols indicates that higher alcohols and esters were formed from lower alcohols, as documented in Table IV, and therefore our
classification scheme is based upon a stepwise attachment of a C_1
intermediate to a growing alcohol chain. The C_1 intermediate is
assumed to have an intact C-O bond and to be capable of attaching
itself by either the carbon atom or the oxygen atom to the growing
alcohol chain as depicted below.

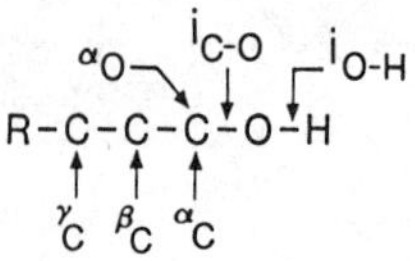

The processes α_C, β_C and γ_C are attachments of the C_1 intermediate
by its carbon at the α, β or γ carbon of the reacting chain, α_O is
the attachment of the C_1 intermediate by its oxygen at the α-carbon,
i_{C-O} and i_{O-H} are insertions of the carbon end of the C_1 intermediate
into the C-O and the O-H bonds of the growing alcohol chain. The
products obtained by these individual processes are shown in the
diagram below with a star label on the carbon atom that originates
from the C_1 intermediate.

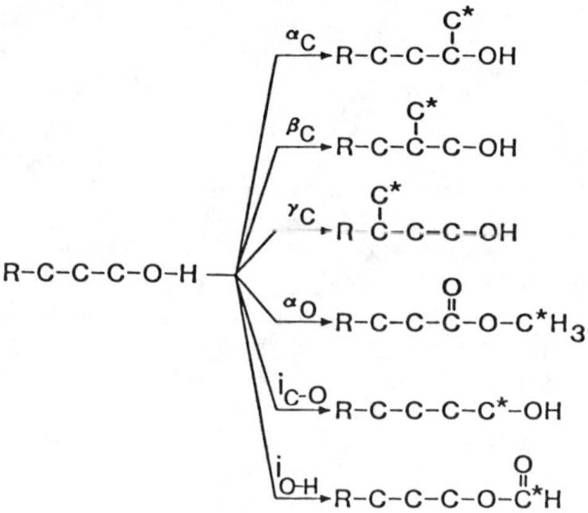

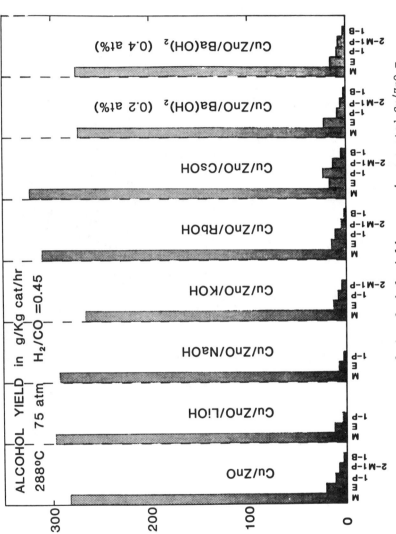

Figure 4. A comparison of the alcohol yields, over impregnated Cu/ZnO = 30/70 mol% catalyst, at 288 C, 75 atm and a total gas flow of 8.0 Z(STP)/hr over 2.45 g of the reduced catalyst. M = methanol, E = ethanol, l–P = 1–propanol, 2–MI–P = 2–methyl 1–propanol, and 1–B = 1–butanol.

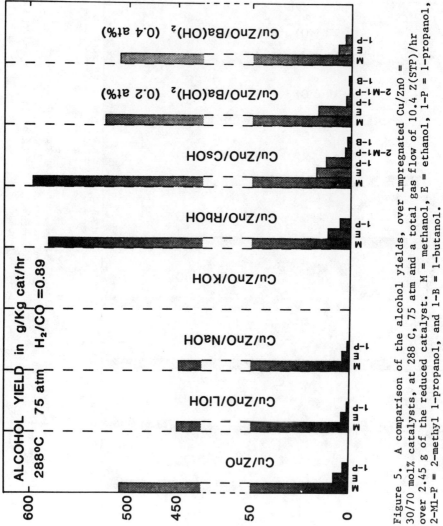

Figure 5. A comparison of the alcohol yields, over impregnated Cu/ZnO catalysts, at 288 C, 75 atm and a total gas flow of 10.4 Z(STP)/hr over 2.45 g of the reduced catalyst. M = methanol, E = ethanol, 1-P = 1-propanol, 2-M1-P = 2-methyl 1-propanol, and 1-B = 1-butanol. 30/70 mol% catalysts, Cu/ZnO = 0.89.

Table IV.

The Effect of Injection of Alcohols on Product Composition

Temperature 288°C; Pressure 75 atm; Feed: H_2/CO = 0.45   Total flow rate of CO/H_2 = 8.0 liters(STP)/hr over 2.45 g of reduced catalyst.   Selectivities are given in carbon atom percent.

Catalyst	Cu/ZnO/Ba(OH)$_2$ (0.2 at% Ba)	Cu/ZnO/Ba(OH)$_2$ (0.2 at% Ba)	Cu/ZnO/NaOH[a]	Cu/ZnO/NaOH[a]
Alcohol added	None	1-Propanol	None	2-Propanol
Mol/hr in feed	0.0000	0.0040	0.0000	0.0039
		Selectivity %		
Methane	3.01	3.95	3.60	1.79
Ethane	0.37	0.47	<0.05	<0.05
Propane	<0.05	<0.05	<0.05	2.77
Methanol	78.57	53.48	93.70	41.72
Ethanol	6.77	3.49	<0.1	1.09
2-Propanol	<0.1	<0.1	<0.1	25.64
1-Propanol	3.38	14.65	<0.1	0.91
2-Methyl 1-Propanol	2.26	11.40	<0.1	<0.1
1-Butanol	1.13	2.56	<0.1	1.79
2-Butanol	<0.1	<0.1	<0.1	13.27
Pentanols	<0.1	6.74	<0.1	10.26
Propanaldehyde	<0.1	0.47	<0.1	<0.1
Methyl formate	1.88	1.16	2.70	0.77
Methyl acetate	2.63	1.16	<0.1	<0.1
Methyl propionate	<0.1	0.47	<0.1	<0.1
Ethyl acetate	<0.1	<0.1	<0.1	<0.1

[a] The Cu/ZnO/NaOH catalyst was slightly deactivated prior to the alcohol additions.

Thus, secondary alcohols would be formed by the α_C attachment, branch-ed primary alcohols by β_C and γ_C, linear alcohols by i_{C-O}, methyl es-ters by α_O, and formate esters by i_{O-H}. The product composition found in this study indicates that α_C and γ_C are negligible, that α_O is more effective than i_{O-H} in the ester formation and that β_C domi-nates the synthesis of the C_3^+ products. The orders of efficiency of the C_1 addition and insertion reactions can also be determined from the product composition and are summarized as follows:

$H_2/CO = 0.45$

Cu/ZnO catalyst: $\beta_C > i_{C-O} > \alpha_O >>> i_{O-H}, \alpha_C, \gamma_C \simeq 0$

CsOH/Cu/ZnO catalyst: $\beta_C >> i_{C-O} > \alpha_O >>> i_{O-H}, \alpha_C, \gamma_C \simeq 0$

$H_2/CO = 0.89$

Cu/ZnO catalyst: $i_{C-O} > \alpha_O > \beta_C >>> i_{O-H}, \alpha_C, \gamma_C \simeq 0$

CsOH/Cu/ZnO catalyst: $i_{C-O} \simeq \beta_C > \alpha_O >>> i_{O-H}, \alpha_C, \gamma_C \simeq 0$

It is evident that the increased H_2/CO ratio suppresses the β_C addi-tion while the CsOH promoter enhances it and that the β_C addition requires a stronger base than the α_O addition or the linear growth i_{C-O}.
    The present finding that the α_C addition is negligible is appar-ent, for example, from the absence of isopropanol in our products. Evidently the α_C attachment is overridden by the α_O, β_C and i_{C-O} pro-cesses under our experimental conditions. The lack of α_C addition is in disagreement with earlier reports on the product composition in low alcohol syntheses(2-4). It is suggested that the effective α_C attachment may have been realized in this earlier work by an isomeri-zation of primary alcohols via a dehydration-rehydration mechanism involving acid centers which are absent in our catalysts.
    Each of the growth steps can materialize by several specific mech-anisms which have not been resolved in detail. We shall give exam-ples of plausible mechanisms for the three dominant processes β_C, α_O and i_{C-O} to account for the dependence of the product composition on the H_2/CO ratio and on the basicity of the alkali hydroxide promoter. The β_C attack dominates at low H_2/CO ratios and in the presence of CsOH, indicating that aldol synthesis of aldehyde precursors or pro-ducts of alcohol dehydrogenation is involved. As shown below, only primary or secondary but not tertiary β-carbon is attacked by the β_C process, a feature characteristic of aldol condensation.

$\boxed{\beta_C}$: $RCH_2CH_2CH_2OH \rightleftharpoons RCH_2CH_2CHO + H_2$

$RCH_2CH_2CHO + CsOH \rightleftharpoons RCH_2\overset{\ominus}{C}HCHO + H_2O$
$\overset{+}{(Cs)}$

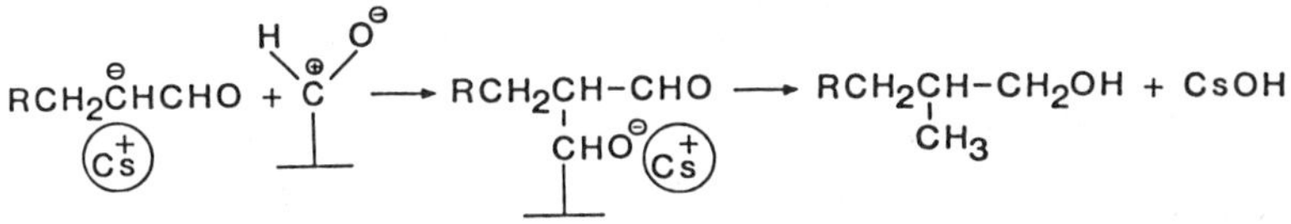

The insertion i_{C-O} may proceed by the mechanism proposed by Natta and coworkers (3) i.e., by the reaction of surface alkoxides with CO as shown below:

$\boxed{i_{C-O}}$:
$$RCH_2O^{\ominus} + CO \longrightarrow RCH_2COO^{\ominus}$$

$$RCH_2COO^{\ominus} + 2H_2 \longrightarrow RCH_2CH_2OH + {}^{\ominus}OH$$

Since this reaction requires hydrogen, it will occur at higher rates when the H_2/CO ratio is increased, as observed.

The methyl ester-forming attack α_O could be due to the Canizzaro-type condensation of two aldehydes, to the reaction of surface carboxylates with formyl or formaldehyde, or to the reaction of surface methoxide with an aldehyde with a hydride elimination. To be consistent with the experimental observations, we shall give an example for the last mechanism because it requires the fewest number of steps, utilizes methoxide which may be favored in hydrogen rich synthesis gas and does not require a strong base catalyst.

$\boxed{\alpha_O}$:

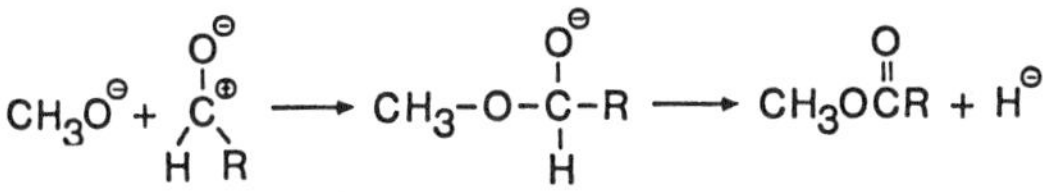

Methyl formate is the product of this reaction when R = H. If methyl formate were the result of the insertion i_{O-H}, one would expect this reaction to produce also higher formate esters, contrary to the finding that only methyl esters were produced.

The principal difference between the aldol addition β_C, the insertion i_{C-O}, and the α_O addition is that the first of these three reaction types requires a strong base catalyst while the remaining two utilize alkoxides which are merely heterolytically dissociated alcohols and are expected to be common surface intermediates in a wide range of H_2/CO ratios and of surface basicity. It is therefore evident that the surface concentrations of the alkali promoters and the H_2/CO ratio can be used as nearly independent variables to achieve selectivities given in Figures 4 and 5.

The selectivity pattern controlled by the processes β_C, i_{C-O} and α_O has an inherent limitation, however, which is apparent from the following consideration. If one wishes to suppress the formation of higher linear alcohols, the insertion i_{C-O} must be minimized but as a consequence also the first C-C bond forming reaction from C_1 to C_2 will be suppressed since neither β_C nor α_O can be involved in this reaction. Hence the suppression of i_{C-O} results in selective methanol synthesis no matter how effective the catalyst may be for the β_C addition. On the other hand, when the i_{C-O} growth mechanism does operate, the β_C addition must be significantly faster in order that the alcohol synthesis may be kinetically restricted so that only low alcohols are formed. If one takes the ratio of (1-butanol:1-propanol) in the product as a crude measure of i_{C-O} and the ratio of (2-methyl-1-propanol: 1-propanol) as a measure of β_C, then it appears from Figures 4 and 5 that the effect of alkali, particularly CsOH, has been to increase the $\beta_C:i_{C-O}$ ratio in comparison to the undoped Cu/ZnO catalyst.

Appendix

Calculations of the surface concentrations of alkali and barium from XPS intensities

The model for the calculation of surface concentrations utilized here is that of Dreiling (11). The measured photoelectron intensity I_i^B due to species i that is evenly distributed over a specimen of thickness t_i is given by

$$I_i^B = K\sigma_i X_i^B g\lambda_i (1-\exp[-t_i/g\lambda_i])/E_i \qquad (A-1)$$

where K is an instrumental constant, σ_i is the photoionization cross section, X_i^B is the atomic volume concentration of element i, g is the escape angular factor, and λ_i is the escape depth of photoelectrons of kinetic energy E_i in the specimen. The intensity I_i^B may be attenuated by a surface overlayer of thickness t_j containing species j by a factor $\exp(-t_j/g\lambda_i)$, which for very thin layers approximates as $(1-t_j/g\lambda_i)$.

Define the atomic surface concentration $X_j^S = X_j^B t_j^o$ where t_j^o is the thickness of a monolayer, and take effective $t_j = t_j^o X_j^S/X_i^S$. Then the attentuated photoelectron signal of species i passing through an overlayer of thickness t_j is given by

$$I_i = K\sigma_i X_i^B g\lambda_i (1-t_j^o X_j^S/(g\lambda_i X_i^S))/E_i \qquad (A-2)$$

where we have taken into account large $t_i/g\lambda_i$. The intensity I_j of photoelectrons from species j in the thin overlayer is given by

$$I_j = K\sigma_j X_j^B t_j^o/E_j = K\sigma_j X_j^S/E_j \qquad (A-3)$$

which follows from (A-1) for small $t_j^o/g\lambda_j$. The relative intensities of the overlayer species j and the bulk species i are, from (A-2) and (A-3),

$$\frac{I_j}{I_i} = \frac{\sigma_j}{\sigma_i}\ \frac{x_j^S}{x_i^S}\ \frac{E_i}{E_j}\ \frac{t_i^o}{g\lambda_i}\ /\,(1 - t_j^o x_j^S /(x_i^S g\lambda_i)) \qquad (A-4)$$

where we used $x_i^B t_i^o = x_i^S$.

The ratio of surface concentrations follows from (A-4).

$$\frac{x_j^S}{x_i^S} = \frac{I_j}{I_i}\ \frac{g\lambda_i}{t_i^o}\ /\ (\frac{\sigma_j E_i}{\sigma_i E_j} + \frac{t_j^o}{t_i^o}\ \frac{I_j}{I_i}) \qquad (A-5)$$

Equation (A-5) permits the intensity I_i of the bulk species to be used as an internal reference for the determination of the surface concentration of the species j that is present in the overlayer only.

In the present work interest centers upon overlayers of alkali compounds on the Cu/ZnO catalyst. The Zn $2p_{\frac{3}{2}}$ photoelectron intensity is taken as the internal reference signal and hence the species i is Zn and species j is an alkali ion. The escape depth λ_{Zn} was estimated from a relation given by Chang (12),

$$\lambda_{Zn} = 0.2\sqrt{E_{Zn}}\,t_{Zn}^o \qquad (A-6)$$

where t_{Zn}^o = 0.2824 nm was taken to be the spacing between the (10$\bar{1}$0) planes of ZnO (13) and E_{Zn} = 232.1 eV, yielding λ_{Zn} = 0.86 nm. Following Dreiling (11), g = 0.75 was used. Other data used in the calculations of the alkali surface concentrations are given in Table A-I.

Table A-I

XPS Data for Photoelectrons of Elements Analyzed in the Present Work

Element i	Photoelectron	E_i (eV)	$\sigma_i{}^a$
Li	1s	1198.1	0.0593
Na	1s	182.2	7.99
K	$2p_{\frac{3}{2}}$	960.7	2.67
Rb	$3d_{\frac{3}{2}}$, $3d_{\frac{5}{2}}$	1143.1	4.44[b]
Cs	$3d_{\frac{5}{2}}$	529.6	22.93
Ba	$3d_{\frac{5}{2}}$	473.6	24.75
Zn	$2p_{\frac{3}{2}}$	232.1	18.01

a
From reference (14).
[b]Sum of photoionization cross sections for the Rb $3d_{\frac{3}{2}}$ and $3d_{\frac{5}{2}}$ photoelectrons. The experimental intensity (cf. Table II) is the integrated intensity of these two photoelectron emissions.

The alkali monolayer thickness t_A^o was taken equal to the sum of the diameters of the OH^- group, 0.272 nm, and of the alkali ion, 0.12 nm for Li^+, 0.19 for Na^+, 0.266 for K^+, 0.296 for Rb^+, 0.338 for Cs^+, and 0.27 for Ba^{++}, yielding $t_{Li}^o = 0.392$, $t_{Na}^o = 0.462$, $t_K^o = 0.538$, $t_{Rb}^o = 0.568$, $t_{Cs}^o = 0.610$ and $t_{Ba}^o = 0.542$ nm. Using these values of t_A^o, g, λ_{Zn} and t_{Zn}^o as noted above, the measured intensity ratios (I_A/I_{Zn}) shown in column 2 of Table II, as well as the parameters listed in Table A-I, the surface concentration ratios (X_A^S/X_{Zn}^S) were calculated from equation (A-5) and summarized in column 3 of Table II.

Acknowledgments

This work was supported by the U.S. Department of Energy (Subcontract No. XX-2-02173 under prime contract No. EG-77-C-01-4092).

References Cited

1. Morgan, G. T., Hardy, D. V. N. and Procter, R. H. J., Soc. Chem. Ind. Trans. and Comm. 51, 1T (1932).
2. Graves, G. D., Ind. and Eng. Chem. 1381 (1931).
3. Natta, G., Colombo, U. and Pasquon, I., "Catalysis," Reinhold, New York, NY, 5, p 131 (1957).
4. Smith, K. J., and Anderson, R. B., "The Higher Alcohol Synthesis over Promoted Methanol Catalysts," presented at the 8th Can. Symp. on Catalysis, May 26-29, 1982, U. of Waterloo, Ont., Canada.
5. Herman, R. G., Klier, K., Simmons, G. W., Finn, B. P., Bulko, J. B. and Kobylinski, T. P., J. Catal. 56, 407 (1979).
6. Mehta, S., Simmons, G. W., Klier, K. and Herman, R. G., J. Catal. 57 (1979).
7. Buch, Von P., Bärnighausen, H., Acta Cryst., B24, 1705 (1968).
8. The systematic errors are likely to originate, in the order of decreasing magnitude, from (a) the uncertainties in λ_{Zn} (if Seah's (9) λ_{Zn} = 1.4 nm were used in the present calculations, there would ensue unrealistic surface alkali concentrations higher by relative 30% than the total amount of the dopant alkali introduced into the system; (b) the values of S_{Zn}; (c) the average escape angular factor g; (d) the estimated thickness t_A^o, and (e) the possibly uneven distribution of the alkali between the surface and the bulk.
9. Seah, M. P., Dench, W. A., Surf. Interface Analysis 1(1), 2(1979).
10. Klier, K., Chatikavanij, V., Herman, R. G., and Simmons, G. W., J. Catal. 74, 343 (1982).
11. Dreiling, M. J., Surface Sci. 71, 231-246 (1978).
12. Chang, C. C., Surface Sci. 48, 9 (1975).
13. Bulko, J. B., Thesis, Lehigh University, 1980.
14. Scofield, S. H., J. Electron Spectroscopy 8, 129 (1976).

RECEIVED October 26, 1984

Water Gas Shift over Magnetite-Based Catalysts

Nature of Active Sites for Adsorption and Catalysis

CARL R. F. LUND[1], JOSEPH E. KUBSH[2], and J. A. DUMESIC[3]

[1] Exxon Research & Engineering Company, Annandale, NJ 08801
[2] Davison Chemical Division-Research, W. R. Grace & Company, Columbia, MD 21044
[3] Department of Chemical Engineering, University of Wisconsin, Madison, WI 53706

Recent studies of the adsorptive and catalytic properties of magnetite (Fe_3O_4) are discussed with respect to
the water-gas shift reaction (WGS) in this review arti-
cle. Proposed mechanisms are examined, and the properties of the active sites required for these mechanisms
are considered. Kinetic relaxation measurements of the
rates of surface oxidation and reduction by WGS reactants and products suggest that a primary WGS pathway
involves successive oxidation and reduction of the magnetite surface. This regenerative mechanism takes place
over a small fraction (ca. 10%) of the catalyst surface.
Surface coverages by CO and CO_2 were determined volu-
metrically and gravimetrically at WGS temperatures
(e.g., 660 K) through the use of CO/CO_2 gas mixtures.
These results suggested that coordinatively unsaturated
iron cations were the sites for CO adsorption, while
adsorption of CO_2 was associated with surface oxygen
species. This was confirmed directly by temperature
programmed desorption studies of magnetite surfaces
under ultra-high vacuum conditions. Studies of a series
of silica-supported magnetite catalysts suggested that
the total extent of adsorption from CO/CO_2 gas mixtures
was proportional to the number of active sites for WGS.
In contrast, the nitric oxide uptake at room temperature
was found to be equal to the BET monolayer uptake of
magnetite. Thus, the ratio of the CO/CO_2 uptake to the
NO uptake provides a measure of the fraction of the
magnetite surface which is active for WGS. Finally,
catalytic effects of solid-state substitutions in magnetite are discussed with respect to the geometric properties of the WGS sites. It is suggested that octahedral-
ly-coordinated iron cations are important for the WGS
reaction.

0097–6156/85/0279–0313$07.50/0
© 1985 American Chemical Society

Magnetite, Fe_3O_4, is the major constituent and the active component in industrial, high temperature (ca. 650 K) water-gas shift (WGS) catalysts (1). The kinetics of the WGS reaction,

$$CO + H_2O \; \rightleftarrows \; CO_2 + H_2 \qquad\qquad (1)$$

have been studied extensively, and several mechanistic interpretations have been developed (1,2). Considerably less information is available concerning the nature of the active site for the reaction. However, recent research in the area has focused upon identifying the surface features of Fe_3O_4 which are responsible for its activity in this reaction.

The solid state structure of magnetite, a spinel(3), contains iron cations in two different oxidation states (Fe^{2+} and Fe^{3+}) and in two lattice sites of different coordination (octahedral and tetrahedral); therefore, the catalytic surface of this material may be expected to provide a variety of possible sites capable of acting as adsorption or reaction centers. Also, it has been demonstrated that substitution of other cations for iron can significantly alter the catalytic activity for WGS (4,5).

The nature of the surface sites for WGS on Fe_3O_4 is the subject of this short review. To begin, the solid state properties of magnetite will be briefly detailed along with proposed reaction pathways for WGS on magnetite-based catalysts. These pathways are discussed with respect to both their kinetic implications and the catalyst surface characteristics which may facilitate these proposed mechanisms. A more detailed investigation of the oxygen transfer characteristics of a magnetite-based catalyst is then presented and discussed with respect to proposed oxidation/reduction pathways for WGS. Chemisorption experiments are then detailed which provide information about the total available magnetite surface area and the density of sites involved in the redox reaction pathway. Finally, solid state substitutions in the magnetite lattice are related to their effect on WGS activity.

Magnetite Structural and Catalytic Properties

Magnetite exists in the spinel structure which can be represented by the formula $(Fe^{3+})[Fe^{2+},Fe^{3+}]O_4$, where the parentheses denote cations in tetrahedral lattice sites, and the brackets denote cations in octahedral lattice sites (3). Figure 1 is a representation of the idealized spinel structure (note that the structure has been extended in the [001] direction for clarity). The oxygen anions form a cubic close-packed framework in which there are 2 tetrahedral vacancies and 1 octahedral vacancy per oxygen anion. From the above formula, it can be seen that one-eighth of the tetrahedral sites and one-half of the octahedral sites are occupied by iron cations. The ordered occupation of octahedral sites shown in Figure 1 facilitates electron hopping between ferrous and ferric cations at temperatures above 119 K(3). As a result, the oxidation state of these octahedral cations can be considered to be +2.5.

Unfortunately, at present, while the bulk structure of magnetite is well understood, little is known about the surface structure of magnetite-based catalysts. In general, the surface may be nonstoichiometric (Fe_3O_{4-x}) with iron cations in both octahedral and tetrahedral sites (6).

The literature pertaining to the catalytic properties of magnetite focuses primarily on the water-gas shift reaction. A number of reaction kinetics studies have been reported in which WGS reaction pathways have been proposed (1,2,7-18). In short, two types of mechanisms have been put forward, these being the adsorptive and regenerative mechanisms. In the adsorptive pathway, reactants adsorb on the surface where they react to form surface intermediates, followed by decomposition to products and desorption from the surface (12-18). Support for this adsorptive mechanism has been provided by tracer studies and apparent stoichiometric number analyses. Two such adsorptive mechanisms consistent with experimental observations are shown below.

$$CO_{(g)} \rightarrow CO_{(ads)} \quad (2a)$$

$$H_2O_{(g)} \rightarrow 2H_{(ads)} + O_{(ads)} \quad (2b)$$

$$CO_{(ads)} + O_{(ads)} \rightarrow CO_{2(ads)} \quad (2c)$$

$$CO_{2(ads)} \rightarrow CO_{2(g)} \quad (2d)$$

$$2H_{(ads)} \rightarrow H_{2(g)} \quad (2e)$$

or

$$CO_{(g)} \rightarrow CO_{(ads)} \quad (3a)$$

$$H_2O_{(g)} \rightarrow OH_{(ads)} + H_{(ads)} \quad (3b)$$

$$CO_{(ads)} + OH_{(ads)} \rightarrow HCOO_{(ads)} \quad (3c)$$

$$HCOO_{(ads)} \rightarrow CO_{2(ads)} + H_{(ads)} \quad (3d)$$

$$CO_{2(ads)} \rightarrow CO_{2(g)} \quad (3e)$$

$$2H_{(ads)} \rightarrow H_{2(g)} \quad (3f)$$

General statements concerning the nature of the adsorption sites required by such adsorptive mechanisms can be suggested based upon the chemical nature of CO, CO_2, H_2O, and H_2 and the manner in which these species have been shown to interact with metal oxide surfaces.

Carbon monoxide, a soft base, is expected to interact with a soft acidic surface site (19). The octahedral iron cations (+2.5 average oxidation state) are the softer of the acid sites on magnetite and may be expected to provide CO adsorption sites. The initial interaction should result in a carbonyl surface species, and such species have been observed by infrared spectroscopy (20-22).

This adsorption mode may undergo an activated transformation, interacting with surface oxygen to form a bidentate carbonate. In contrast, steam is a hard base and prefers to adsorb on hard acid sites. Coordinatively unsaturated Fe^{3+} cations should provide such hard acid sites. The resulting interaction may initially involve a single coordinative bond between the oxygen of water and the cation, but depending on temperature and surface hydroxyl concentration, dissociation of H_2O into surface hydroxyls may occur. This dissociation requires a pair-site consisting of a hard cation and an adjacent oxygen anion.

Carbon dioxide is a hard acid, and as such, it is expected to initially adsorb on hard base sites, presumably oxygen anions and surface hydroxyl groups. However, the adsorbed carbon dioxide can be as carboxylate or carbonate species, and the carbonate may be monodentate or bidentate:

```
    0    0              0    0                    0
     \..·/               \  /                     |
       C                   C                       C
       |                   |                      / \
    M-0-M-0-M           0-M-0-M-0              M-0-M-0-M

                        Monodentate              Bidentate
    Carboxylate          Carbonate               Carbonate
```

The existence of such species has been confirmed by IR spectroscopy for CO_2 interactions with several metal oxide surfaces.(23-26) Finally, hydrogen can adsorb either heterolytically (H+ on an anion and H- on a cation) or reductively (forming two hydroxyl groups) on oxides (23). Depending upon relative surface populations, desorption is possible in each of these fashions. For heterolytic adsorption, pair-sites are again indicated.

The existence of surface intermediates resulting from the interaction of two adsorbates, as postulated, for example, in Step 3c above, has also been confirmed on several metal oxide surfaces (22,29-30). Formate species have been observed by infrared spectroscopy following interaction of CO with a hydroxylated oxide surface. The decomposition of formic acid on a variety of oxides, including iron oxide, was also shown to depend on surface acidity and basicity, with acidic surfaces forming CO and H_2O (WGS reactants) and basic surfaces forming CO_2 and H_2 (WGS products) (31-36). It is clear that adsorptive mechanisms involving formate species are dependent on the acid/base character of the magnetite surface. In general, it appears that both acidic and basic properties are desired to make formate species from CO and H_2O and then to decompose these species to CO_2 and H_2.

In the second class of WGS mechanisms (i.e., regenerative processes), the magnetite catalyst serves as an oxygen transfer agent. In such regenerative mechanisms, oxygen in the catalyst surface and perhaps even the catalyst bulk participates in the reaction. This can be represented by the following equations:

$$CO + *\text{-}0 \underset{k_{CO_2}}{\overset{k_{CO}}{\rightleftarrows}} CO_2 + * \tag{4}$$

$$H_2O + * \underset{k_{H_2}}{\overset{k_{H_2O}}{\rightleftarrows}} H_2 + *\text{-}0 \tag{5}$$

where $*$ denotes a surface site. In this simple form, the regenerative mechanism merely requires a cation site which is capable of reversible oxidation/reduction. It is also possible that the two steps above are simplifications of the actual processes. For example, weakly adsorbed species may alternately oxidize and reduce anion-cation pair-sites, and the transformation to an adsorbed, activated intermediate may be rate-limiting. In this case, the requirements upon the sites are stricter and similar to those described above for the adsorption of CO, CO_2, H_2O and H_2.

The regenerative mechanism fits well with the observed catalytic properties of Fe_3O_4. The octahedral iron cations, which undergo rapid electron hopping, would appear to be the natural cation sites for the reversible oxidation/reduction required by the mechanism, as will be discussed later. Indeed, assuming the sites to be exponentially distributed with respect to oxygen affinity, a rate expression can be derived which fits measured kinetic data well (7). Additional support for this mechanism comes from the work of Boreskov et al.(9,10) who showed that the rate of WGS corresponds to the rate at which H_2O oxidizes and CO reduces the surface of magnetite. This comparison, however, was not carried out over a range of reactant and product partial pressures.

Oxygen Transfer Properties of Magnetite

Recently, the simplified regenerative mechanism has been tested in a more rigorous manner (37). Kinetic relaxation measurements were carried out for each of the two reactions of the regenerative mechanism, equations 4 and 5, using an in situ gravimetric technique. A chromia-promoted magnetite catalyst was equilibrated in either CO/CO_2 or H_2O/H_2 at 637 K. Either the gas phase composition or the temperature was then perturbed, and the kinetics of relaxation to the new equilibrium state were monitored. It can be shown that under these conditions, integration of the kinetic expression gives the following relationship:

$$\frac{n-n_e}{n^0 - n_e} = \exp \{Sr_e[\frac{d}{dx} \ell n (\frac{a_{0-*}}{a_*})_{x_e}]t\} \tag{6}$$

where n is the number of sites which contain oxygen at a given time, the subscript e denotes the value measured after equilibrium is established, the superscript O denotes an arbitrary reference state, S is the catalyst surface area, r_e is the equilibrium exchange rate, and t is the time. The variables a_{o-*} and a_* are the thermodynamic activities of surface oxygen and anion vacancies, respectively. The extent of surface oxygen removal, x, is defined by the equation:

$$x = n^o - n \tag{7}$$

A related quantity, θ_x, is the value of x normalized to the BET monolayer. The chemical activities of surface oxidized sites, a_{o-*} and surface reduced sites, a_* are defined at equilibrium by the equation:

$$\frac{a_{o-*}}{a_*} = K_{CO_2} \left(\frac{P_{CO_2}}{P_{CO}}\right) = K_{CO_2} K_{WGS} \left(\frac{P_{H_2O}}{P_{H_2}}\right) \tag{8}$$

here K_i is the equilibrium constant for either reaction 4 or reaction 5 above, written with species i as the reactant.

Figure 2 shows the experimentally determined extent of oxygen removal as a function of the activity ratio, measured via perturbations of either CO or CO_2 in their mixtures, or H_2 in H_2/H_2O mixtures. It should be noted that both CO and CO_2 adsorb appreciably on the catalyst under the conditions of these experiments, and corresponding corrections to observed weight changes were made in determining x. (A further discussion of CO and CO_2 adsorption on magnetite-based catalysts is presented in the following section.) From this plot, the derivative in equation 6 can be evaluated and thus r_e can be measured from kinetic relaxation experiments. The value of r_e measured at a number of pressure ratios (both CO/CO_2 and H_2/H_2O) can then be used to determine the forward and reverse rate constants of reactions 4 and 5.

Rate constants for these reactions were determined assuming that equations 4 and 5 are mechanistic steps. This analysis also assumes that the surface is uniform. With these assumptions, r_e is related to θ_x according to the expression:

$$\frac{r_e}{P_i} = k_i \left(\theta_{o-*}^{sat} - \theta_x^o\right) - k_i \theta_x \quad (i = CO \text{ or } H_2) \tag{9}$$

for the forward rate of reaction 4 and the reverse rate of reaction 5, and according to

$$\frac{r_e}{P_i} = k_i \theta_x^o + k_i \theta_x \quad (i = CO_2 \text{ or } H_2O) \tag{10}$$

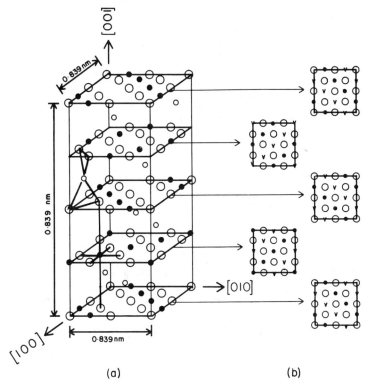

Figure 1. The crystal structure of magnetite: (a) expanded in [001] direction (b) overview of individual anion layers. Large open circles represent oxygen anions; small open circles, tetrahedral cations; small filled circles, octahedral cations; and v's, vacant octahedral sites.

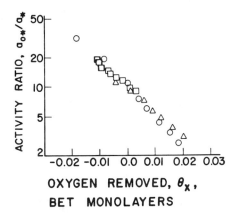

OXYGEN REMOVED, θ_x,
BET MONOLAYERS

Figure 2. Surface oxygen activity ratio as a function of surface oxygen removal measured via gas phase perturbations of (O) CO, (□) CO_2 and (Δ) H_2 pressure. Reproduced with permission from Ref. 37. Copyright 1982, AIChE.

for the remaining two rates. Here k_i refers to the rate constants of the mechanistic steps 4 and 5 and θ^o_x is the value of θ_x at the reference state. Figures 3 and 4 show appropriate plots of equations 9 and 10 from which the rate constants can be determined. Table I presents the rate constants so determined and the values of θ^o_x. In this table, the subscripts refer to the species which is the reactant for a particular rate constant, "sat" refers to the condition where the active sites are all in the oxidized form.

The validity of the measured values of the four rate constants was checked in two ways. First the following relation should be obeyed:

$$\frac{k_{CO} \cdot k_{H_2O}}{k_{CO_2} \cdot k_{H_2}} = K_{WGS} \tag{11}$$

Using the experimentally determined values of the rate constants, K_{WGS} was calculated to be 25.4 by equation 11. The actual value of the equilibrium constant at 637 K is 17.6, and this agreement is acceptable. The second check was performed by predicting both the rate and the kinetic rate expression for WGS and comparing these predictions to reported values.

For a uniform surface, the regenerative mechanism predicts the following kinetic expression for the forward rate of the WGS reaction:

$$r_{WGS} = \frac{k_{CO} k_{CO_2} \theta^{sat}_{ox} P_{CO} P_{H_2O}}{k_{CO} P_{CO} + k_{H_2O} P_{H_2O} + k_{CO_2} P_{CO_2} + k_{H_2} P_{H_2}} \tag{12}$$

Reaction rates predicted by equation 12 were then compared to the rates given by the empirically determined equation:

$$r_{WGS} = 2.0 \times 10^{15} \frac{P_{CO}^{0.9} P_{H_2O}^{0.25}}{P_{CO_2}^{0.6}} \tag{13}$$

This equation was determined using a catalyst similar to that used in the gravimetric study (1). The predicted rate of 2.1×10^{15} m^{-2} s^{-1} (for the case in which the partial pressures of both reactants and products are set equal to 10 kPa) agreed well with the experimental value of 7.3×10^{15} m^{-2} s^{-1}. Furthermore, Figure 5 compares the experimentally determined effects of changing each of the constituent partial pressures from equation 13 to the effects predicted

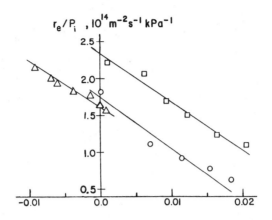

OXYGEN REMOVED, θ_x, BET MONOLAYERS

Figure 3. r_e/P_i versus extent of surface oxygen removal at 637 K for i = CO (○); P_{CO_2} = 19.4 kPa; P_{CO} varied for i = CO (△); P_{CO} = 1.99 kPa; P_{CO_2} varied, and for i = H_2 (□); P_{H_2O} = 2.18 kPa; P_{H_2} varied. Reproduced with permission from Ref. 37. Copyright 1982, AIChE.

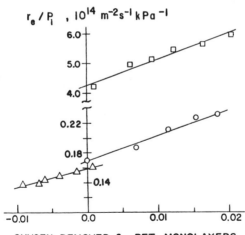

OXYGEN REMOVED, θ_x, BET MONOLAYERS

Figure 4. r_e/Pi versus extent of surface oxygen removal at 637 K for i = CO_2 (○); P_{CO_2} = 19.4 kPa; P_{CO} varied, for i = CO_2 (△); P_{CO} = 1.99 kPa; P_{CO_2} varied, and for i = H_2O (□); P_{H_2O} = 2.18 kPa, P_{H_2} varied. Reproduced with permission from Ref. 37. Copyright 1982, AIChE.

Table I

Summary of Kinetic Parameters[1]

	k_{CO} (m/s)	k_{CO_2} (m/s)	k_{H_2} (m/s)	k_{H_2O} (m/s)	θ_{0*}^{sat}	θ_x^0
CO Pulses	5.6×10^{-8}	3.2×10^{-9}	--	--	0.082	0.049
CO_2 Pulses	4.7×10^{-8}	2.1×10^{-9}	--	--	0.092	0.058
H_2 Pulses	--	--	5.5×10^{-8}	7.0×10^{-8}	0.086	0.050
Average	5.2×10^{-8}	$2.6 \times .10^{-9}$	5.5×10^{-8}	7.0×10^{-8}	0.087	0.052

Source: Reproduced with permission from Ref. 37. Copyright 1982, American Institute of Chemical Engineers.

by equation 12. Qualitatively, the predictions are in agreement with the observed behavior.

These results provide convincing evidence that the regenerative mechanism is an important pathway for WGS over magnetite. The differences between the predicted kinetic behavior and the observed kinetics may be due to any of several possibilities, including a contribution from an adsorptive mechanism or nonuniformity of the catalyst surface. This study also indicates that the active portion of the catalyst surface for the regenerative pathway may be quite small (e.g., ca. 10% of the BET monolayer). Since this study indicated that the reactant gases chemisorb to a significant extent on magnetite surfaces, the adsorption properties of Fe_3O_4 may provide further insight into the details of the active sites, as discussed below.

The Titration of Magnetite Surface Sites

The previous discussions have suggested that anion-cation pair-sites are important actors in both the adsorptive and catalytic characteristics of magnetite. These pair-sites can serve as sites for CO_2, H_2, or H_2O adsorption, and may participate in a regenerative pathway for WGS. We consider first the titration of coordinatively unsaturated iron cations, these cations being potential members of anion-cation pair-sites. Nitric oxide has been reported to adsorb at room temperature on Fe_3O_4 in one-to-one correspondence with the surface iron cations,(38) and thus is useful in this respect. Although the adsorption of NO on unsupported Fe_3O_4 has been shown to be dependent upon sample pretreatment, a standard titration procedure has been identified (39). Specifically, after magnetite samples were reduced in a CO/CO_2 mixture which was chosen on thermodynamic grounds to ensure that Fe_3O_4 was the iron phase, the sample was evacuated at 650 K for 1 h before cooling to 273 K and initiating NO chemisorption. (It was shown that prolonged evacuations at 650 K resulted in apparent NO uptakes which exceeded as much as 3 BET monolayers.) An NO adsorption isotherm was then collected at 273 K. The sample was subsequently evacuated for 0.5 h at 273 K, after which a second isotherm was collected. Extrapolation of the high pressure portions of the two isotherms to zero pressure and subtraction yielded on NO uptake equal to the N_2 uptake in the BET monolayer.

The titration of Fe_3O_4 surfaces with NO is especially useful for supported magnetite (40). A series of silica-supported magnetite catalysts were prepared with various Fe_3O_4 loadings. The standard NO titration scheme presented above was used to measure selectively the magnetite surface areas, from which it was possible to calculate the average Fe_3O_4 particle size for each catalyst preparation. The particle size was then measured by X-ray diffraction line broadening and by low and high field magnetization methods. Agreement between all the methods was satisfactory, as indicated in Table II.

While the adsorption of nitric oxide is very useful for the measurement of magnetite surface areas, it is not necessarily true that this molecule titrates the active sites for the water-gas shift reaction. (This point will be discussed later in this paper.) For this reason, recent studies have focused on the adsorptive properties of magnetite for other molecules, and in particular, on the

Table II

Comparison of Magnetite Particle Size Determinations for Silica-Supported Samples[2]

Sample wt.% Fe_3O_4	Particle Size (nm)				
	NO adsorption		Magnetization		X-ray line-broadening
	Volumetric	Gravimetric	Low Field	High Field	
4.5%	6.5	6.1	7.3	6.2	9.1
9.2%	7.4	--	7.4	6.2	9.6
18.3%	6.6	--	7.7	7.1	9.4

Reproduced with permission from Ref. 40. Copyright 1981, Reaction Kinetics and Catalysis Letters.

adsorption of water-gas shift reactants and products at reaction temperatures. The most detailed study to date has involved the adsorption of CO and CO_2 from CO/CO_2 gas mixtures (41), the composition of these mixtures chosen such that magnetite was the thermodynamically stable phase of iron. Indeed, it is important to note that magnetite would be reduced to metallic iron in pure CO, while it would be oxidized to Fe_2O_3 in pure CO_2 at water-gas shift temperatures.

Three different CO/CO_2 ratios were studied at three different temperatures over chromia-promoted magnetite. Figure 6 presents Langmuir isotherms for the different gas ratios at 637 K. From the small changes in composition of the gas phase upon adsorption, the individual amounts of CO and CO_2 adsorbed were measured independently. Figures 7 and 8 show Langmuir isotherms for CO and CO_2 individually measured in this manner at 637 K in the three CO/CO_2 gas phase compositions. The adsorption behavior shown in these figures was demonstrated to be reversible.

In addition to the observation that the total uptake was only about 20% of the BET monolayer, in agreement with the previously discussed gravimetric studies, the data showed that when the CO_2 pressure was increased at constant CO pressure, the amount of adsorbed CO decreased. Similarly, increasing the pressure of CO decreased the amount of adsorbed CO_2. These results are consistent with adsorption on anion-cation pair-sites, where CO adsorbs on a cation and interacts with a neighboring anion, and where CO_2 adsorbs as a bidentate carbonate species. For competitive adsorption on a fixed number of surface sites, the coverages are given by the following expressions:

$$\theta_{CO} = \frac{K_{CO} P_{CO} \theta_T^{sat}}{1 + K_{CO} P_{CO} + K_{CO_2} P_{CO_2}} \tag{14}$$

$$\theta_{CO_2} = \frac{K_{CO_2} P_{CO_2} \theta_T^{sat}}{1 + K_{CO} P_{CO} + K_{CO_2} P_{CO_2}} \tag{15}$$

where K_{CO} and K_{CO_2} are equilibrium adsorption constants and θ_T^{sat} is the fraction of the BET monolayer composed of adsorption sites. In the case where $K_{CO} \approx K_{CO_2}$, expressions 14 and 15 may be reduced to yield:

$$\frac{P_{CO}}{\theta_{CO}} = \frac{P_{CO_2}}{\theta_{CO_2}} = \frac{1}{K_T \theta_T^{sat}} + \frac{P_T}{\theta_T^{sat}} \tag{16}$$

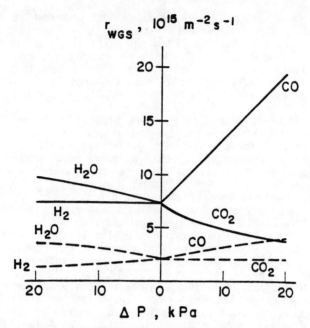

Figure 5. Measured (solid lines) effect of pressure changes on WGS rate compared to effect predicted by oxygen transfer (regenerative) mechanism (dashed lines). ($\Delta P = 0$ corresponds to all P_i = 10 kPa). Reproduced with permission from Ref. 37. Copyright 1982, AIChE.

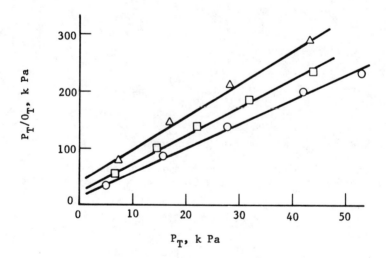

Figure 6. Langmuir isotherms for total CO_2/CO adsorption at (O) - 613 K, (□) - 637 K, and (△) - 663 K. Reproduced with permission from Ref. 41. Copyright 1981, Academic Press.

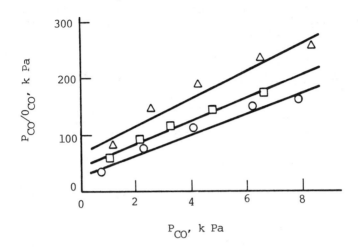

Figure 7. Langmuir isotherms for CO adsorption in $CO_2/CO = 5.45$ mixture at (0) - 613 K, (□) - 637 K, and (Δ) - 663 K. Reproduced with permission from Ref. 41. Copyright 1981, Academic Press.

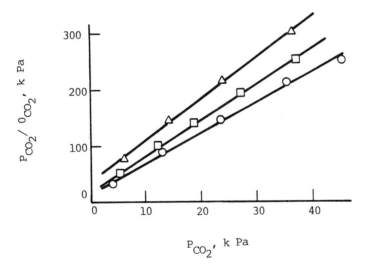

Figure 8. Langmuir isotherm for CO_2 adsorption in $CO_2/CO = 5.45$ mixture at (0) - 613 K, (□) - 637 K, and (Δ) - 663 K. Reproduced with permission from Ref. 41. Copyright 1981, Academic Press.

Figure 9 shows that when plotted appropriately, the isotherms for CO and CO_2 measured with different feed ratios superimpose, consistent with the competitive model described by isotherm, equation 16.

The idea of competitive adsorption on pair-sites has also been used to describe the interaction of H_2 and H_2O with magnetite (42). When isotherms were collected using H_2/H_2O mixtures following the same approach as discussed above, it was found that the data fit a model where H_2 adsorbed dissociatively and H_2O adsorbed associatively, with both species competing for pair-sites. These studies were conducted at water-gas shift reaction temperatures (e.g., 650 K) and as for the adsorption of CO and CO_2, only a fraction of the magnetite surface was capable of adsorbing H_2 and H_2O.

While the competitive model adequately describes the CO_2/CO adsorption data, it is possible that the relationship between CO and CO_2 coverages observed for the magnetite-based catalyst may be due to variations in the oxidation state of the surface. In the previous discussion of the oxygen transfer characteristics of this same catalyst, it was shown that the surface oxygen content was dependent on the CO_2/CO gas phase ratio (Figure 2). The fraction of the surface which participated in redox reactions was about 10% of the BET monolayer. In view of the aforementioned acid/base properties of CO_2 and CO, it is reasonable (41) to correlate θ_{CO_2} with θ_{O-*} (a surface oxygen anion) and θ_{CO} with θ_* (a coordinatively unsaturated cation). Table III summarizes this acid/base-type correlation. Saturation coverages by CO do, in fact, correspond directly to θ_*, while changes in CO_2 saturation coverage correlate with changes in θ_{O-*} for the three P_{CO_2}/P_{CO} ratios examined. A two-site adsorption model was then developed in which the coverages of CO and CO_2 are described by the following isotherm expressions:

$$\frac{\theta_{CO}}{\theta_*} = \frac{K_{CO}P_{CO}}{1 + K_{CO}P_{CO}} \qquad (17)$$

$$\frac{\theta_{CO_2}}{\theta_{O-*} + \phi_0} = \frac{K_{CO_2}P_{CO_2}}{1 + K_{CO_2}P_{CO_2}} \qquad (18)$$

The parameter ϕ_0 was included to account for values of θ_{CO_2} larger than θ_{O-*}, and therefore, represents surface oxygen species capable of adsorbing CO_2 but not capable of participating in the oxygen transfer process.

In view of the above discussion, the following qualitative model may be suggested to describe the regenerative mechanism over magnetite-based catalysts:

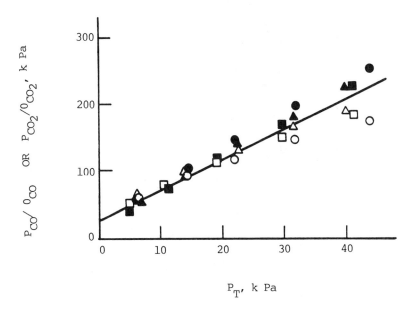

Figure 9. Langmuir isotherms for competitive adsorption at 637 K for CO_2/CO mixtures of (O) - 5.45, (Δ) - 2.37 and (□) - 1.56. Reproduced with permission from Ref. 41. Copyright 1981, Academic Press.

Table III

Summary of Adsorption Site Densities on Chromia-Promoted Magnetite in CO_2/CO Gas Mixtures at 637 K[1]

P_{CO_2}/P_{CO}	Θ_*	Θ_{CO} at Saturation	$\dfrac{\Theta_{CO}^{sat}}{\Theta_*}$	Θ_{O^*}	Θ_{CO_2} at Saturation	$\dfrac{\Delta\Theta_{CO_2}^{sat}}{\Delta\Theta_{O^*}}$
5.45	0.060	0.061	1.02	0.027	0.156	1.31
2.37	0.073	0.080	1.10	0.014	0.139	1.38
1.56	0.086	0.095	1.10	0.001	0.121	--

Source: Reproduced with permission from Ref. 41. Copyright 1981, Academic Press Inc.

$$
\begin{array}{c}
O \\
C \\
|
\end{array}
$$
$$
CO + [O - M] \rightleftharpoons [O - M]
$$

$$
\begin{array}{c}
O \\
C \\
|
\end{array}
$$
$$
[O - M] \rightleftharpoons CO_2 + [\square - M]
$$

$$
\begin{array}{cc}
H & H \\
| & |
\end{array}
$$
$$
H_2O + [\square - M] \rightleftharpoons [O - M]
$$

$$
\begin{array}{cc}
H & H \\
| & |
\end{array}
$$
$$
[O - M] \rightleftharpoons H_2 + [O - M]
$$

where [O - M] is an anion-cation pair-site and $\square$ represents an anion vacancy.

In the first step, CO adsorbs weakly on a coordinatively unsaturated metal cation of an anion-cation pair-site. This weakly adsorbed molecule then reacts with surface oxygen to form CO_2 and an anion vacancy. In the third step, the oxygen atom of water fills the anion vacancy via dissociative adsorption of H_2O. Finally, desorption of H_2 takes place in the fourth step, regenerating an anion-cation pair-site. In view of the arguments presented by Kubsh and Dumesic (37), it is suggested that the above reactions involving surface oxygen and anion vacancies take place with weakly adsorbed water-gas shift reactants and products, such that these reactions can be described by Eley-Rideal kinetics instead of Langmuir-Hinshelwood kinetics.

It is possible, however, that available pair-sites may be blocked by the adsorption of CO_2 or H_2O. For example, CO_2 may adsorb on these sites, forming bidentate carbonate species. Although such adsorbed species may suppress the regenerative mechanism over anion-cation pair-sites, these species may be important for adsorptive WGS mechanisms. The general conclusion, however, is that the adsorption isotherms for CO and CO_2 and for H_2 and H_2O are useful for probing the pair-sites necessary for WGS.

The above ideas that anion-cation pair sites are the surface sites for CO and CO_2 adsorption on magnetite was verified directly by Udovic and Dumesic (43). These authors prepared films of magnetite on polycrystalline iron foils and varied the oxidation state of the surface by vacuum-annealing at different temperatures. In short, it was shown by Auger electron spectroscopy and X-ray photoelectron spectroscopy that the surface became more reduced (i.e., Fe/O and Fe^{2+}/Fe^{3+} atomic ratios both increased) as the vacuum-annealing temperature was increased. Temperature programmed desorption experiments, under ultra-high vacuum conditions, were then performed. The saturation coverage by CO increased as the surface became more reduced, indicating that adsorbed CO was interacting with coordinatively unsaturated cations, as suggested above. In contrast, the saturation coverage by CO_2 decreased as the surface became more reduced. This result would imply that the number of pair-sites was limited not by the number of surface iron cations, but by the number of reactive oxygen species on the surface. The observation of isotopic exchange of ^{18}O between CO_2 and the surface confirmed that the CO_2 was present as bidentate carbonate species. Finally, the saturation coverages by CO_2 and CO were significantly less than one monolayer. This is consistent with the above findings that only a small fraction of the surface is capable of adsorbing these species.

Additional insight into the nature of the adsorption sites for CO_2 was obtained by carrying out temperature programmed desorption after coadsorption of CO_2 and NO. Specifically, it has been noted above that NO adsorbs on iron cations, and this was further confirmed by noting that the saturation coverage by NO increased as the surface of magnetite became more reduced. Importantly, it was then observed that exposure of the magnetite surface to NO caused a decrease in the amount of CO_2 which could be subsequently adsorbed on the surface. Thus, iron cations must be associated with the sites for CO_2 adsorption. Since it had also been shown that CO_2 adsorption is associated with surface oxygen, it was concluded that CO_2 adsorption takes place on anion-cation pair-sites. (This would, in fact, be expected for formation of bidentate carbonate species). It has recently been proposed by Tinkle and Dumesic (44) that this mode of adsorption may not follow the Langmuir isotherm.

Solid State Probes of the Active WGS Sites

The effects of solid state alterations of the magnetite structure on the catalytic activity for WGS provide additional insight into the nature of the active sites. While gravimetric and chemisorptive studies provided a chemical picture of the active sites, a geometric or crystallographic description was lacking. Solid state probes of the active sites have supplied information on this aspect of the mechanism.

It has been shown that the addition of lead to a chromia-promoted magnetite WGS catalyst enhances the activity for WGS (4). A study of the solid state changes which occur upon this substitution was made to probe the active sites of the catalyst. Through a combination of oxidation studies, Mossbauer spectroscopy, and X-ray diffraction line broadening, a model for the catalyst was developed. It was concluded that Pb was present as Pb^{4+} at tetrahedral sites. The Pb substitution resulted in the expansion of the tetrahedral sites, contraction of the octahedral sites, and the oxidation of some Fe^{2+} to Fe^{3+}. The resulting octahedral cations became more covalent in nature, and since the octahedral cations have been reported to be the active sites for CO oxidation over ferrites

(45-48), it was speculated that this increased covalency was responsible for the enhanced activity.

In another study, the effect of silica incorporation into the Fe_3O_4 lattice was studied (5,49,50). A 20% Fe_3O_4 on silica catalyst was prepared using conventional techniques. It was found that while direct oxidation of the catalyst at 800 K produced the expected α-Fe_2O_3, if the catalyst was previously reduced in CO/CO_2 to produce magnetite, then subsequent oxidation resulted in the formation of γ-Fe_2O_3. Figure 10 shows Mössbauer spectra of this catalyst after various thermal treatments. In these spectra, the central doublets were demonstrated to be a result of small iron oxide particles which were superparamagnetic at the conditions where the spectrum was recorded. The suppression of the γ-Fe_2O_3 to α-Fe_2O_3 transition is characteristic of the substitution of foreign cations into the magnetite lattice (4,51).

When a series of silica supported-magnetite catalysts of varying iron oxide particle size were investigated, it was determined that Si substitutes into the magnetite lattice according to the following reaction(49):

$$2SiO_2 + 2(Fe^{3+})[Fe^{2+},Fe^{3+}]O_4 \rightarrow$$
$$(Si_{2,\square}^{4+})[Fe_2^{2+},Fe_4^{3+}]O_{12} \qquad (19)$$

Additionally, it was proposed that reaction 19 occurred only in the outermost 3-4 atomic layers of the magnetite crystallites. The Mössbauer spectrum of the catalyst in the reduced form agreed with this substitution. The spectral parameters of the tetrahedral cations were unaffected by the substitution, whereas the isomer shift and magnetic hyperfine field of the octahedral cations decreased. Also, the line width of the octahedral cations increased relative to an unsubstituted catalyst. Finally, the spectral area ratio of the iron cations in the tetrahedral to octahedral sublattices decreased.

The effect of Si substitution on the turnover frequency for WGS is shown in Figure 11. The turnover frequencies plotted in this figure were based on the magnetite surface area as determined by the NO chemisorption technique. The turnover frequencies shown for unsupported Fe_3O_4 indicate that the factor of 10^3 decline in activity for the silica-supported catalysts is not a particle size effect, but instead is a consequence of the substitution of Si into the lattice. However, when the adsorption of CO/CO_2 at 663 K was used to titrate the surface sites instead of NO, the resulting turnover frequencies were essentially constant as shown in Figure 12. Accordingly, the CO/CO_2 mixture apparently titrates the sites active for WGS. Clearly, the number of active sites is decreased markedly as the particle size decreases in the silica-substituted magnetite catalysts.

The substitution of Si into the magnetite lattice near the surface causes the effective charge of the octahedral cations to increase from its nominal value of 2.5+. Furthermore, because the substitution appears to occur in the outermost surface layers only, the perturbation is greatest for the smallest particles, because in

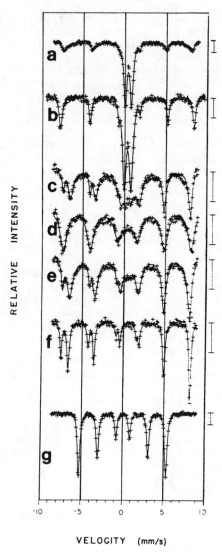

VELOCITY (mm/s)

Figure 10.   Room temperature Mössbauer spectra of 20% Fe_3O_4/SiO_2 catalyst (a) untreated, (b) oxidized (773K), (c) reduced in CO_2/CO (658K), (d) reoxidized (773K), and (e) rereduced (663K) (f) unsupported Fe_3O_4. (g) metallic Fe.   Reproduced from Ref. 49.   Copyright 1981, ACS.

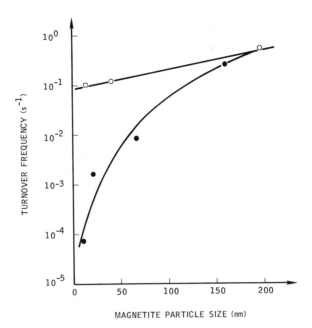

Figure 11. Turnover frequency versus particle size for unsupported (open) and SiO_2-supported (solid) Fe_3O_4. Reproduced with permission from Ref. 5. Copyright 1982, Academic Press.

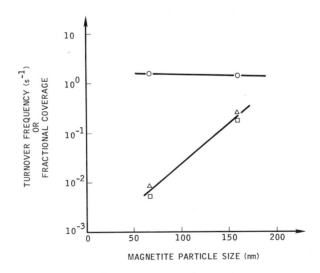

Figure 12. Comparison of turnover frequencies based upon NO (△) and CO/CO_2 (O) adsorption. (□) - ratio of NO to CO/CO_2 uptakes. Reproduced with permission from Ref. 5. Copyright 1982, Academic Press.

SOLID STATE CHEMISTRY IN CATALYSIS

these particles the substituted layers comprise a more significant fraction of the total particle. Because the tetrahedral sites in Si-substituted magnetite do not contain Fe cations and yet the catalytic activity is maintained for the larger magnetite particles, it can be suggested that the tetrahedral sites are not responsible for the WGS reaction. Since according to the regenerative mechanism, the active sites must be capable of both receiving and donating oxygen, the octahedral cations are the natural sites for this reaction, as discussed previously. Upon substitution by Si, these cations become more ferric in nature, thereby perturbing the balance between Fe^{2+} and Fe^{3+} responsible for catalytic activity.

In closing, it is important to note that the CO/CO_2 adsorption technique effectively titrates the active sites for WGS on magnetite catalysts which differ in activity by over an order of magnitude. Nitric oxide on the other hand titrates all of the surface cation sites and is unaffected by Si-substitution. Indeed, NO is known to chemisorb strongly on iron oxides and may even be able to reconstruct the surface. Thus, the combined use of NO and CO/CO_2 adsorption provides information about the total magnetite surface area and fraction of the magnetite surface which is active for the WGS reaction.

Conclusions

This review has focused on recent research directed toward characterization of the active sites for water-gas shift over magnetite-based catalysts. The reaction can be described by a regenerative mechanism wherein gas phase or weakly adsorbed CO reduces anion sites and steam oxidizes the resultant surface oxygen vacancies. Kinetic relaxation techniques indicate this to be a primary pathway. The sites which participate in this reaction comprise only about 10% of the BET monolayer, and these sites can be titrated using CO/CO_2 adsorption at 663 K. In contrast, the total cation site density is effectively titrated with NO at 273 K. In fact, the ratio of the extent of CO/CO_2 adsorption to the extent of NO adsorption provides a measure of the fraction of the magnetite surface which is active for water-gas shift.

Solid state substitution of Si in the magnetite structure occurs only near the surface and at the tetrahedral sites. On large crystallites this substitution alters the catalytic activity slightly, indicating that the tetrahedral sites are not catalytically significant. As the crystallite size decreases, the activity declines in Si-substituted magnetite, and concomitantly the ferric character of the octahedral cations increases. Thus, the regenerative mechanism appears to occur at sites with octahedral iron cations which have an effective 2.5+ charge.

Acknowledgment

The authors thank Martha Tinkle for critical reading of the manuscript and helpful discussion and comments.

Literature Cited

1. Bohlbro, H., "An Investigation on the Kinetics of the Conver-
    sion of Carbon Monoxide with Water Vapour Over Iron Oxide
    Based Catalysts," (2nd ed). Gjellerup, Copenhagen, 1969.
2. Happel, J., Catal. Rev. 1972, 6, 221.
3. Verwey, E. J. W., and Heilmann, E. L., J. Chem. Phys. 1947,
    15, 174.
4. Topsøe, H., and Boudart, M., J. Catal. 1973, 31, 346.
5. Lund, C. R. F., and Dumesic, J. A., J. Catal 1982, 76, 93.
6. Udovic, T. J., Ph.D Dissertation, Univ. of Wisconsin 1982.
7. Shchbrya, G. G., Morozov, N. M., and Temkin, M. I., Kinet.
    Catal. 1965, 6, 955.
8. Wagner, C., Adv. Catal. 1970, 20, 323.
9. Boreskov, G. K., Yur'eva, T. M., and Sergeeva, A. S., Kinet.
    Catal. 1970, 11, 1230.
10. Boreskov, G. K., Kinet. Catal. 1980, 11, 312.
11. Stotz, S., Ber. Bunsenges. Phys. Chem. 1966, 70, 37.
12. Kaneko, Y., and Oki, S., J. Res. Inst. Catal. Hokkaido Univ.
    1965, 13, 55.
13. Kaneko, Y., and Oki, S., J. Res. Inst. Catal. Hokkaido Univ.,
    1965, 13, 169.
14. Kaneko, Y., and Oki, S., J. Res. Inst. Catal. Hokkaido Univ.
    1967, 15, 185.
15. Oki, S., Happel, J. Hnatow, M., and Kaneko, Y., Proc. Int.
    Congr. Catal, 5th, 1, (J. Hightower, editor), 173.
16. Oki, S., and Mezaki, R., J. Phys. Chem. 1973, 77, 447.
17. Oki, S., and Mezaki, R., J. Phys. Chem. 1973, 77, 1601.
18. Mezaki, R., and Oki, S., J. Catal. 1973, 30, 488.
19. Jensen, W. B., Chem. Rev. 1978, 78, 1.
20. Kasatkina, L. A., Nekipelov, V. N. and Zhivotenki, N. N.,
    Kinet. Katal. 1973, 14, 304.
21. Kozub, G. M., Voroshilov, I. G., Roev, L. M., and Rusov, M.
    T., Kinet. Katal. 1976, 17, 903.
22. Rubene, N. A., Davydov, A. A., Kravtsov, A. V., Usheva, N. V.,
    and Smol'yaninov, S. I. Kinet. Katal. 1976, 17, 400.
23. Burwell, R. L., Jr., NBS Special Publication 455, 1970, 155.
24. Knözinger H., Adv. Catal. 1976, 25, 184.
25. Ai, M., J. Catal. 1978, 54, 223.
26. Morterra, C., Ghiotti, G., Boccuzzi, F. and Coluccia, S. J.
    Catal. 1978, 51, 299.
27. Mars, P., Z. Physik. Chem. (Frankfurt) 1959, 22, 309.
28. Ueno, A., Onishi, T. and Tamaru, K., Trans. Faraday Soc. 1970,
    66, 756.
29. Amenomiya, Y., Applied Spectroscopy 1978, 32, 484.
30. Deluzarche, A., Kieffer, R., and Papadopoulos, M., C.R. Hebd.
    Seances Acad. Sci., Ser. C 1978, 287, 25.
31. Ai, M., J. Catal. 1977, 49, 305.
32. Ai, M., J. Catal. 1977, 49, 313.
33. Ai, M., J. Catal. 1978, 50, 291.
34. Ai, M., J. Catal. 1978, 52, 16.
35. Ai, M., J. Catal. 1978, 54, 223.
36. Ai, M., J. Catal. 1978, 54, 426.
37. Kubsh, J. E., and Dumesic, J. A., AICHE J. 1982, 28, 793.

338   SOLID STATE CHEMISTRY IN CATALYSIS

38. Otto, K. and Shelef, M. J. Catal. 1970, 18, 184.
39. Lund, C. R. F., Schorfheide, J. J., and Dumesic, J. A., J.
     Catal. 1979, 57, 105.
40. Kubsh, J. E., Lund, C. R. F., Chen. Y., and Dumesic, J. A.,
     React. Kinet. Catal. Lett. 1981, 17, 115.
41. Kubsh, J. E., Chen, Y., and Dumesic, J. A., J. Catal. 1981,
     71, 192.
42. Tinkle, M., and Dumesic, J. A., J. Phys. Chem., in press.
43. Udovic, T. J. and Dumesic, J. A., J. Catal. in press.
44. Tinkle, M., and Dumesic, J. A., J. Phys. Chem. 1983, 87, 3557.
45. Krishnamurthy, T. R., Viswanathan, B., and Sastri, M.V.C. J.
     Res. Inst. Catal. Hokkaido Univ. 1977, 24, 219.
46. Schwab, G.-M., Roth, E., Grintzos, C., and Mavrakis, N.
     "Structure and Properties of Solid Surfaces." University of
     Chicago Press, Chicago 1953.
47. Popovskii, V. V., Boreskov, G. K., Dzevetski, Z., Muzykantov,
     V. S. and Shul'meister, T. T. Kinet. Katal. 1971, 12, 979.
48. Wolski, W. Roczniki Chemii. 1960, 34, 309.
49. Lund, C. R. F., and Dumesic, J. A., J. Phys. Chem. 1981, 85,
     3175.
50. Lund, C. R. F., and Dumesic, J. A., J. Phys. Chem. 1982, 86,
     130.
51. DeBoer, F. E., and Selwood, P. W., J. Amer. Chem. Soc. 1954,
     76, 3365.

RECEIVED October 10, 1984

Secondary Ion Mass Spectrometry Studies of the Structure and Reactivity of Carbon on Ruthenium(001)

L. L. LAUDERBACK[1] and W. N. DELGASS

School of Chemical Engineering, Purdue University, West Lafayette, IN 47907

Most of the ethylene that interacts with an Ru(001) surface at 323 K produces a nondesorbable carbon layer.
Thermal desorption of CO, Auger electron spectroscopy,
and temperature programmed oxidation all show that the
carbon layer 1) is immobile below 550 K 2) forms a more
densely packed surface phase at temperatures of 550-
1150 K and 3) dissolves into the bulk at 1350 K. SIMS
measurements of isotope mixing in the C_2^- ions confirm
formation of dense-phase (graphitic) islands after heating
the carbon layer to 923 K. SIMS spectra also demonstrate
that at 520 K, CO dissociates on Ru(001). The oxygen-
free carbon layer that forms behaves similarly to the
carbon from ethylene. Both SIMS and thermal desorption
results show no positive interaction between adsorbed CO
and D_2 but significant attraction between D_2 and C
formed by CO dissociation.

In a recent investigation (1) we showed, using secondary ion mass
spectrometry (SIMS) and thermal desorption spectroscopy (TDS), that
the interaction of ethylene with Ru(001) at 323 K is accompanied by
substantial dissociation and subsequent desorption of hydrogen giving
rise to an adlayer consisting mainly of non-desorbable carbon along
with small amounts of dissociated hydrogen and various hydrocarbon
species. Most importantly, it was shown that the hydrocarbon-
containing secondary ions seen in SIMS could be directly related to
hydrocarbon species on the surface and that the high temperature
desorption peak for C_2H_4 results from associative desorption of C_2H_2.
In this paper we extend our previous study by investigating in more
detail the nature of the non-desorbable carbon adlayer formed by
C_2H_4 or CO interaction with Ru(001). SIMS, CO thermal desorption
spectroscopy, temperature programmed oxidation (TPO) and Auger elec-
ton spectroscopy (AES) measurements are used to characterize the
structure and reactivity of the carbon adlayer and to develop SIMS
as a tool for analysis of the chemistry and structure of surface
layers.

[1]Current address: Department of Chemical Engineering, University of Colorado, Boulder,
CO 80309

0097-6156/85/0279-0339$06.00/0

Experimental

All experiments were carried out in an ion-pumped, stainless steel
UHV bell jar with a base pressure of 1x10^{-10} Torr. In SIMS experiments, primary Ar$^+$ ions were generated by a Riber CI 50 ion gun,
and secondary ions were detected with a Riber Q156 quadrupole mass
spectrometer equipped with a 45° energy prefilter. The mass spectrometer is also equipped with an ionization filament for residual gas
analysis and thermal desorption measurements. All experiments were
performed with 5 keV Ar$^+$ ions impinging on the sample surface at a
45° polar angle measured from the surface normal. The primary ion
current density was 5x10^{-8} amps/cm2.

     Auger electron spectroscopy measurements were carried out using
a PHI, single-pass, cylinderical mirror analyzer and an off-axis
electron gun. The primary electron energy was 3.5 keV and the primary electron current was 9μA in all experiments.

     The Ru single crystal was oriented by Laue x-ray back-scattering
to within 1° of the Ru(001) plane, cut by a diamond saw and mechan-
ically polished. After being etched in hot aqua-regia for about 15
min., the crystal was spot welded to two tantalum heating wires which
were connected to two stainless steel electrodes on a sample manip-
ulator. The temperature was monitored by a Pt/Pt-10%Rh thermocouple
which was spot welded to the back of the crystal. In this configuration temperatures up to 1700 K could be routinely achieved. The surface cleaning procedure, which was similar to that used by Madey et
al. (2), involved many heating and cooling cycles up to 1600 K in 5x
10^{-7} Torr of oxygen, followed by heating in vacuum 2-5 times to 1700
K to remove surface oxygen. Surface cleanliness was verified by SIMS
and AES.

Results and Discussion

Carbon Layers From Ethylene. Carbon coverage as a function of ethylene exposure at 323 K has been determined previously by recording the
amount of CO desorbed during TPO (1). Half monolayer (M.L.) coverage by carbon requires about 2 Langmuirs (L) of C_2H_4, while 1.1
monolayer coverage corresponds to 15 L. Molecular species removed
from the surface by TDS after 15 L exposure to C_2H_4 amount to only
1% of a monolayer. We describe below studies of the residual carbon
layer both in its initial state and after heating to elevated temperatures.

     CO Thermal Desorption. On a clean Ru(001) surface, 9 L of CO
induces saturation coverage by molecular CO. When this dose of CO
is applied to a surface preexposed to C_2H_4 at 323 K, the CO uptake is
diminished but not completely blocked by the carbon layer. The CO
uptake is still 90% of the saturation value when the carbon coverage
is 1/4 M.L. and falls to 1/4 of the saturation value after 15 L preexposure to ethylene.

     In order to probe the effect of temperature on the structure of
the carbon layer, carbon was first deposited by exposure to C_2H_4.
The sample was then heated to the desired temperature at a rate of
6 K/sec and then cooled to 323 K before exposure to 9.0 L of CO.
Figure 1 shows the CO TDS spectra for a 15 L C_2H_4 preexposure. The
TDS heating rate was 6 K/sec. The 1145 K spectrum in this figure is

identical in shape to that from the clean surface, but the CO coverage is about 1/2 the saturation value. A significant increase in
CO coverage occurs after annealing to 763 K and above.

Analysis of the CO desorption spectra, and others obtained at
lower carbon coverage, in terms of results for CO adsorption on the
clean surface (2-5), leads to several conclusions. At low carbon
coverages the effect of the carbon is slight. This corresponds to a
CO TPD spectrum similar to that found for a clean surface but
decreased in intensity and with a broadened high temperature peak.
High carbon coverage blocks some sites and weakens the remainder.
This effect shifts the CO TPD spectrum to lower desorption temperatures. The effects of annealing the carbon layers are first seen at
763 K, where carbon mobility allows a rearrangement of the carbon
layer to accommodate more CO, as shown in Figure 1. The 1145 K spectrum suggests that the carbon layer rearrangement has produced some
clean Ru(001) which desorbs CO in its characteristic fashion. Since
TPO shows that the amount of carbon remaining on the surface is almost unchanged after annealing at 980 K and is still 80% of the original value at 1145 K, the carbon layer appears to rearrange into a
more dense phase at temperatures above 763 K.

Temperature Programmed Oxidation. These measurements characterize both the amount and chemical nature of the carbon on the
surface. After a surface is exposed to ethylene and pretreated as
desired, it receives a 6 L dose of O_2 at 323 K. The TPO spectrum is
the CO desorption signal at a 6 K/sec programming rate. CO_2 accounts
for less than 1% of the oxidation, so the CO signal accounts for
essentially all of the carbon removed. O_2 dosing is repeated until
no further CO is evolved during heating. SIMS results show that all
carbon has been removed from the surface at the TPO end point.

TPO spectra as a function of annealing temperature are shown in
Figure 2 for a carbon coverage of about 0.25 monolayers, corresponding to an ethylene exposure of 1 L. One TPO cycle removes all the
carbon at this coverage. A qualitative examination of these spectra
reveals that the low temperature peak, A, does not change position
with carbon coverage and is consistent with the first order evolution
of CO from immobile C - O neighbors. The sharp peak, B, occurs in
a temperature region where carbon transformation to the dense phase
and oxygen disordering (6) occur. Most interesting is the evolution
of a high temperature peak beginning at an annealing temperature of
663 K. The gradual decline of the low temperature peak and growth
of the high temperature peak as the annealing temperature increases
from 663 K to 1145 K correlates well with the conversion of the low
density carbon phase to the high density carbon phase seen in this
region in the CO TDS spectra. The loss of carbon after 1323 K and
1573 K annealing (Figure 2) indicates dissolution of carbon into the
bulk.

Auger Electron Spectroscopy. The dominant Ru MNN and carbon
KLL peaks overlap strongly. A carbon fine structure peak at 249 eV
is partially resolved, however. Up to an annealing temperature of
about 600 K, this peak position from 1.1 monolayers of carbon stays
constant, but it falls to 245.5 eV after annealing at 600 to 900 K.
This change again corresponds to the temperature region in which
transformation to the dense carbon phase occurs. The shift is
consistent with assignment of the dense phase to a graphitic structure (7).

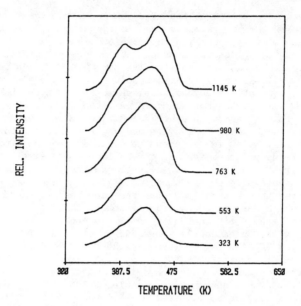

Figure 1. Thermal desorption of CO following a 15 L C$_2$H$_4$ exposure at 323 K, heating to the temperature indicated, cooling to 323 K, and exposing to 9.0 L CO.

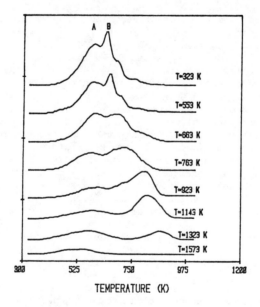

Figure 2. Temperature programmed oxidation spectra (CO signal) following a 1.0 L C$_2$H$_4$ dose, heating to the temperatures indicated, cooling to 323 K, and exposing to 6.0 L O$_2$.

Secondary Ion Mass Spectrometry. The surface structural changes
implied by the data discussed so far influence a variety of SIMS ions.
We discuss here a SIMS experiment designed specifically to examine
effects of temperature on carbon mobility and the formation of dense
phase carbon islands on the surface by monitoring carbon isotope
mixing. The low mass negative ion spectrum of 0.5 monolayers of ^{12}C
deposited from $^{12}C_2H_4$ at 435 K shows peaks at 12, 16, and 24 amu with
only minor contributions at 13, 25, and 26. The higher exposure
temperature was used in these experiments to eliminate the molecular
species on the surface since fragments from those species complicate
interpretation of the spectra. The strong peaks at 12 and 24 correspond to $^{12}C^-$ and $^{12}C_2^-$. The small impurities at 25 and 26 are
$^{12}C_2H^-$ and $^{12}C_2H_2^-$. The $^{16}O^-$ peak at 16 is strong even at the very low
oxygen coverages present because of the high SIMS sensitivity for
oxygen. Most importantly, these data show that the background due to
hydrocarbon contamination is low in the 25, 26 amu region. The
isotope mixing experiments were conducted as follows. In Case A, 2 L
of $^{12}C_2H_4$ and 3 L of $^{13}C_2H_4$ were adsorbed sequentially at 435° K to
give equal coverage of carbon from each. The surface was then annealed by heating to a desired temperature at 6 K/sec and cooling to 435
K. In Case B, the sample was exposed to 2 L of $^{12}C_2H_4$ at 435 K,
annealed at 923 K, cooled to 435 K, exposed to 3 L of $^{13}C_2H_4$, and
annealed again to the temperature corresponding to that for Case A.
Since proximity is essential for cluster ion formation in SIMS (8,9),
the mobility and structure of the carbon layer is reflected in the
isotopic composition of the C_2^- ions. With the equal coverages of ^{12}C
and ^{13}C used in these experiments, a random distribution of carbon
would yield relative intensity of 1:2:1 for masses 24:25:26.
Complete isolation would yield 2:0:2 for the corresponding intensities. Figure 3 displays the $I_{25}/(I_{24} + I_{26})$ intensity ratios as a
function of annealing temperature for both Cases A and B. Examples
of the C_2 data are shown in the inset.
     For Case A, the peak area ratio of 25 to (24 + 26) goes from
0.75 at 435 K to nearly 1 at 818 K. The substantial isotope mixing
at low temperature shows that molecular emission of intact C_2 from
the original ethylene molecule does not occur. It is interesting,
however, that the isotope mixing is not complete at the lower annealing temperatures. This suggests that proximity of the parent carbons
is preserved at low temperature and supports the conclusion that the
carbon layer is immobile at temperatures below about 550 K.
     The much lower extent of atomic mixing at 435 K for case B
relative to case A clearly demonstrates that annealing to 923 K
alters the distribution of carbon on the surface to prevent close
proximity of much of the annealed carbon with the additional carbon
subsequently deposited. This is consistent with previously described
results indicating the formation of high density carbon islands upon
annealing above 763 K since in that case the annealed carbon can be
in close proximity to the carbon deposited subsequent to annealing
only along island edges. The increase in the peak area ratio of 25
to (24 + 26) from 0.4 at 435 K to a plateau of 0.6 at 670 to 923 K
corresponds to the carbon phase transition previously discussed.
     The rapid increase in the extent of mixing to the statistically
complete level as the annealing temperature increases from 1000 K to
about 1450 K corresponds to the temperature region where carbon
diffuses into the bulk. This suggests that during annealing at these

high temperatures the high density carbon phase becomes unstable,
allowing individual carbon atoms to migrate independently over the
surface as well as to diffuse into the bulk thereby increasing the
extent of isotopic randomization in both the low and high density
carbon phases.

Carbon Layers from CO

CO Dissociation. It is well established in the literature and
implicit in the data already discussed that CO adsorbs molecularly on
Ru(001) at temperatures below 350 K. Dissociation of CO is to be
expected at higher temperatures, however, and is thought to be a key
step in catalytic methanation (10-12). Figure 4 shows direct SIMS
evidence for CO dissociation at 520 K. Spectrum A corresponds to
molecular adsorption and shows the $RuCO^+$ and Ru_2CO^+ ions characteristic of molecular CO on the surface (13). Spectrum B shows that a
long CO exposure at 520 K deposits carbon, as indicated by Ru_4C^+ and
Ru_2C^+, but leaves essentially no molecular CO since the $RuCO^+$ signal
is negligible. The absence of molecular CO at this temperature is as
expected from Figure 1. The absence of RuO^+ in spectrum B shows that
oxygen does not accumulate in the surface layer and suggests that it
is removed either by reaction with CO to form CO_2 or by diffusion
into the bulk. Reaction probabilities for formation of carbon from
CO were found to be 4.9×10^{-4}, 3.7×10^{-4} and 2.1×10^{-4} at 435 K, 520 K
and 815 K respectively. A coverage of nearly 1/2 monolayer was found
after 480 L exposure to CO at 435 K.

Temperature programmed oxidation experiments showed this carbon
layer to be similar to that formed from C_2H_4 at equivalent coverages.
Annealing to 663 K produced the high density carbon phase (14).

Interaction with D_2. Thermal desorption spectra for coadsorbed
CO and D_2 confirmed literature findings (15,16) that CO displaces D_2
from the surface and TDS peak positions do not shift when both gases
are present on the surface. SIMS spectra also show no D/CO combination ions under these conditions. As shown in Figure 5, however, the
D_2 TDS spectrum is clearly shifted to higher temperature by the presence of surface carbon from dissociated CO. The CO preadsorption
was done at 425 K to eliminate the presence of molecular CO on the
surface. The heating rate for TDS of D_2 was 65 K/sec.

These results indicate that the presence of surface carbon
creates a new deuterium adsorption state having a significantly
higher binding energy than that for the clean surface. The relatively narrow and symmetric desorption peaks and the invariance of
the peak maximum with carbon coverage, furthermore, indicate that
only a single new state is created by the presence of surface carbon,
as opposed to a distribution of adsorption states of different binding energies. This suggests the possibility that the new state may
be due to the formation of a specific deutero-carbon complex having
a well defined C-D binding energy. This seems plausible since if the
new adsorption state were created by modification of metal adsorption
sites by a through-metal type electronic interaction, then one would
generally expect a distribution of binding states due to a non-
uniform local distribution of atoms which would change with carbon
coverage.

The existence of strong C-D interaction is supported by SIMS
data which show the presence of $RuCD^+$ and the loss of this ion in
concert with the thermal desorption of D_2 (14).

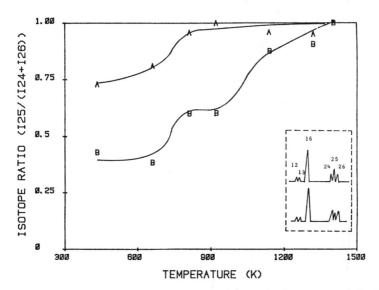

Figure 3. Intensity of the 25 amu peak divided by the sum of the intensities of the 24 and 26 amu peaks, indicating the degree of isotope mixing in SIMS of the C_2^- ion as a function of annealing for Case A, sequential adsorption of equal amounts of $^{12}C_2H_4$ and $^{13}C_2H_4$ and Case B, $^{12}C_2H_4$ annealed to 923 K followed by addition of $^{13}C_2H_4$. Inset indicates negative ion SIMS spectra for Case A (upper) and Case B (lower) after annealing at 663 K.

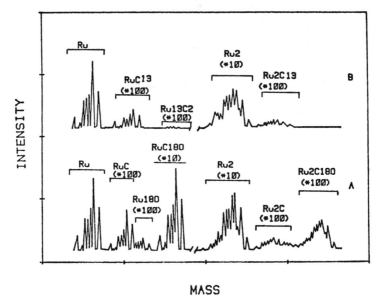

MASS

Figure 4. Positive ion SIMS spectra following: A) 9.0 L $C^{18}O$ at 320 K B) 480 L ^{13}CO exposure at 520 K.

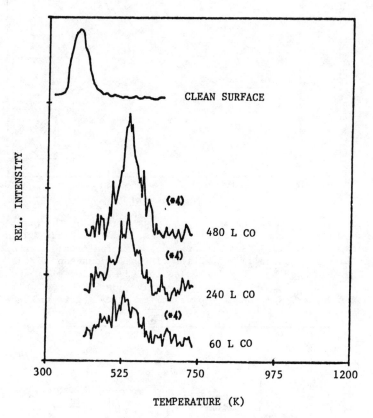

Figure 5.   Thermal desorption of D_2 from Ru(001) and Ru(001) pre-exposed at 520 K to the ^{13}CO doses indicated. The exposure to D_2 was 720 L at 425 K.

Conclusions

The combination of TDS, TPO, SIMS and AES shows that carbon deposited on Ru(001) from C_2H_4 becomes mobile at about 550 K and transforms into a more dense, perhaps graphitic, phase at temperatures above 550 K. SIMS studies of isotopic exchange in the C_2 ion were particularly effective in demonstrating carbon mobility and dense-phase island formation. SIMS also demonstrated that CO adsorbs dissociatively on Ru(001) at 520 K. This carbon phase formed by dissociation also showed the dense phase transformation after heating. D_2 was found not to interact with CO, but to have a positive interaction with surface carbon. This last finding is in good agreement with current mechanistic explanations of catalytic methanation on ruthenium. Overall, the results show both the importance of surface-phase chemistry in catalysis and the utility of SIMS for studies of surface structure.

Acknowledgments

We are grateful for support of this work by NSF Grants CHE 78-08728, CPE 7911597 and DMR 77-23798 and by the Amoco Oil Company.

Literature Cited

1. Lauderback, L.L. and Delgass, W.N., in Proceedings of the Symposium on Catalytic Materials: Relationship Between Structure and Reactivity, American Chemical Soc., 1984, 248, 21.
2. Madey, T.E. and Menzel, D., Japan J. Appl. Phys. Suppl., 1974, 2, Pt.2, 229.
3. Williams, E.D. and Weinberg, W.H., Surf. Sci., 1979, 82, 93.
4. Thomas, G.E. and Weinberg, W.H., J. Chem. Phys., 1979, 80, 954.
5. Pfnur, H., Menzel, D., Hoffman, F.M., Ortega, A., and Bradshaw, A.M., Surf. Sci., 1980, 93, 431.
6. Madey, T.E., Engelhardt, H.A., Surf. Sci., 1975, 48, 304.
7. Bonzel, H.P. and Krebs, H.J., Surf. Sci., 1980, 91, 494.
8. Winograd, N., Garrison, B.J., and Harrison, D.E., J. Chem. Phys., 1980, 73, 3473.
9. Garrison, B.J., Winograd, N., and Harrison, D.E., Phys. Rev. B, 1978, 18, 6000.
10. Rabo, J.A., Risch, A.P., and Poutsma, J.L., J. Catal., 1978, 53, 295.
11. Low, G.G. and Bell, A.T., J. Catal., 1979, 57, 397.
12. Ekerdt, J.G. and Bell, A.T., J. Catal., 1979, 58, 170.
13. Lauderback, L.L. and Delgass, W.N., Phys. Rev. B, 1982, 266, 5258.
14. Lauderback, L.L. and Delgass, W.N., J. Catal., to be submitted.
15. Peebles, D.E., Schreifels, J.A., and White, J.M., Surf. Sci., 1982, 116, 117.
16. Fischer, G.B., Madey, T.E., and Yates, J.T., J. Vacuum Sci. & Techol., 1978, 15, 543.

RECEIVED October 4, 1984

INDEXES

Author Index

Subject Index

Production by Hilary Kanter
Indexing by Susan Robinson
Jacket design by Pamela Lewis

Elements typeset by Hot Type Ltd., Washington, D.C.
Printed and bound by Maple Press Co., York, Pa.